Periodic Table of the Elements with the Gmelin System Numbers

1 H 2																1 H 2	2 He 1
3 Li 20	4 Be 26											5 B 13	6 C 14	7 N 4	8 O 3	9 F 5	10 Ne 1
11 Na 21	12 Mg 27											13 Al 35	14 Si 15	15 P 16	16 S 9	17 Cl 6	18 Ar 1
19 * K 22	20 Ca 28	21 Sc 39	22 Ti 41	23 V 48	24 Cr 52	25 Mn 56	26 Fe 59	27 Co 58	28 Ni 57	29 Cu 60	30 Zn 32	31 Ga 36	32 Ge 45	33 As 17	34 Se 10	35 Br 7	36 Kr 1
37 Rb 24	38 Sr 29	39 Y 39	40 Zr 42	41 Nb 49	42 Mo 53	43 Tc 69	44 Ru 63	45 Rh 64	46 Pd 65	47 Ag 61	48 Cd 33	49 In 37	50 Sn 46	51 Sb 18	52 Te 11	53 I 8	54 Xe 1
55 Cs 25	56 Ba 30	57** La 39	72 Hf 43	73 Ta 50	74 W 54	75 Re 70	76 Os 66	77 Ir 67	78 Pt 68	79 Au 62	80 Hg 34	81 Tl 38	82 Pb 47	83 Bi 19	84 Po 12	85 At 8a	86 Rn 1
87 Fr 25a	88 Ra 31	89*** Ac 40	104 71	105 71													* NH_4 23

**Lanthanides 39	58 Ce	59 Pr	60 Nd	61 Pm	62 Sm	63 Eu	64 Gd	65 Tb	66 Dy	67 Ho	68 Er	69 Tm	70 Yb	71 Lu
***Actinides	90 Th 44	91 Pa 51	92 U 55	93 Np 71	94 Pu 71	95 Am 71	96 Cm 71	97 Bk 71	98 Cf 71	99 Es 71	100 Fm 71	101 Md 71	102 No 71	103 Lr 71

A Key to the Gmelin System is given on the Inside Back Cover

Gmelin Handbook of Inorganic Chemistry

8th Edition

Gmelin Handbook of Inorganic Chemistry

8th Edition

Gmelin Handbuch der Anorganischen Chemie

Achte, völlig neu bearbeitete Auflage

Prepared and issued by
Gmelin-Institut für Anorganische Chemie
der Max-Planck-Gesellschaft
zur Förderung der Wissenschaften
Director: Ekkehard Fluck

Founded by
Leopold Gmelin

8th Edition
8th Edition begun under the auspices of the Deutsche Chemische Gesellschaft by R. J. Meyer

Continued by
E.H.E. Pietsch and A. Kotowski, and by Margot Becke-Goehring

Springer-Verlag Berlin Heidelberg GmbH 1986

STAFF GMELIN HANDBOOK

D. Barthel, Dr. N. Baumann, Dr. W. Behrendt, Dr. L. Berg, Dipl.-Chem. E. Best, Dipl.-Ing. V. A. Chavizon, E. Cloos, Dipl.-Phys. G. Czack, I. Deim, Dipl.-Chem. H. Demmer, R. Dowideit, Dipl.-Chem. M. Drößmar, U. Duggen, M. Engels, Dr. H.-J. Fachmann, Dr. J. Faust, Dr. W.-D. Fleischmann, V. Frick, G. Funk, Dipl.-Ing. N. Gagel, E. Gerhardt, Dr. U. W. Gerwarth, M.-L. Gerwien, C. Gorr, Dipl.-Phys. D. Gras, Dipl.-Bibl. W. Grieser, H. Hartwig, Dipl.-Min. H. Hein, G. Heinrich-Sterzel, H.-P. Hente, H. W. Herold, U. Hettwer, Dr. I. Hinz, Dr. W. Hoffmann, Dipl.-Chem. K. Holzapfel, Dr. S. Jäger, Dr. R. Jotter, Dr. J. von Jouanne, Dipl.-Chem. W. Karl, H.-G. Karrenberg, Dipl.-Phys. H. Keller-Rudek, A. Klusch, Dipl.-Phys. E. Koch, Dipl.-Chem. K. Koeber, Dipl.-Chem. H. Köttelwesch, R. Kolb, E. Kranz, Dipl.-Chem. I. Kreuzbichler, Dr. A. Kubny, Dr. P. Kuhn, Dr. W. Kurtz, M. Langer, Dr. A. Leonard, Dipl.-Chem. H. List, H. Mathis, E. Meinhard, K. Meyer, Dr. M. Mirbach, Dr. U. Neu-Becker, K. Nöring, Dipl.-Chem. R. Nohl, Dipl.-Min. U. Nohl, Dr. W. Petz, I. Rangnow, Dipl.-Phys. H.-J. Richter-Ditten, Dipl.-Chem. H. Rieger, B. Riegert, E. Rieth, A. Rosenberger, E. Rudolph, G. Rudolph, Dipl.-Chem. S. Ruprecht, V. Schlicht, Dipl.-Chem. D. Schneider, Dipl.-Min. P. Schubert, A. Schwärzel, Dipl.-Ing. H.M. Somer, E. Sommer, M. Teichmann, Dr. W. Töpper, Dipl.-Ing. H. Vanecek, Dipl.-Chem. P. Velić, Dipl.-Ing. U. Vetter, H.-M. Wagner, Dipl.-Phys. J. Wagner, R. Wagner, Dr. E. Warkentin, Dr. C. Weber, Dr. B. Wöbke, K. Wolff, U. Ziegler

STAFF GMELIN ONLINE DATA SYSTEM

Dr. L. Kießling — Dr. R. Froböse, Dr. B. Roth

Dipl.-Min. M.-B. Cinarz, Dr. J. Fippinger, Dr. G. Friedrich, Dr. B. Fröhlich, I. Güllich, M. Klöffler, R. Lohmeyer, Dr. M. Pauluth, Dr. H.-J. Schröder, Dr. K. Schücke

Organometallic Compounds in the Gmelin Handbook

The following listing indicates in which volumes these compounds are discussed or are referred to:

Ag	Silber B5 (1975)
Au	Organogold Compounds (1980)
Bi	Bismut-Organische Verbindungen (1977)
Co	Kobalt-Organische Verbindungen 1 (1973), 2 (1973), Kobalt Erg.-Bd. A (1961), B1 (1963), B2 (1964)
Cr	Chrom-Organische Verbindungen (1971)
Cu	Organocopper Compounds 1 (1985), 2 (1983)
Fe	Eisen-Organische Verbindungen A1 (1974), A2 (1977), A3 (1978), A4 (1980), A5 (1981), A6 (1977), A7 (1980), Organoiron Compounds A8 (1986), B1 (partly in English; 1976), B2 (1978), Eisen-Organische Verbindungen B3 (partly in English; 1979), B4 (1978), B5 (1978), Organoiron Compounds B6 (1981), B7 (1981), B8 (1985), B9 (1985), B10 (1986) present volume, B11 (1983), B12 (1984), Eisen-Organische Verbindungen C1 (1979), C2 (1979), Organoiron Compounds C3 (1980), C4 (1981), C5 (1981), C7 (1986), and Eisen B (1929-1932)
Hf	Organohafnium Compounds (1973)
Nb	Niob B4 (1973)
Ni	Nickel-Organische Verbindungen 1 (1975), 2 (1974), Register (1975), Nickel B3 (1966), C1 (1968), C2 (1969)
Np, Pu	Transurane C (partly in English; 1972)
Pt	Platin C (1939) and D (1957)
Ru	Ruthenium Erg.-Bd. (1970)
Sb	Organoantimony Compounds 1 (1981), 2 (1981), 3 (1982), 4 (1986) *present volume*
Sc, Y, La to Lu	D6 (1983)
Sn	Zinn-Organische Verbindungen 1 (1975), 2 (1975), 3 (1976), 4 (1976), 5 (1978), 6 (1979), Organotin Compounds 7 (1980), 8 (1981), 9 (1982), 10 (1983), 11 (1984), 12 (1985), 13 (1986)
Ta	Tantal B2 (1971)
Ti	Titan-Organische Verbindungen 1 (1977), 2 (1980) 3 (1984), 4 and Register (1984)
U	Uranium Suppl. Vol. E2 (1980)
V	Vanadium-Organische Verbindungen (1971), Vanadium B (1967)
Zr	Organozirconium Compounds (1973)

Gmelin Handbook of Inorganic Chemistry

8th Edition

Sb Organoantimony Compounds

Part 4

Compounds of Pentavalent Antimony
with Three Sb-C Bonds

With 19 illustrations

AUTHOR Markus Wieber, Anorganisches Institut, Universität Würzburg

FORMULA INDEX Edgar Rudolph, Gmelin-Institut, Frankfurt am Main

EDITOR Marlies Mirbach, Gmelin-Institut, Frankfurt am Main

CHIEF EDITOR Ulrich Krüerke, Gmelin-Institut, Frankfurt am Main

Springer-Verlag Berlin Heidelberg GmbH 1986

LITERATURE CLOSING DATE: END 1983

Library of Congress Catalog Card Number: Agr 25-1383

ISBN 978-3-662-06311-8 ISBN 978-3-662-06309-5 (eBook)
DOI 10.1007/978-3-662-06309-5

Originally published by Springer-Verlag Berlin Heidelberg New York in 1986
Softcover reprint of the hardcover 8th edition 1986

Preface

This fourth volume on organoantimony compounds describes pentavalent antimony compounds of the type R_3SbX_2 and $R_3Sb{=}X$. The R denotes an organic group bonded by carbon to the antimony atom. X represents a group, inorganic or organic, which is bonded to the antimony by an atom other than carbon. The X atoms in R_3SbX_2 may be part of a ring system. In the case of bidentate X ligands like O^{2-}, S^{2-}, SO_4^{2-}, CO_3^{2-}, and others, the compounds are placed with the mononuclear R_3SbX_2 compounds.

I once again thank Dr. Margot Becke and Dr. Ekkehard Fluck for the stimulus that led to this book. To the editor of the former volumes, Dr. Hubert Bitterer† gratitude and memory are due. I especially thank Drs. Ulrich Krüerke and Marlies Mirbach for editing this volume, Mrs. Ursula Hettwer for systematically arranging the compounds, and Mr. Edgar Rudolph for preparing the index. Last but not least I thank my wife Sigrid for putting my handwritten manuscript into legible form.

Gramschatz, Altes Forsthaus, June 1986 — Markus Wieber

Explanations, Abbreviations, and Units

Many compounds in this volume are presented in tables in which abbreviations are used and the units are omitted for the sake of conciseness. This necessitates the following clarification.

Temperatures are given in °C, otherwise K stands for Kelvin. Abbreviations used with temperatures are m.p. for melting point, b.p. for boiling point, dec. for decomposition, and subl. for sublimation. Terms like 80°/0.1 mean the boiling or sublimation point at a pressure of 0.1 Torr. **Densities** d are given in g/cm^3. d_c and d_m distinguish calculated and measured values, respectively.

NMR represents **nuclear magnetic resonance.** Chemical shifts are given as δ values in ppm and positive to low field from the following reference substances: $Si(CH_3)_4$ for ^1H and ^{13}C, $BF_3 \cdot O(C_2H_5)_2$ for ^{11}B, $CFCl_3$ for ^{19}F, and H_3PO_4 for ^{31}P. Multiplicities of the signals are abbreviated as s, d, t, q (singlet to quartet), quint, sext, sept (quintet to septet), and m (multiplet); terms like dd (double doublet) and t's (triplets) are also used. Assignments referring to labelled structural formulas are given in the form C-4, H-3,5. Coupling constants J in Hz appear usually in parenthese behind the δ value, along with the multiplicity and the assignment, and refer to the respective nucleus. If a more precise designation is necessary, they are given as, e.g., $^nJ(C,H)$ or J(1,3) referring to labelled formulas.

Nuclear quadrupole resonance is abbreviated NQR, with the transitions in MHz.

Mössbauer spectra are represented by ^{121}Sb-γ; the isomer shift δ (vs. $Ba^{121}SnO_3$ or $^{121}SnO_2$ at room temperature), the quadrupole splitting Δ, and the width Γ are given in mm/s; the experimental error has generally been omitted.

Optical spectra are labelled as IR (infrared), R (Raman), and UV (electronic spectrum including the visible region). IR bands and Raman lines are given in cm^{-1}; the assigned bands are usually labelled with the symbols ν for stretching vibration and δ for deformation vibration. Intensities occur in parentheses either in the common qualitative terms (s, m, w, vs, etc.) or as numerical relative intensities. The UV absorption maxima, λ_{max}, are given in nm followed by the extinction coefficient ε ($L \cdot cm^{-1} \cdot mol^{-1}$) or log ε in parentheses; sh means shoulder.

Solvents or the **physical state** of the sample and the temperature (in °C or K) are given in parentheses immediately after the spectral symbol, e.g., R (solid), ^{13}C NMR (C_6D_6, 50 °C), or at the end of the data if spectra for various media are reported. Common solvents are given by their formula (C_6H_{12}=cyclohexane) except THF and HMPT, which represent tetrahydrofuran and hexamethylphosphoric triamide, respectively.

The data of **mass spectra,** abbreviated MS, are given as m/e, relative intensity in parentheses, and fragment ions in brackets; $[M]^+$ is the molecular ion.

References, usually quoted in the last column, are occasionally also placed in other columns if statements from different sources must be distinguished.

Figures give only selected parameters. Barred bond lengths (in Å) or angles are mean values for parameters of the same type.

Table of Contents

Organoantimony Compounds

2 Organoantimony Compounds with Pentavalent Antimony

2.5 Organoantimony Compounds with Three Sb-C Bonds

2.5.1 Mononuclear Compounds

2.5.1.1 Compounds of the R_3SbX_2 Type

General Remarks. In contrast to previously described substances ("Organoantimony Compounds" Parts 1 to 3) the compounds of this volume are relatively stable. Unless mentioned otherwise they are usually obtained as colorless crystals, which are not sensitive towards dry air. They hydrolyze slowly in the presence of moisture. Most of them are soluble in aprotic polar solvents.

2.5.1.1.1 Triorganoantimony Difluorides

2.5.1.1.1.1 R_3SbF_2 Compounds with R = Alkyl and Alkenyl

$(CH_3)_3SbF_2$

Trimethylantimony difluoride is prepared by reacting a hot aqueous solution of $(CH_3)_3SbBr_2$ with AgF in the same solvent. After removal of the AgBr, the filtrate is evaporated to dryness, and the product recrystallized from ethanol (yield 42%) [1]. Aqueous HF may also be used as a reaction medium. The compound is soluble without undergoing solvolysis in water, methanol, and chloroform [2]. $(CH_3)_3SbF_2$ may also be obtained by reaction of $(CH_3)_3SbCl_2$ with AsF_3 [16].

IR spectra of the compound were measured and discussed in [1 to 7]. Raman spectra are published in [3, 4, 6]. The complete IR and Raman spectra of the solid compound are shown in Table 1 [6]. From these data it is concluded that the molecule has a nonionic, five-coordinated trigonal bipyramidal structure [1 to 7]. Normal coordinate calculations were performed using the modified Urey-Bradley force field [3, 7], the modified general valence force field [7], and Redington's approach. In the latter case, force constants, mean amplitudes of vibration, and rotation distortion constants have also been calculated [8]. Using general valence force field constants, mean amplitudes of vibration, generalized mean square amplitudes, shrinkage constants, Coriolis coupling coefficients, and centrifugal distortion constants have been evaluated and compared with those of the other trimethylantimony dihalides [9]. A correlation between fundamental frequencies of the dihalides and atomic mass, electronegativity of the halogen, and moment of inertia of the molecule has been made [10].

The ^1H NMR spectrum in $CDCl_3$ shows a triplet at $\delta = 1.75$ ppm with a coupling constant of J = 5 Hz at −32 °C. At higher temperatures (30.5 °C, 70 °C) the coupling of the fluorine atoms with the protons is no longer observed; therefore, an intermolecular exchange is concluded. A ^{19}F NMR in $CDCl_3$ at 30.5 °C shows a resonance at 106.1 ppm. Mixing with equal amounts of $(CH_3)_3SbX_2$ (X = Cl, Br, I) in $CDCl_3$ gives $(CH_3)_3SbFX$ (X = Cl, Br, I), as measured by ^1H and ^{19}F NMR [11]. Equilibrium constants were measured for these redistribution products by NMR spectroscopy [12, 13]. Halide exchange reactions with $(C_6H_5)_3SbX_2$ (X = Cl, Br) are also described [13]. Some data are given in this volume on pp. 37, 80, and 99.

References on p. 5

Table 1
IR and Raman Vibrations (in cm^{-1}) of Solid $(CH_3)_3SbF_2$ [6].

IR	Raman	assignment	IR	Raman	assignment
3017 mw	3033 mw	ν_{as} CH	590 vs	592 ms	ν_5, ν_{as} SbC
2930 mw	2928 ms	ν_s CH		546 s	ν_1, ν_s SbC
2858 w			543 mw		
2780 vw			526 mw		
2477 vw			484 s		ν_3, ν_{as} SbF
2414 vw				465 w	ν_2, ν_s SbF
2381 vvw			450 mw		
1775 vvw			356 vw		
1728 w			423 vw		2 ν_6
1642 ms			292 vw		
1403 ms		δ_{as} CH	284 vw		
1240 w	1232 w	δ_s CH	274 vw		
1086 s			256 w		
1225 vw	1205 w		228 vw	229 vvw	
1125 m		$\nu_1+\nu_3$		245 w	ν_8, $\varrho\,(CH_3)_3SbF_2$
1065 vw		$\nu_2+\nu_5$	215 sh		
870 vs		ϱ CH_3	210 vs	213 vw	ν_6, δSbC out-of-plane
			146 mw	145 m	ν_7, δSbC in-plane

A correlation of ^1H NMR chemical shifts in $(CH_3)_3SbX_1X_2$ (X_1, X_2 = F, Cl, Br, I) with inverse ionization potentials of the ligands is established and discussed in [14]. Hartree-Fock-Slater LCAO calculations of the Mössbauer parameters were made and compared with other antimony and organoantimony compounds [15].

An X-ray crystal structure analysis was done. Crystals were obtained by sublimation. The compound crystallizes as a monoclinic solid in the space group $P2_1/c-C^5_{2h}$ (No. 14) with a = 965.1(1), b = 785.8(1), c = 804.5(1) pm, β = 115.53(1)°, V = 618.97 × 10^6 pm^3, Z = 4, d_c = 2.198 g/cm^3, and d_m = 2.14 g/cm^3. The structure was determined from 2726 independent reflections. The final refinement is R = 0.031. The molecule has nearly an ideal trigonal bipyramidal shape with the fluorine atoms in axial and the methyl groups in equatorial positions. The angle F-Sb-F is 178.3°. The structure, the main bond distances, and angles are given in **Fig. 1** [16].

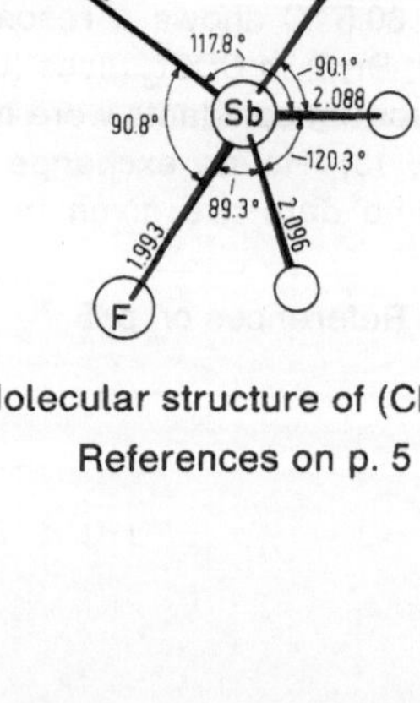

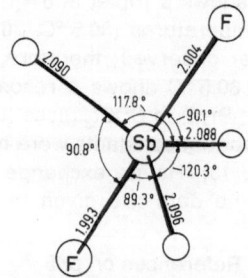

Fig. 1. Molecular structure of $(CH_3)_3SbF_2$ [16].

References on p. 5

$(C_2H_5)_3SbF_2$

A preparation of the compound by reacting the corresponding dichloride or dibromide with AgF in methanol or acetonitrile is mentioned without details. The low frequency IR and Raman spectra are very similar in the solid state, in C_6H_6, and cyclohexane solutions. The bands are assigned on the basis of a C_{3v} symmetry of the molecule and compared with those of other R_3SbX_2 (R = alkyl, X = halogen) compounds. IR: 555 ($\nu_{as}Sb(CC)_3$), 490 ($\nu_{as}SbF_2$), ca. 240 ($\varrho F_2Sb(CC)_3$), ca. 220 (δSbCC) cm^{-1}; Raman: 557 (dp, $\nu_{as}Sb(CC)_3$), 513 (p, $\nu_sSb(CC)_3$), 475 (p, $\nu_s SbF_2$), 280 (p, δSbCC) cm^{-1} [5].

$(i\text{-}C_3H_7)_3SbF_2$

The compound is prepared like the previous one. IR: 482 ($\nu_{as} SbF_2$), ca. 470 (sh, ν_sSbF_2); 410, 300 ($\delta SbCC_2$) cm^{-1}; Raman: 534 (dp, $\nu_{as}Sb(CC_2)_3$), 500 (p, $\nu_s Sb(CC_2)_3$), 472 (p, ν_sSbF_2); 410, 265 (p, $\delta SbCC_2$) cm^{-1} [5].

$(C_4H_9)_3SbF_2$

Tributylantimony difluoride is prepared by galvanostatic electrolysis at 20 °C under argon atmosphere on a platinum anode. The anolyte is a solution of tributylantimony and $[N(C_2H_5)_4]BF_4$ in acetonitrile. The current density was 2.5 mÅ/cm^2. At the end of the electrolysis the solvent is removed in vacuum, and the products are extracted from the residue with ether. After removal of the ether, the residue is washed with water, and then vacuum distilled. Yield 95%; b.p. at 5 Torr 132 to 135 °C; n_D^{20} = 1.4741; IR: 1180, 1095, 715 (butyl), 640 (SbC), 500 (SbF) cm^{-1}; ^{19}F NMR (vs. CF_3CO_2H): δ = 49 ppm [17].

$(C_5H_{11})_3SbF_2$

The compound is prepared like the previous one from $Sb(C_5H_{11})_3$. Yield 93%; b.p. 151 to 152 °C at 2 Torr; n_D^{20} = 1.4738; IR: 1465, 1180, 1105, 710 (C_5H_{11}), 640 (SbC), 500 (SbF) cm^{-1}; ^{19}F NMR (vs. CF_3CO_2H): δ = 48 ppm [17].

$(CF_3CH_2)_3SbF_2$

The compound is prepared either by reacting CF_2=CH_2 with SbF_5 at 50 °C or by passing CF_2=CH_2 through a solution of SbF_5 in liquid SO_2. After pouring the latter mixture on ice, one obtains white crystals in a yield of 20% which melt at 72 °C after recrystallization from CCl_4. The ^1H NMR shows a quartet at δ = 3.34 ppm (vs. $(CH_3)_6Si_2O$) with a coupling constant of J(C, H) = 12.2 Hz. The ^{19}F NMR spectrum has a triplet at δ = −21.5 ppm (vs. CF_3CO_2H) with J(C, H) = 11.2 Hz. The mass spectrum shows peaks for $[M-F]^+$, $[M-CF_3CH_2]^+$, and $[CF_3CH_2SbF_3]^+$ [18].

$(C_6H_5CH_2)_3SbF_2$

The compound is prepared by treating a solution of the respective dichloride in acetonitrile with an excess of solid AgF. The slurry is stirred for 20 h just below its boiling point. After filtration, the filtrate is evaporated to dryness and the residue is extracted with boiling petroleum ether (65 to 110 °C). The hot filtered extract is cooled yielding white needles which are recrystallized from petroleum ether with a yield of 43%, m.p. 113 to 115 °C [12].

IR [5, 19] and Raman spectra were measured in solid state and in benzene solution. From the assignments of fundamental vibrations, see Table 2, a C_{3v} (C_{3h}) symmetry of the molecule with a slight distortion of the trigonal bipyramidal geometry in the solid state is proposed [19].

References on p. 5

Table 2
Fundamental IR and Raman Vibrations (in cm^{-1}) of $(C_6H_5CH_2)_3SbF_2$ [19].

$C_6H_5CH_2Sb$-skeleton IR in C_6H_6 (Nujol) Raman in C_6H_6 (Nujol)	assignment	$(CC)_3SbF_2$-skeleton IR in C_6H_6 (Nujol) Raman in C_6H_6 (Nujol)	assignment C_{3v}, C_{3h}
620 w, 540 w (620 m, 547 s)	δC_6H_5	578 vw (579 w)	$\nu_s Sb(CC)_3$
622, 553, 544 p (622, 545 sh)	in-plane	583 p (583)	A_1, A′
457 s (463, 448 s)	δC_6H_5	ca. 485 sh	$\nu_s SbF_2$
442 p, 456 sh (466, 447)	out-of-plane	485 p (500)	A_1, A′
—	δ CCSb	498 vs (495 s)	$\nu_{as} SbF_2$
—(335)		ca. 500 sh	A_1, A″
222 w, ca. 205 vw (222 s)	δ CC	—	$\delta Sb(CC)_3$
ca. 220 sh, 205 p (217)	in-plane	—(203)	A_1, A″
—	δ CC	608 w (609 w, 601 w)	$\nu_{as} Sb(CC)_3$
ca. 110	out-of-plane	611 dp (613, 605)	E, E′
		—	$\delta Sb(CC)_3$
		169, 156	2E, 2E′
		240 (247)	δSbF_2
		—(189)	E, E′
		—	$\delta F_2Sb(CC)_3$
		—(242, 246 sh)	2E, 2E″

The ^1H NMR spectrum of tribenzylantimony difluoride in $CDCl_3$ solution at −60 °C shows a triplet at δ=3.26 ppm, and the ^{19}F NMR shows a multiplet at δ=126.6 ppm under the same conditions [12]. These signals collapse at room temperature due to intermolecular exchange yielding signals at δ=3.06 (CH_2) and 73 (F) ppm [19]. Equilibrium constants for redistribution reactions with $(C_6H_5CH_2)_3SbX_2$ (X=Cl, Br, I) at 35 °C in $CHCl_3$ to give $(C_6H_5CH_2)_3SbFX$ [12, 13] and for exchange reactions with $(CH_3)_3SbX_2$ (X=Cl, Br, I) [13] were established by ^1H NMR and ^{19}F NMR spectroscopy.

$(CH_2{=}CH)_3SbF_2$

A solution of 2 g KF · 2 H_2O in 3 mL H_2O is added to a 3 mL ethanol solution containing 3 g $(CH_2CH)_3SbBr_2$. An additional 3 mL H_2O and 1 mL C_2H_5OH are added, and the solution is then extracted with four 5 mL portions of $CHCl_3$. After drying with K_2SO_4 and removal of the solvent, the residue is distilled at 10 Torr; 77% yield is obtained at a boiling point of 90 to 93 °C with $n_D^{20}=1.598$ [20]. Another method for the preparation of the compound from $(CH_2CH)_3SbCl_2$ and KF or CsF is mentioned without details. The boiling point is given as 40 to 41 °C at 0.1 Torr.

IR and Raman spectra for the solid compound were measured and assigned, see Table 3. According to these data, the compound has a trigonal bipyramidal structure probably with slightly twisted vinyl-groups [21].

Table 3
IR and Raman Vibrations (in cm−1) of Solid $(CH_2{=}CH)_3SbF_2$ [21].

IR	Raman (intensity)	assignment
1593 w-m	1598 p (24)	ν C=C (A′ + E′)
1385 s	1388 p (30)	$\delta_s CH_2$ (A′ + E′)
1234 s	1247 p (34), 1232 dp (1)	δ CH in-plane (A′ + E′)
	993 p (3)	τCH_2
983 vs, 965 sh		ωCH_2
574 s	585 s, 576 dp (7)	$\nu_{as} SbC_3$ (E′)
	544 p (100)	$\nu_s SbC_3$ (A′)
530 s-m	534 p (9)	δ HCC out-of-plane (A′ + E′)
513 vs		$\nu_{as} SbF_2$ (A″)
	497 p (19)	$\nu_s SbF_2$ (A′)
347 s-m	352 sh, dp	δ SbCC (E′)
	316 p (69)	δ SbCC (A′)
	330 dp (6)	δ CSbF (E″)
	280 p (13)	δSbF_2 (A′)

The ^1H NMR spectrum of the compound is very complex; for details and the calculation of parameters, see original text. In the ^{13}C NMR spectrum (in C_6D_6 at 25 to 30 °C), chemical shifts at $\delta = 132.2$ (C-1) and 139.8 (C-2) ppm with coupling constants of J(C-1, F) = 20 and J(C-2, F) = 6.3 Hz were found [21].

$(cis\text{-}CH_3CH{=}CH)_3SbF_2$

A solution of 2 g $KF \cdot 2\,H_2O$ in 3 mL H_2O is added to a solution of 3 g of the corresponding dibromide in 3 mL ethanol. The oil which forms is dissolved in 10 mL of a 1:1 mixture of ethanol and water. The solution is extracted with four 5 mL portions of $CHCl_3$. After drying with K_2SO_4 the chloroform is driven off, and the residue is distilled at 2 Torr. Yield 74%; b.p. 108 to 113 °C at 2 Torr; $n_D^{20} = 1.5185$ [20].

$(trans\text{-}CH_3CH{=}CH)_3SbF_2$

The compound is prepared like the previous one. Yield 76%; b.p. 105 to 107 °C at 2 Torr; $n_D^{20} = 1.5090$ [20].

$(CH_2{=}CCH_3)_3SbF_2$

The compound is prepared like the two previous ones. Yield 76%; b.p. 80 °C at 2 Torr; $n_D^{20} = 1.4990$ [20].

References:

[1] G. G. Long, G. O. Doak, L. D. Freedman (J. Am. Chem. Soc. **86** [1964] 209/13).
[2] H. C. Clark, R. G. Goel (Inorg. Chem. **5** [1966] 998/3).
[3] R. G. Goel, E. Maslowsky Jr., C. V. Senoff (Inorg. Chem. **10** [1971] 2572/7).
[4] R. G. Goel, E. Maslowsky Jr., C. V. Senoff (Inorg. Nucl. Chem. Letters **6** [1970] 833/5).
[5] L. Verdonck, G. P. Van der Kelen (Spectrochim. Acta A **31** [1975] 1707/11).
[6] B. A. Nevett, A. Perry (Spectrochim. Acta A **33** [1977] 755/60).
[7] B. A. Nevett, A. Perry (J. Mol. Spectrosc. **66** [1977] 331).

[8] R. Namasivayam, S. Viswanathan (Bull. Soc. Chim. Belges **87** [1978] 733/6).
[9] A. Natarajan, K. Chockalingam (Indian J. Pure Appl. Phys. **19** [1981] 672/5; C.A. **95** [1981] No. 105806).
[10] B. A. Nevett, A. Perry (J. Organometal. Chem. **71** [1974] 399/402).

[11] G. G. Long, C. G. Moreland, G. O. Doak, M. Miller (Inorg. Chem. **5** [1966] 1358/61).
[12] C. G. Moreland, M. H. O'Brien, C. E. Douthit, G. G. Long (Inorg. Chem. **7** [1968] 834/6).
[13] C. G. Moreland, G. G. Long (Inorg. Nucl. Chem. Letters **8** [1972] 347/51).
[14] T. Schaefer, F. Hruska, H. M. Hutton (Can. J. Chem. **45** [1967] 3143/51).
[15] W. Ravenek, J. W. M. Jacobs, A. Van der Avoird (Chem. Phys. **78** [1983] 391/404; C.A. **99** [1983] No. 130904).
[16] W. Schwarz, H. J. Guder (Z. Anorg. Allgem. Chem. **444** [1978] 105/11).
[17] E. V. Nikitin, A. A. Kazakova, O. V. Parakin, Yu. M. Kargin (Zh. Obshch. Khim. **52** [1982] 2027/9; J. Gen. Chem. [USSR] **52** [1982] 1802/4).
[18] G. G. Belen'kii, Y. L. Kopaevich, L. S. German, I. L. Knunyants (Izv. Akad. Nauk SSSR Ser. Khim. **1972** 983; Bull. Acad. Sci. USSR Div. Chem. Sci. **1972** 950).
[19] L. Verdonck, G. P. Van der Kelen (Spectrochim. Acta A **29** [1973] 1675/80).
[20] A. N. Nesmeyanov, A. E. Borisov, N. V. Novikova (Izv. Akad. Nauk SSSR Ser. Khim. **1964** 1202/9).

[21] K. Sille, J. Weidlein, A. Haaland (Spectrochim. Acta A **38** [1982] 475/82; C.A. **97** [1982] No. 92450).

2.5.1.1.1.2 R_3SbF_2 Compounds with R = Aryl

$(C_6H_5)_3SbF_2$

Preparation. Triphenylantimony difluoride can be prepared by reacting $Sb(C_6H_5)_3$ with 30% H_2O_2 and 40% HF in aqueous solution by stirring in a PTFE beaker just below boiling for a few hours. After filtration, the residue is washed with water and dried in vacuum, yielding 97% of the compound which melts at 117 °C [1]. Direct fluorination of $Sb(C_6H_5)_3$, dissolved in $CFCl_3$, by a F_2/Ar mixture (1:4) under a −90 °C circulation cooling system leads to 88% of the compound, which melts at 116 °C when recrystallized from n-hexane [2]. The compound can also be prepared with fluorinating agents like $C_6H_5IF_2$ [3] or XeF_2 [4] in CH_2Cl_2, 30 min at reflux, or $(C_6H_5)_2SF_2$ in $CDCl_3$, two days at room temperature [5]. In the first two cases, the yields are 70 or 95% [3, 4]. Recrystallization from benzene/hexane [3] or CH_3OH [4] gives melting points of 140 to 142 °C and 121 to 122 °C, respectively. Undecafluoropiperidine and $Sb(C_6H_5)_3$, probably in petroleum ether, give about 70% of the title compound and nonafluoro-3,4,5,6-tetrahydropyridine [6]. $(C_6H_5)_3SbF_2$ is also prepared by a galvanostatic electrolysis of a mixture of $Sb(C_6H_5)_3$ and $[N(C_2H_5)_4]BF_4$ in acetonitrile at 20 °C under an argon atmosphere. After workup of the mixture one obtains 96% of the compound which melts at 110 to 113 °C, recrystallized from ethanol [7]. Metathesis reactions for the preparation of the title compound have also been reported, see Table 4. Friedel-Crafts chemistry of benzene with SbF_5 in different solvents as C_6H_6, CCl_2FCClF_2, AsF_3, and CCl_4, at different reaction times, yields 4 to 20% of $(C_6H_5)_3SbF_2$ [15].

The compound can be purified from traces of free Sb by chromatography on Wofatit L-150 or Wofatit SBS-400 with acetone as eluent and subsequent recrystallization from acetone [9].

Properties. IR and Raman spectra are published and discussed in [13, 16 to 18]. A complete listing of the IR and Raman bands between 4000 and 40 cm^{-1} is given. The bands are assigned according to Whiffen's nomenclature [18]. The characteristic SbF and SbC

Table 4
Metathesis Reactions for the Preparation of $(C_6H_5)_3SbF_2$.

starting materials	reaction conditions (yield in %), m.p. in °C	Ref.
$(C_6H_5)_3SbCl_2$ in $C_2H_5OH + KF$ in H_2O/C_2H_5OH	1 h reflux (80), m.p. 115°	[8]
$(C_6H_5)_3SbCl_2$ in $(CH_3)_2CO + KF$ in H_2O	m.p. 115°	[9]
$(C_6H_5)_3SbCl_2$ in $HC(O)N(CH_3)_2 + KF$ in H_2O	2 to 3 h heating (95 crude)	[10]
$(C_6H_5)_3SbCl_2$ + aqueous HF	extraction with $CHCl_3$ (95)	[11]
$(C_6H_5)_3SbCl_2 + KF \cdot 2\,H_2O$	m.p. 115° from n-heptane	[12]
$(C_6H_5)_3SbCl_2 + AgF$ in H_2O/C_2H_5OH	m.p. 115 to 116°	[13]
$(C_6H_5)_3SbCl_2 + AgSCF_3$ in CH_3CN	1 h at 50° (77)	[14]
$(C_6H_5)_3SbO$ in $CH_2Cl_2 + SF_4$	−10° (94), m.p. 120 to 122°	[14]
$(C_6H_5)_3Sb(O_2CCF_3)_2$ in $CH_2Cl_2 + SF_4$	−10° (92)	[14]

stretching vibrations for the solid are at 508 vs (ν_{as}SbF) [18], 491 w (ν_sSbF), 292 (ν_{as}SbC), and 244 (ν_sSbC) [17] cm^{-1} in the IR spectrum, and at 507 vw (ν_{as}SbF) and 488 m [18] (485 s [17]) (ν_sSbF) cm^{-1} in the Raman spectrum.

The ^1H NMR spectrum shows signals at $\delta = 7.85$ (H-3,4,5) and 8.55 (H-2,6) ppm in CCl_4 at 30 °C or at $\delta = 7.52$ (H-3,4,5) and 8.17 (H-2,6) ppm in $CDCl_3$ or $(CD_3)_2CO$ [12]. ^{19}F NMR resonances are given as 153.0 ppm (in CCl_4) [15]; 151.0(s) [5] or 153.2(s) ppm [2] (in $CDCl_3$), or 57.8 ± 0.1 (vs. CF_3CO_2H) ppm [19, 20]. ^{13}C NMR spectra in $CDCl_3$ were measured and show resonances at $\delta = 129.44$ (C-3), 132.06 (C-4), 134.07 (C-1), 135.18 (C-2) ppm [19] or $\delta = 129.6$ (s, C-3), 132.2 (s, C-4), 134.3 (t, C-1; J = 15.4), 135.4 (t, C-2; J = 4.5) ppm [2]. Coupling constants are J(C-2, H) = 165.9, J(C-3, H) = 163.0, J(C-4, H) = 160.9 Hz, and absolute coupling constants $^nJ(^{19}F, ^{13}C)$ for n = 2 to 5 are 15.9, 34.9, 41.3, and 50.7 ± 0.2 Hz, respectively [19].

NQR measurements at 300 K lead to following resonances: $\nu(^{121}Sb)$: 5/2 ⇄ 3/2 at 184.25 (20), 3/2 ⇄ 1/2 at 92.592 (25); $\nu(^{123}Sb)$: 7/2 ⇄ 5/2 at 167.89 (2), 5/2 ⇄ 3/2 at 111.69 (8), 3/2 ⇄ 1/2 at 56.924 (10) MHz; $e^2Qq/h = 603.91$ for ^{121}Sb, and 769.90 MHz for ^{123}Sb. From the small asymmetry parameter $\eta = 0.065$, a linear F-Sb-F arrangement is suggested [21].

^{121}Sb Mössbauer spectra have been published several times. The experimental results are summarized in Table 5. From the experimental and from calculated coupling constants [25], the electric field gradient [26], and the Sb^V orbital population [27], a trigonal bipyramidal geometry for the molecule is concluded.

Table 5
^{121}Sb Mössbauer Spectra of $(C_6H_5)_3SbF_2$.

source	T in K	isomeric shift δ in mm/s	referred to	quadrupole coupling constant e^2qQ in mm/s	line width Γ in mm/s	Ref.
$^{121}SnO_2$	80	-4.9 ± 0.25	source	-17 ± 2	3.0 ± 0.5	[22]
		-5.5 ± 0.3	source	–	–	[28]
$Ca^{121}SnO_3$	4	-4.69 ± 0.03	source	-22 ± 0.4	2.66 ± 0.06	[22, 23]
$^{121}SnO_2$	80	4.1 ± 0.02	InSb	-21 ± 2	–	[24]

References on p. 10

The compound is soluble in nonpolar solvents [15]. The solubility of the compound is 1.6 in benzene, 0.8 in CCl_4, and 0.2 mol/L in $CHCl_3$ at 25 °C. The distribution coefficient between these solvents and water is 5×10^4, 3×10^4, and 7.5×10^3, respectively. Hydrolysis constants in benzene are determined by $\log K_1 = 7 \pm 1$ and $\log K_2 = 4.5 \pm 1$ [11]. The dipole moment in benzene at 25 °C is 0.72 D [29].

Reactions. $(C_6H_5)_3SbF_2$ reacts with BCl_3 in CH_2Cl_2 to form $(C_6H_5)_3SbCl_2$ in 98% yield [14]. Dissolved in benzene, it reacts with C_6H_5MgBr in ether at room temperature to give $Sb(C_6H_5)_5$ in good yields [10, 12]. For exchange reactions with $(C_6H_5)_3SbX_2$ (X = Cl, Br, I) to give $(C_6H_5)_3SbFX$ (X = Cl, Br, I), studied by ^{1}H NMR spectroscopy, see [30, 31] and the other halides in this volume, pp. 37, 80, and 99. The kinetics of the isotope exchange in the system $^{125}Sb(C_6H_5)_3/(C_6H_5)_3SbF_2/C_3H_5OH$ was investigated [42]. Solid $(C_6H_5)_3SbF_2$ was irradiated with fast and thermal neutrons (n, γ-reactions). The chemical effect and the degree of isotopic enrichment was determined [9, 32].

Uses. $(C_6H_5)_3SbF_2$ is claimed to be useful as a retarding agent for the burning of epoxy resins [33] and unsaturated polyester resins [34]. Concerning flammability of epoxy resins in the presence of the title compound and other triphenylantimony halides, see [35]. The compound is claimed as a cocatalyst together with Lewis bases for the polymerization of epoxides [37].

$(4\text{-}FC_6H_4)_3SbF_2$

The compound is prepared in a Friedel-Crafts reaction from SbF_5/HF or SbF_5/AsF_3 with an excess of fluorobenzene when stirred for several hours at room temperature. The mixture is then quenched with aqueous methanol or pyridine. The remaining residue is extracted with CCl_4. After evaporation of the solvent, the product remains. Yield 4.8%; melting point 77 to 78 °C from methanol; ^{1}H NMR in CCl_4: $\delta = 7.6$ (H-3,5), 8.52 (H-2,6) ppm; ^{19}F NMR in CCl_4: $\delta = 105.7$ (FC), 149.8 (FSb) ppm [15].

$(3,4\text{-}F_2C_6H_3)_3SbF_2$

The compound is prepared like the previous one, but with 1,2-difluorobenzene. Oily residue identified by NMR; ^{1}H NMR in CCl_4: $\delta = 7.82$ (H-5), 8.33 (H-2), 8.53 (H-6) ppm; ^{19}F NMR in CCl_4: $\delta = 129.2$, 133.1 (FC), 150.1 (FSb) ppm [15].

$(C_6F_5)_3SbF_2$

The compound can be prepared by reaction of $Sb(C_6F_5)_3$ with elemental fluorine at low temperature in the liquid phase. No details are given [41].

$(4\text{-}ClC_6H_4)_3SbF_2$

$(4\text{-}ClC_6H_4)_3SbCl_2$, dissolved in ethanol, is treated with stoichiometric amounts of KF in aqueous ethanol. After one hour reflux the mixture is concentrated and the compound crystallizes in a yield of 80%. It melts at 115.5 to 116 °C [8]. A similar reaction of the chloride with $KF \cdot 2H_2O$ is mentioned. The product melts at 116 °C recrystallized from heptane. The ^{1}H NMR spectrum (in $CDCl_3$ or $(CD_3)_2CO$) shows resonances at $\delta = 7.54$ (H-3,5) and 8.10 (H-2) ppm [12].

The compound, dissolved in benzene or 1,2-dimethoxyethane, reacts with ethereal Grignard solutions of C_6H_5MgBr, $4\text{-}ClC_6H_4MgBr$, or $4\text{-}CH_3C_6H_4MgBr$ at room temperature to yield the corresponding pentaarylantimony compounds. With $(N(CH_3)_2)_3PO$, a hexacoordinated adduct is formed [12].

References on p. 10

$(4\text{-}BrC_6H_4)_3SbF_2$

The compound is prepared by reacting the corresponding dichloride, dissolved in ethanol, with KF in aqueous ethanol. One hour reflux and concentration of the mixture gives a precipitate in 72% yield with a melting point of 149 to 149.5 °C [8].

$(4\text{-}(CH_3)_2NC_6H_4)_3SbF_2$

The corresponding dibromide and KF are refluxed for two hours in a water/ethanol (1:5) mixture. Subsequent addition of water gives a precipitate which is recrystallized from $CHCl_3$. It is thus obtained as a $CHCl_3$ adduct. This substance, upon treatment in a high vacuum, forms the compound in 72% yield. It decomposes at 210 to 215 °C. ^1H NMR spectrum in $CDCl_3$: $\delta=3.52$ (NCH_3), 7.28 (H-3,5), 8.44 (H-2,6) ppm; J(H-2,3)+J(H-2,5)=9 Hz [38].

$(2\text{-}CH_3C_6H_4)_3SbF_2$

The compound is prepared by reaction of the corresponding dichloride, dissolved in ethanol, with stoichiometric amounts of KF in aqueous ethanol, with one hour of reflux. Concentration of the mixture gives crystals in 76% yield, which melt at 195.5 °C [8]. It is also obtained by treating $(2\text{-}CH_3C_6H_4)_3SbCl_2$ with 1 to 2 N aqueous HF for 10 min and extraction of the mixture with $CHCl_3$. After evaporation of the solvent, 95% of the product remains. The solubility of the compound is 4.5×10^{-2} in C_6H_6, 3×10^{-2} in CCl_4, and 0.15 mol/L in $CHCl_3$. The distribution coefficients between these organic phases and H_2O are 10^3, 6×10^2, and 3×10^3, respectively [11].

$(3\text{-}CH_3C_6H_4)_3SbF_2$

The compound is prepared like the previous one from the corresponding dichloride with HF. Yield 85%, m.p. 109 °C [8].

$(4\text{-}CH_3C_6H_4)_3SbF_2$

The compound can be prepared by the same procedures as described for the ortho compound. The yields range from 75 [8] to 95% [11]. Other methods of preparation are the reaction of $(4\text{-}CH_3C_6H_4)_3SbO$ with 48% aqueous HF in excess acetone and precipitation of the product with cold water [39] or the reaction of the same oxide with SF_4 in CH_2Cl_2 in a quartz tube at −10 °C (86% yield). Fluorination of $Sb(C_6H_4CH_3\text{-}4)_3$ with $C_6H_5IF_2$ in CH_2Cl_2 with 30 min reflux yields 57% of the compound. Fluorination with XeF_2 at −30 °C under similar conditions forms 94% of the title compound [3]. The title compound is also formed in yields between 3 and 8% in a Friedel-Crafts reaction of SbF_5, SbF_5/HF, or SbF_5/AsF_3 with toluene [15].

Melting points of the crystals are given as 92 °C [39], 118 °C [8, 40], 118 to 120 °C from hexane [3], and 119 °C from heptane [12]. The solubility in $CHCl_3$ is 1.5×10^{-2} mol/L, and the distribution coefficient between this solvent and water is 1.5×10^3. Hydrolysis constants in $CHCl_3$ were found to be $\log K_1=7.5$ and $\log K_2=5$ [11]. ^1H NMR spectra in CCl_4 show signals at $\delta=2.82$ (CH_3), 7.72 (H-3,5), and 8.48 (H-2,6) ppm [15]. In $CDCl_3$ the values are $\delta=2.38$ (CH_3); 7.32, 8.02 (C_6H_5) ppm [12, 39] with a coupling constant of J=8 Hz [39]. The ^{19}F NMR resonance of the compound in CCl_4 is $\delta=154.7$ ppm [15].

$(4\text{-}CH_3C_6H_4)_3SbF_2$ crystallizes as a monoclinic solid with a=26.8, b=10.4, c=22.1 Å; $\beta=112°$; Z=12 in the space group C_{2h}^6-C2/c (No. 15) or C_s^4-Cc (No. 9). The experimental density is 1.47 g/cm³ [8, 40].

References on p. 10

$(4\text{-}CH_3C_6H_4)_3SbF_2$, dissolved in benzene, reacts with ethereal Grignard solutions of $4\text{-}CH_3C_6H_4MgBr$ or $4\text{-}ClC_6H_4MgBr$ at room temperature to form the expected pentaaryl-antimony compounds [12].

$(3\text{-}F\text{-}4\text{-}CH_3C_6H_3)_3SbF_2$

The compound may be prepared as an oil in poor yields from 2-fluorotoluene with HF/SbF_5. Workup of the Friedel-Crafts mixture with pyridine and extraction with CCl_4 gives, after evaporation of the solvent, the compound identified by NMR spectroscopy. ^1H NMR in CCl_4: $\delta = 2.81$ (CH_3), 7.60 (H-5), 8.39 (H-2), and 8.45 (H-6) ppm. ^{19}F NMR in acetone: $\delta = 101.3$ (CF) and 152.3 (SbF) ppm [15].

$(2,4,6\text{-}(CH_3)_3C_6H_2)_3SbF_2$

Trimesitylantimony difluoride is obtained by galvanostatic electrolysis at 20 °C under argon atmosphere from trimesitylstibine and $[N(C_2H_5)_4]BF_4$ in a 2:1 mixture of CH_3CN and C_6H_6 as anolyte. At the end of the electrolysis the solvents are removed in a vacuum; the residue is washed with water and recrystallized from ethanol with addition of some CH_2Cl_2. The yield is 96% of crystals which melt at 223 to 225 °C [7].

$(1\text{-}C_{10}H_7)_3SbF_2$

Trinaphthylantimony difluoride is obtained from the corresponding dichloride by reaction with KF in aqueous ethanol. Refluxing for one hour gives a yield of 90% [8]. Similarly, one obtains the compound by extracting a mixture of $(C_{10}H_7)_3SbCl_2$ and aqueous HF with $CHCl_3$. After evaporation of the solvent, 95% of the compound remains [11]. It melts at 279 to 280 °C [8]. The solubility in mol/L is 10^{-2} in C_6H_6, 3.5×10^{-3} in CCl_4, and 6.5×10^{-2} in $CHCl_3$ [11].

References:

[1] M. Cartwright, A. A. Woolf (J. Fluorine Chem. **19** [1981] 101/22; C.A. **96** [1982] No. 41633).
[2] I. Ruppert, V. Bastian (Angew. Chem. **90** [1978] 226/7).
[3] V. I. Popov, N. V. Kondratenko (Zh. Obshch. Khim. **46** [1976] 2597/601; J. Gen. Chem. [USSR] **46** [1976] 2477/80).
[4] L. M. Yagupol'skii, V. I. Popov, N. V. Kondratenko, B. L. Korsunskii, N. N. Aleinikov (Zh. Org. Khim. **11** [1975] 459/60; J. Org. Chem [USSR] **11** [1975] 454/5).
[5] I. Ruppert (Chem. Ber. **112** [1979] 3023/30).
[6] R. E. Banks, R. N. Haszeldine, R. Hatton (Tetrahedron Letters **1967** 3993/6).
[7] E. V. Nikitin, A. A. Kazakova, O. V. Parakin, Yu. M. Kargin (Zh. Obshch. Khim. **52** [1982] 2027/9; J. Gen. Chem. [USSR] **52** [1982] 1802/4).
[8] V. P. Glushkova, T. V. Talalaeva, Z. P. Razmanova, G. S. Zhdanov, K. A. Kocheshkov (Sb. Statei Obshch. Khim. **2** [1953] 992/6).
[9] G. Grossmann, A. Winzer (Isotopen Tech. **2** [1962] 193/8; C.A. **58** [1963] 2051).
[10] G. Doleschall, N. A. Nesmeyanov, O. A. Reutov (J. Organometal. Chem. **30** [1971] 369/75).
[11] M. Benmalek, H. Chermette, C. Martelet, D. Sandino, J. Tousset (J. Organometal. Chem. **67** [1974] 53/9).
[12] B. Raynier, B. Waegell, R. Commandeur, H. Mathais (Nouv. J. Chim. **3** [1979] 393/401).
[13] G. O. Doak, G.G. Long, L.D. Freedman (J. Organometal. Chem. **4** [1965] 82/91).
[14] L. M. Yagupol'skii, N. V. Kondratenko, V. I. Popov (Zh. Obshch. Khim. **46** [1976] 620/3; J. Gen Chem. [USSR] **46** [1976] 618/21).

[15] G. A. Olah, P. Schilling, I. M. Gross (J. Am. Chem. Soc. **96** [1974] 876/83).
[16] R. G. Goel, E. Maslowsky Jr., C. V. Senoff (Inorg. Chem. **10** [1971] 2572/7).
[17] R. G. Goel, E. Maslowsky Jr., C. V. Senoff (Inorg. Nucl. Chem. Letters **6** [1970] 833/5).
[18] B. A. Nevett, A. Perry (Spectrochim. Acta A **33** [1977] 755/60).
[19] J. Havranek, A. Lycka (Sb. Ved. Pr. Vys. Sk. Chemickotechnol. Pardubice **43** [1980] 123/7; C.A. **95** [1981] No. 212464).
[20] E. L. Muetterties, W. Mahler, K. J. Packer, R. Schmutzler (Inorg. Chem. **3** [1964] 1298/1303).

[21] T. B. Brill, G. G. Long (Inorg. Chem. 9 [1970] 1980/5).
[22] G. G. Long, J. G. Stevens, R. J. Tullbane, L. H. Bowen (J. Am. Chem. Soc. **92** [1970] 4230/5).
[23] J. G. Stevens, S. L. Ruby (Phys. Letters A **32** [1970] 91/2).
[24] S. E. Gukasyan, V. P. Gor'kov, P. N. Zaikin, V. S. Shpinel (Zh. Strukt. Khim. **14** [1973] 650/5; J. Struct. Chem. [USSR] **14** [1973] 603/7).
[25] G. M. Bancroft, V. G. K. Das, T. K. Sham, M. G. Clark (J. Chem. Soc. Dalton Trans. **1976** 643/54).
[26] J. N. R. Ruddick, J. R. Sams, J. C. Scott (Inorg. Chem. **13** [1974] 1503/7).
[27] L. H. Bowen, G. G. Long (Inorg. Chem. **15** [1976] 1039/44).
[28] S. E. Gukasyan, V. S. Shpinel (Phys. Status Solidi **29** [1968] 49/52).
[29] L. M. Kataeva, Yu. V. Rydvanskii, N. I. Trofimova (Zh. Fiz. Khim. **50** [1976] 814/5; Russ. J. Phys. Chem. **50** [1976] 486/7).
[30] C. G. Moreland, M. H. O'Brien, C. E. Douthit, G. G. Long (Inorg. Chem. **7** [1968] 834/6).

[31] C. G. Moreland, G. G. Long (Inorg. Nucl. Chem. Letters **8** [1972] 347/51).
[32] V. D. Nefedov, L. N. Evtikheev (Zh. Fiz. Khim. **30** [1956] 2090/2; C.A. **1957** 7182).
[33] J. Havranek, J. Mleziva (Angew. Makromol. Chem. **84** [1980] 105/17).
[34] J. Havranek, J. Muller, J. Mleziva (Sb. Ved. Pr. Vys. Sk. Chemickotechnol. Pardubice **42** [1980] 123/32; C.A. **94** [1981] No. 209645).
[35] J. Havranek (Sb. Dokl. 1st Nats. Konf. Mladite Nauchni Rab. Spets. Neft Khim., Burgas, Bulg., 1976 [1977], pp. 152/9; C.A. **93** [1980] No. 187219).
[36] A. Ninagawa, H. Matsuda, R. Nomura (Kenkyu Hokoku-Asahi Garasu Kogyo Gijutsu Shoreikai **40** [1982] 141/7; C.A. **99** [1982] No. 88609).
[38] J. M. Keck, G. Klar (Z. Naturforsch. **27b** [1972] 596/9).
[39] G. L. Kuykendall, J. L. Mills (J. Organometal. Chem. **118** [1976] 123/8).
[40] G. S. Zhdanov, Z. P. Razmanova (Dokl. Akad. Nauk SSSR **72** [1950] 1055/7).

[41] R. Kasemann, G. Klein, D. Naumann (J. Fluorine Chem. **29** [1985] 99).
[42] N. I. Trofimova, V. E. Zhuravlev, E. N. Sinotova, N. E. Shchepina, M. V. Moshkovskaya (Tr. Estestvennonauch. Inst. Perm. Univ. **13** [1975] 187/93; C.A. **86** [1977] No. 88696).

2.5.1.1.2 Triorganoantimony Dichlorides

2.5.1.1.2.1 R_3SbCl_2 Compounds with R = Alkyl and Alkenyl

2.5.1.1.2.1.1 Trimethylantimony Dichloride $(CH_3)_3SbCl_2$

Preparation. Trimethylantimony dichloride was first prepared by H. Landolt in 1861. He reacted $Sb(CH_3)_3$ with Cl_2 in the absence of solvent, or in CS_2 solution, or $(CH_3)_3SbO$ with aqueous HCl to form small hexagonal crystals which were recrystallized from water or ethanol [1]. Similarly the reaction of trimethylstibine with chlorine may be performed in ether [2 to 4] or in ether/CCl_4 mixtures [5]. The yields of the compound are between 60 and 85% [4, 5]. The reaction of $Sb(CH_3)_3$ with PCl_3, PCl_5, $SbCl_3$, or $SbCl_5$ without solvent at 0 °C in a vacuum system leads quantitatively to the title compound with reduction of the phosphorus and antimony chlorides to the elements [6]. A yield of 16% of the compound besides 71% $[(CH_3)_3SbCl]NH$ is obtained by reacting $Sb(CH_3)_3$ with NH_2Cl in ether [7]. $(CH_3)_3Sb(OH)_2$ and PCl_5 form the substance upon warming [8]. $(CH_3)_3SbS$ and organotin-halides, for example $(C_6H_5)_3SnCl$, react in acetone/methanol to form 88% $(CH_3)_3SbCl_2$ and $[(C_6H_5)_3Sn]_2S$ [9]. With $InCl_3$, the same educt reacts in ether and, after six hours reflux, forms 25% of the compound besides InSCl and In_2S_3 [10]. $(CH_3)_3Sb(OSi(CH_3)_3)_2$ and $SOCl_2$ yield $(CH_3)_3SiCl$, SO_2, and $(CH_3)_3SbCl_2$ [11, 55]. $(CH_3)_3Sb(OSi(CH_3)_3)_2$ and HCl yield $(CH_3)_3SbCl_2$ and $(CH_3)_3SiOSi(CH_3)_3$ [55]. From $(CH_3)_3Sb(ON(CF_3)_2)_2$ and HCl, the title compound and $(CF_3)_2NOH$ are obtained quantitatively [12, 81]. In a metathetic reaction of $(CH_3)_3SbX$ ($X = 2\text{-}OC_6H_4CH{=}NC_6H_4O\text{-}2$) with $(CH_3)_2SbCl_3$ or $(C_6H_5)_2SbCl_3$ the title compound is formed besides $(CH_3)_2SbClX$ or $(C_6H_5)_2SbClX$ [13, 82]. A reaction of $(CH_3)_3Sb(O_2CCCl_3)_2$ with $P(C_6H_5)_3$ and cyclopentadiene leads to 79% of the compound, 92% $(C_6H_5)_3PO$, and 35% 7,7-dichlorobicyclo [3.2.0]hept-2-en-6-one [44].

Properties. Trimethylantimony dichloride melts at 228.2 to 229 °C after recrystallization from acetone [6]. It decomposes at 230 °C [8]. IR spectra are described and discussed in several publications [5, 17 to 20]. Complete IR and Raman data with assignments are shown in Table 6 [21]. Normal coordinate analyses were made using a Urey-Bradley force field [17, 22], general valence force field [19, 20, 22], orbital valence force field [22], and Redington's approach [23] assuming in all cases a D_{3h} symmetry of the molecule. For the resulting force field constants, calculated frequencies, and discussions concerning the structure of the molecule see [17 to 24]. Correlations between fundamental frequencies and atomic mass, electronegativity of the halogen, and moment of inertia of the molecule are established in [25]. In addition to general force field constants, mean amplitudes of vibration, generalized mean square amplitudes, shrinkage constants, Coriolis coupling coefficients, and centrifugal distortion constants were evaluated in [26].

Table 6
Complete IR and Raman Vibrations (in cm^{-1}) of Solid $(CH_3)_3SbCl_2$ [21].

IR	Raman	assignment	IR	Raman	assignment
3000 m	3024 m, br	$\nu_{as}CH$		538 vvs	ν_1, ν_sSbC
2921 w	2930 vs	ν_sCH	540 vw		
2858 vw			530 vw		
2775 vw	2790 vw		452 w	444 vvw	
2416 w	2424 vw			385 vvw	$\nu_5-\nu_2$
2354 mw	2357 vvw		333 vvw	342 vvw	
2333 mw	2340 vw				
1784 w			302 vw	307 vvw	$2\nu_6$
1740 w			282 vs, br		ν_3, $\nu_{as}SbCl$
1624 mw	1625 vvw			272 vvs	ν_2, ν_sSbCl
1401 ms	1401 vvw	$\delta_{as}CH$	236 vvw		
	1240 m	δ_sCH	229 vvw	228 vvw	
1212 w	1213 mw			208 m	ν_8, $\varrho(CH_3)_3SbCl_2$
1096 vw		$\nu_1+\nu_5$	188 vs		ν_4, δSbC out-of-plane
1018 vvw			158 ms	173 s	ν_6, $\delta SbCl$
877 vs		ϱCH_3	130 vs	108 s	ν_7, δSbC in-plane
575 vs	579 s	ν_5, $\nu_{as}SbC$		52 s	lattice vibration

References on p. 18

A ^1H NMR spectrum recorded in $CDCl_3$ solution at 20 °C gives a chemical shift of δ = 2.3 ppm [13]. In $CHCl_3$ at 30 °C the value is δ = 2.28 ppm, and at −65 °C δ = 2.36 ppm [27]. In $CDCl_3$ the following values were found: δ = 2.38 at −32 °C, 2.34 at 30.5 °C, and 2.31 ppm at 70 °C [28]. For ^1H NMR spectra of the compound in mixtures with the other trimethylantimony dihalides see [28]. A correlation of the ^1H NMR chemical shift with the inverse ionization potential is made in [29].

NQR spectra are measured and discussed on basis of a D_{3h} symmetry of the molecule [30, 31]. At 300 K the following resonances (in MHz) were found: $\nu(^{121}Sb)$: 5/2 ⇌ 3/2 at 198.61 (70), 3/2 ⇌ 1/2 at 99.327 (70); $\nu(^{123}Sb)$: 7/2 ⇌ 5/2 at 180.95 (10), 5/2 ⇌ 3/2 at 120.63 (15), 3/2 ⇌ 1/2 at 60.313 (60); e^2Qq/h = 662.18 for ^{121}Sb (see also [83]), 884.43 MHz for ^{123}Sb; η = 0.00; $\nu(^{35}Cl)$: 14.045 (20), $\nu(^{37}Cl)$: 11.065 (4) [31]; e^2Qq/h = 28.0918 (20) MHz for Cl [83].

Mössbauer resonances for ^{121}Sb were recorded at 80 K vs. $^{121}SnO_2$ leading to $\delta = -5.7 \pm 0.2$, $e^2qQ = -24 \pm 1$, and line width $\Gamma = 3.0 \pm 0.2$ mm/s [32]. At 4 K vs. $Ca^{121}SnO_3$, the values are: $\delta = -6.11 \pm 0.01$, $e^2qQ = -24 \pm 0.2$, and $\Gamma = 2.74 \pm 0.02$ mm/s [32, 33]. Hartee-Fock-Slater LCAO calculations of these parameters were performed [34] in connection with the additive electric field gradient model [35]. For the calculation of the e^2qQ value see [36], and for the correlation of this value with that of $[(CH_3)_3SnCl_2]^-$ see [37].

A photoelectron (PE) spectrum of the compound in the gas phase was measured and assigned on the basis of a qualitative D_{3h} model and by comparison with the PE data of $Sb(CH_3)_3$ and $SbCl_3$. The maxima are at 9.8 sh, 10.22, 10.65, 12.5, and 14.65 eV [38].

An X-ray powder pattern is shown in [41], see also [65]. An X-ray structure determination of the compound was published in 1938. It crystallizes in a hexagonal space group $P\bar{6}2c - D^4_{3h}$ (No. 190) with two molecules per unit cell. Cell dimensions are a = 7.27 and c = 8.44 Å. The structure was determined as a trigonal bipyramid with the chlorines in axial positions at a distance of 2.49 Å from the Sb atom [42].

Mass spectra of the compound were measured [39, 40]. No molecular ion could be found. The highest fragment is $[(CH_3)_2SbCl_2]^+$ with a relative intensity of 100. Further fragments were found at m/e = 201 (60), 186 (60), 171 (40), 208 (55), 166 (5), and 151 (20). A fragmentation scheme is given in [40].

For conductivity measurements in CH_3CN and in H_2O see [14]. Polarographic half wave potentials at 25 °C vs. saturated calomel were detected in 1 N HCl as −0.78 V, at pH = 1 as −0.76 V, in 1 N NH_4OH as −1.24 V [15]. Polarograms of H_2O/C_2H_5OH solutions are published in [16].

Reactions and Uses. The reactions of trimethylantimony dichloride are summarized in Tables 7 and 8. About the fungitoxicity of the compound see [43].

$(CH_3)_3SbCl_2 \cdot SbCl_3$

Equimolar amounts of $(CH_3)_3SbCl_2$ and $SbCl_3$ are dissolved in boiling CH_2Cl_2. Upon cooling the solution, the compound crystallizes in a yield of 62%. For the X-ray structure determination, crystals were obtained by mixing cold saturated solutions of $(CH_3)_3SbCl_2$ and $SbCl_3$ in CH_2Cl_2 and slowly evaporating the solvent at room temperature in an inert atmosphere [71].

References on p. 18

Table 7
Reactions of $(CH_3)_3SbCl_2$.

reactants	reaction conditions	products (yield in %)	Ref.
—	heating to 220 °C at 760 Torr	$Sb(CH_3)_3$ (15), $(CH_3)_2SbCl$ (75)	[40]
—	heating under CO_2 at 600 Torr	$(CH_3)_2SbCl$, CH_3Cl	[45, 46]
Na	ratio 1:7 in liquid NH_3, addition of $(ClCH_2)_2$	$[(CH_3)_2Sb]_2$ (79)	[47]
Na	ratio 1:4 in liquid NH_3, addition of $Cl(CH_2)_nCl$ (n = 3 to 6)	$[(CH_3)_2Sb]_2(CH_2)_n$ (100)	[48]
H_2O	in hot H_2O treated with ion exchange material M 500 [84] or Amberlite IR 4B [5]	$(CH_3)_3Sb(OH)_2$ (85 [5])	[5, 84]
$HN_3 + NaN_3$	in C_6H_6, 12 h at 20 °C	$(CH_3)_3Sb(N_3)_2$	[49]
NaN_3	in C_6H_6	$(CH_3)_3Sb(N_3)_2$	[50]
NaN_3	ratio 1:4 in $(ClCH_2)_2$, 24 h at 20 °C	$(CH_3)_3Sb(N_3)_2$ (87)	[51]
KSCN	in C_2H_5OH	$(CH_3)_3Sb(SCN)_2$	[1]
NaSCN or AgSCN	in CH_3OH or $(CH_3)_2O$	$(CH_3)_3Sb(SCN)_2$	[50]
AgNCO	in ether, 24 h at 20 °C	$(CH_3)_3Sb(NCO)_2$	[49, 50]
$SbCl_5$	in CH_2Cl_2 at 20 °C	$[(CH_3)_3SbCl]SbCl_6$	[52]
$AgClO_4$ + D, D = $(CH_3)_2SO$, $(C_6H_5)_2SO$, $(C_6H_5)_3PO$, $(C_6H_5)_3AsO$	ratio 1:2:2 in C_6H_6, 2 h	$[(CH_3)_3SbD_2](ClO_4)_2$	[53]
$NaOSi(CH_3)_3$	ratio 1:1 in ether, 2 h at 25 °C ratio 1:2 in ether, 2 h at 35 °C	$(CH_3)_3SiOSb(CH_3)_3Cl$ (91) $[(CH_3)_3SiO]_2Sb(CH_3)_3$ (70)	[55]
$NaOC_2H_5$	ratio 1:1 in C_6H_6 or $CHCl_3$	$(CH_3)_3Sb(Cl)OC_2H_5$	[54]
CH_3OH	in $CH_3OH + NH_3$ gas	$(CH_3)_3Sb(OCH_3)_2$	[54]
$NaOC_4H_9$-t	in ether, 2 h at 35 °C	$(CH_3)_3Sb(OC_4H_9\text{-t})_2$ (76)	[55]
$HOOC_4H_9$-t	in $CHCl_3/C_6H_6 + N(CH_3)_3$	$(CH_3)_3Sb(OOC_4H_9\text{-t})_2$ (91)	[56]
HOOR, R = t-C_4H_9, 3,4-dihydro-1H-benzoisopyranyl, $(CH_3)_2C(C_6H_5)$	in C_6H_6	$(CH_3)_3Sb(OOR)_2$	[70]

References on p. 18

HON=CRR′, R, R′ = $(CH_2)_5$; R = CH_3, R′ = C_6H_5	in $C_6H_6 + N(C_2H_5)_3$, 3 h reflux	$(CH_3)_3Sb(ON{=}CRR')_2$	[57]
Na-8-quinolinolate	ratio 1:1 in C_2H_5OH	$(CH_3)_3SbCl(OC_9H_7N)$ (54)	[58]
2-methyl-8-hydroxyquinoline	in $C_6H_6 + NaOCH_3/CH_3OH$, 30 min reflux	$(CH_3)_3SbCl(OC_{10}H_9N)$	[59]
Na-acetylacetonate	in CH_3OH ratio 1:1	$[(CH_3)_3SbCl]_2O$	[60]
2-$NaOC_6H_4CN{=}NC_6H_4ONa$-2′	in CH_3OH	$(CH_3)_3Sb(-OC_6H_4CH{=}NC_6H_4O-)$ (65)	[61]
HOC_6H_4X, X = H, 4-Cl, 4-Br, 2-NO_2, 3-NO_2, 2-CH_3, 2-C_6H_5, 4-C_6H_5	ratio 1:2 in CH_3OH 30 min, $+N(C_2H_5)_3$	$(CH_3)_3Sb(OC_6H_4X)_2$	[62]
$LiC{\equiv}CCH_3$	ratio 1:2 in THF, slow addition at −10 °C	$(CH_3)_3Sb(C{\equiv}CCH_3)_2$ (60)	[66]
$Mg(C_2H_5)_2$	ratio 1:1 in ether, 3 h reflux	$(CH_3)_4Sb(C_2H_5)$ (15) $(CH_3)_3Sb(C_2H_5)_2$ (19) $(CH_3)_2Sb(C_2H_5)_3$ (18)	[67]
CH_3MCl_2 or $(CH_3)_2MCl$ or $(CH_3)_3M$ (M = Ga, In)	in CH_2Cl_2, 1 h at 20 °C	$[(CH_3)_4Sb][(MCl_4)]$, M = Ga $[(CH_3)_4Sb][(M(CH_3)Cl_3)]$, M = In, Ga $[(CH_3)_4Sb][(M(CH_3)_2Cl_2)]$, M = In, Ga	[68] [68, 69] [68]
$In(CH_3)_3$	in excess, in CH_2Cl_2, 1 h at 20 °C	$[(CH_3)_4Sb][(In(CH_3)_3)_2Cl_2In(CH_3)_2]$	[68]
$Ga(C_2H_5)_3$	in CH_2Cl_2 or C_6H_6	$[(CH_3)_3SbC_2H_5][(C_2H_5)_2GaCl_2]$	[66]
$(C_2H_5)GaCl_2$	in CH_2Cl_2 or C_6H_6	$[(CH_3)_3SbC_2H_5]GaCl_4$	[66]
$SbCl_3$	ratio 1:1 in boiling CH_2Cl_2	$(CH_3)_3SbCl_2 \cdot SbCl_3$ (62)	[71]
$Hg(CH_3)_2 + SbCl_3$		$(CH_3)_3SbCl_2 \cdot 3Hg(CH_3)Cl$	[72, 74]
$(CH_3)_3SbS + [(CH_3)_2SnS]_3$	ratio 3:3:1 in $CHCl_3$ at 20 °C	$(CH_3)_2SnCl_2 \cdot (CH_3)_3SbS$ (98)	[9, 73]
$(CH_3)_3Sb(SCOR)_2$, R = CH_3, C_6H_5	ratio 1:1 in $CHCl_3$, 30 min at room temperature	$(CH_3)_3SbCl(SCOR)$ (100)	[65]
RSH, R = CH_3, C_2H_5, $CH_2C_6H_5$, C_6H_5	with excess $N(C_2H_5)_3$ in $(CH_3)_2CO$, 6 h at −30 to −25 °C	$(CH_3)_3Sb(SR)_2$ (74 to 79)	[63]
2,6-diamino-8-purinol	$(CH_3)_3SbCl_2$ in CCl_4 is added to the diamine in water/NaOH at 25 °C, reaction time 30 s	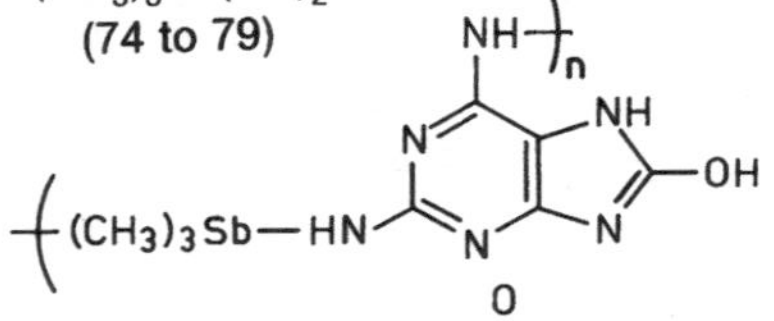 (93)	[64]
4,4′-diaminodiphenylsulfone	as above	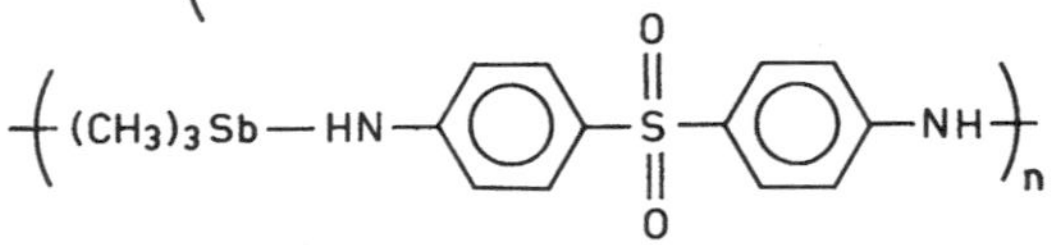	[64]

References on p. 18

Table 8
Exchange and Equilibrium Reactions of $(CH_3)_3SbCl_2$.
(k = rate constants in $L \cdot mol^{-1} \cdot cm^{-1}$, K = equilibrium constants)

reaction; conditions; method of measurement	results	Ref.
exchange reaction with $^{124}Sb(CH_3)_3$ to give $(CH_3)_3{}^{124}SbCl_2$ and $Sb(CH_3)_3$; mass spectroscopy	k = 0.05 at 0 °C k = 0.3 at 10 °C very fast at 40 to 60 °C E_a ca. 5 kcal/mol	[75]
exchange reaction of $(CD_3)_3SbCl_2$ with $Sb(CH_3)_3$ in C_2H_5OH; mass spectroscopy	dominant mechanism for exchange of Sb is electron transfer	[75]
exchange reaction with $^{122/124}SbCl_5$ in 6 N HCl; probably mass spectroscopy	no exchange observed	[76]
exchange reaction with $^{122/124}SbCl_5$ in $CHCl_3$; probably mass spectroscopy	no exchange observed	[76]
$(CH_3)_3SbCl_2 + (CH_3)_3SbF_2 \rightleftharpoons 2(CH_3)_3SbClF$; in $CHCl_3$; ^1H NMR	K = 3.90 ± 0.15 at 35 °C	[78], also [77, 79]
	K = 4.20 ± 0.10 at 60 °C	[78]
$(CH_3)_3SbCl_2 + (CH_3)_3SbBr_2 \rightleftharpoons 2(CH_3)_3SbClBr$; in $CHCl_3$; ^1H NMR	K = 3.28 ± 0.20 at 0 °C	[78]
	K = 3.50 ± 0.15 at 35 °C	[78], also [77,79]
	K = 3.65 ± 0.13 at 60 °C	[78]
$(CH_3)_3SbCl_2 + (CH_3)_3SbI_2 \rightleftharpoons 2(CH)_3SbClI$; in $CHCl_3$; ^1H NMR	K = 2.05 ± 0.07 at 0 °C	[78]
	K = 2.3 ± 0.1 at 32 °C	[77]
	K = 2.25 ± 0.01 at 35 °C	[78]
$(CH_3)_3SbCl_2 + (C_6H_5CH_2)_3SbF_2 \rightleftharpoons (CH_3)_3SbF_2 + (C_6H_5CH_2)_3SbCl_2$; in $CHCl_3$; ^1H NMR	K = 38 ± 6 at 32 °C	[77, 78]
$(CH_3)_3SbCl_2 + (CH_3)_3Sb(NO_3)_2 \rightleftharpoons 2(CH_3)_3Sb(Cl)NO_3$; in $C_6H_5NO_2$; ^1H NMR	K ca. 20 at 32 °C; exchange is not directly observable, overlapped by the following reaction	[79]
$(CH_3)_3SbCl_2 + (CH_3)_3Sb(Cl)NO_3 \overset{k}{\underset{k}{\rightleftharpoons}} (CH_3)_3Sb(Cl)NO_3 + (CH_3)_3SbCl_2$; in $C_6H_5NO_2$; ^1H NMR	k = 46 ± 3 at 20 °C k = 114 ± 9 at 32 °C k = 246 ± 32 at 45 °C k = 314 ± 41 at 55 °C k = 399 ± 51 at 60 °C	[79]
in CH_2Cl_2	k = 60 ± 10 at 64 °C $\Delta H^{\neq} = 10.6 \pm 2.1$ kcal/mol $\Delta S^{\neq} = -16 \pm 6$ cal · $mol^{-1} \cdot K^{-1}$	
$(CH_3)_3SbCl_2 + [(CH_3)_3Sn]_2S \rightleftharpoons 2(CH_3)_3SnCl + (CH_3)_3SbS$; in $CHCl_3$ or CH_2Cl_2; ^1H NMR	K = 7.14 at 20 °C	[80]

see [79] for further exchange reactions involving $(CH_3)_3SbCl_2$

References on p. 18

The Raman spectrum in the solid and in CH_2Cl_2 solution was measured, see Table 9. From this, it is shown that the compound dissociates quantitatively in solution into its two components [71].

Table 9
Raman Vibrations (in cm^{-1}) of $(CH_3)_3SbCl_2 \cdot SbCl_3$ [71].

solid (intensity)	in CH_2Cl_2	assignment
575 (220)		$\nu_{as}SbC_3$
580 (180)	577 (150) dp	$\nu_{as}SbC_3$
527 (1000)	526 (1000) p	ν_sSbC_3
353 (1000)	368 (890) p	ν_sSbCl_3
324 (560)	343 (307) dp	$\nu_{as}SbCl_3$
311 (476)		
280 (sh)		$\nu_{as}SbCl_2$
268 (238)	268 (sh) p	ν_sSbCl_2
240 (155)		$\nu Sb\cdots Cl$
223 (143)		
211 (95)	212 (46) dp	ϱSbC_3
173 (274)	178 (150) dp	δSbC_3
155 (24)	166 (160) dp	δ_sSbCl_3
135 (262)	131 (150) dp	$\delta_{as}SbCl_3$
118 (83)		$\delta SbCl_2$

The compound crystallizes as a monoclinic solid in the space group $P2_1/n\text{-}C_{2h}^5$ (No. 14) with a=803.7, b=1632.5, c=975.9 pm, and β=97.46°; d_m=2.40 and d_c=2.437 g/cm³; Z=4 molecules per unit cell. The structure was solved with 2698 observed reflections and was refined to R=0.055 with anisotropic temperature factors for all atoms. In the crystal, the Sb atom of the $(CH_3)_3SbCl_2$ molecule has a trigonal bipyramidal environment while the Sb^{III} atom is in the center of an octahedron. A projection to the xz plane is shown in **Fig. 2**, p. 18, together with the main distances and angles [71].

References on p. 18

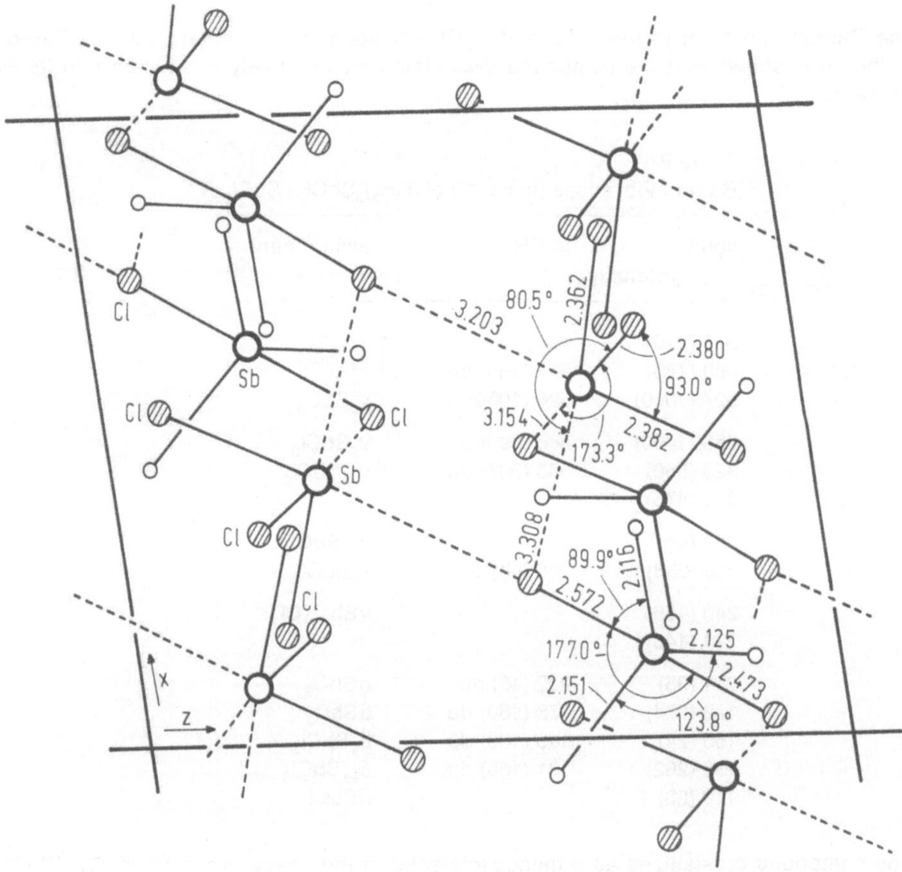

Fig. 2. Projection of the elementary cell of $(CH_3)_3SbCl_2 \cdot SbCl_3$ on the xz plane [71].

References:

[1] H. Landolt (J. Prakt. Chem. **84** [1861] 328/39).
[2] G. T. Morgan, G. R. Davies (Proc. Roy. Soc. [London] A **110** [1926] 523/34).
[3] G. T. Morgan, G. R. Davies (Nature **116** [1925] 499).
[4] G. O. Doak, G. G. Long, M. E. Key (Inorg. Syn. **9** [1967] 92/7).
[5] G. G. Long, G. O. Doak, L. D. Freedman (J. Am. Chem. Soc. **86** [1964] 209/13).
[6] R. R. Holmes, E. F. Bertau (J. Am. Chem. Soc. **80** [1958] 2983/5).
[7] R. L. McKenney, H. H. Sisler (Inorg. Chem. **6** [1967] 1178/82).
[8] G. T. Morgan, V. E. Yarsley (Proc. Roy. Soc. [London] A **110** [1926] 534/7).
[9] M. Shindo, Y. Matsumura, R. Okawara (J. Organometal. Chem. **11** [1968] 299/305).
[10] T. Maeda, G. Yoshida, R. Okawara (J. Organometal. Chem. **44** [1972] 237/41).

[11] H. Schmidbaur, M. Schmidt (Angew. Chem. **73** [1961] 655).
[12] H. G. Ang, W. S. Lien (J. Fluorine Chem. **3** [1973] 235/6).
[13] F. Di Bianca, H. A. Meinema, J. G. Noltes, N. Bertazz, G. C. Stocco, E. Rivarola, R. Barbieri (Atti Accad. Sci. Lettere Arti Palermo I [4] **33** [1973/74] 173/86).
[14] T. M. Lowry, J. H. Simons (Chem. Ber. B **63** [1930] 1595/602).
[15] M. K. Saikina (Uch. Zap. Kaz. Gos. Univ. Khim. **116** [1956] 129/86; C.A. **1957** 7191).

[16] V. F. Toropova, M. K. Saikina (Sb. Statei Obshch. Khim. Akad. Nauk SSSR **1** [1953] 210/5).
[17] R. G. Goel, E. Maslowsky Jr., C. V. Senoff (Inorg. Chem. **10** [1971] 2572/7).
[18] R. G. Goel, E. Maslowsky Jr., C. V. Senoff (Inorg. Nucl. Chem. Letters **6** [1970] 833/5).
[19] C. Woods, G. G. Long (J. Mol. Spectrosc. **38** [1971] 387/95).
[20] C. Woods (Diss. North Carolina State Univ. 1971, pp. 1/100; Diss. Abstr. Intern. B **32** [1972] 3838).

[21] B. A. Nevett, A. Perry (Spectrochim. Acta A **33** [1977] 755/60).
[22] B. A. Nevett, A. Perry (J. Mol. Spectrosc. **66** [1977] 331).
[23] R. Namasivayam, S. Viswanathan (Bull. Soc. Chim. Belges **87** [1978] 733/6).
[24] L. Verdonck, G. P. Van der Kelen (Spectrochim. Acta A **31** [1975] 1707/11).
[25] B. A. Nevett, A. Perry (J. Organometal. Chem. **71** [1974] 399/402).
[26] A. Natarajan, K. Chockalingam (Indian J. Pure Appl. Phys. **19** [1981] 672/5).
[27] W. A. Kustes, C. G. Moreland, G. G. Long (Inorg. Nucl. Chem. Letters **8** [1972] 695/9).
[28] G. G. Long, C. G. Moreland, G. O. Doak, M. Miller (Inorg. Chem. **5** [1966] 1358/61).
[29] T. Schaefer, F. Hruska, H. M. Hutton (Can. J. Chem. **45** [1967] 3143/51).
[30] T. B. Brill, Z. Z. Hugus Jr. (J. Chem. Phys. **53** [1970] 1291/2).

[31] T. B. Brill, G. G. Long (Inorg. Chem. **9** [1970] 1980/5).
[32] G. G. Long, J. G. Stevens, R. J. Tullbane, L. H. Bowen (J. Am. Chem. Soc. **92** [1970] 4230/5).
[33] J. G. Stevens, S. L. Ruby (Phys. Letters A **32** [1970] 91/2).
[34] W. Ravenek, J. W. M. Jacobs, A. Van der Avoird (Chem. Phys. **78** [1983] 391/404).
[35] J. N. R. Ruddick, J. R. Sams, J. C. Scott (Inorg. Chem. **13** [1974] 1503/7).
[36] G. M. Bancroft, V. G. Das, T. K. Sham, M. G. Clark (J. Chem. Soc. Dalton Trans. **1976** 643/54).
[37] G. M. Bancroft, V. G. K. Das, K. D. Butler (J. Chem. Soc. Dalton Trans. **1974** 2355/8).
[38] S. Elbel, H. Tom Dieck (Z. Anorg. Allgem. Chem. **483** [1981] 33/43).
[39] H. Preiss (Z. Anorg. Allgem. Chem. **389** [1972] 280/92).
[40] H. J. Breunig, W. Kanig (Phosphorus Sulfur **12** [1982] 149/59).

[41] J. Otera, R. Okawara (J. Organometal. Chem. **17** [1969] 353/7).
[42] A. F. Wells (Z. Krist. **99** [1938] 367/77).
[43] J. Seifter (J. Am. Chem. Soc. **61** [1939] 530/1).
[44] T. Okada, R. Okawara (J. Organometal. Chem. **42** [1972] 117/21).
[45] G. T. Morgan, G. R. Davies (Proc. Roy. Soc. [London] A **110** [1926] 523/34).
[46] G. T. Morgan, G. R. Davies (Nature **116** [1925] 499).
[47] H. A. Meinema, H. F. Martens,J. G. Noltes (J. Organometal. Chem. **51** [1973] 223/30).
[48] H. A. Meinema, H. F. Martens, J. G. Noltes (J. Organometal. Chem. **110** [1976] 183/93).
[49] R. G. Goel, D. R. Ridley (Inorg. Chem. **13** [1974] 1252/5).
[50] R. G. Goel, D. R. Ridley (Inorg. Nucl. Chem. Letters **7** [1971] 21/3).

[51] A. Schmidt (Chem. Ber. **101** [1968] 3976/80).
[52] A. Schmidt (Chem. Ber. **102** [1969] 380/1).
[53] R. G. Goel,H. S. Prasad (J. Organometal. Chem. **59** [1973] 253/7).
[54] J. Dahlmann, A. Rieche (Chem. Ber. **100** [1967] 1544/9).
[55] H. Schmidbaur, H. S. Arnold, E. Beinhofer (Chem. Ber. **97** [1964] 449/58).
[56] A. Rieche, J. Dahlmann, D. List (Liebigs Ann. Chem. **678** [1964] 167/82).
[57] K. Bajpai, R. C. Srivastava (Syn. Reactiv. Inorg. Metal-Org. Chem. **11** [1981] 7/13).
[58] H. A. Meinema, E. Rivarola, J. G. Noltes (J. Organometal. Chem. **17** [1969] 71/81).
[59] Y. Kawasaki, K. Hashimoto (J. Organometal. Chem. **99** [1975] 107/14).
[60] H. A. Meinema, A. Mackor, J. G. Noltes (J. Organometal. Chem. **37** [1972] 285/95).

[61] F. Di Bianca, E. Rivarola, A. L. Spek, H. A. Meinema, J. G. Noltes (J. Organometal. Chem. **63** [1973] 293/300).
[62] A. Ouchi, M. Nakatani, Y. Takahashi, S. Kitazima, T. Sugihara, M. Matsumoto, T. Uehiro, K. Kitano, K. Kawashima, H. Honda (Sci. Papers Coll. Gen. Educ. Univ. Tokyo **25** [1975] 73/99).
[63] H. Schmidbaur, K. H. Mitschke (Chem. Ber. **104** [1971] 1842/6).
[64] C. E. Carraher Jr., M. D. Naas, D. J. Giron, D. R. Cerutis (J. Macromol. Sci. A **19** [1983] 1101/20).
[65] J. Otera, R. Okawara (J. Organometal. Chem. **17** [1969] 353/7).
[66] N. Tempel, W. Schwarz, J. Weidlein (J. Organometal. Chem. **154** [1978] 21/32).
[67] H. A. Meinema, J. G. Noltes (J. Organometal. Chem. **22** [1970] 653/7).
[68] H. J. Widler, W. Schwarz, H. D. Hausen, J. Weidlein (Z. Anorg. Allgem. Chem. **435** [1977] 179/90).
[69] H. J. Widler, H. D. Hausen, J. Weidlein (Z. Naturforsch. **30 b** [1975] 645/7).
[70] A. Rieche, J. Dahlmann (Ger. 1 155 127 [1960/63]; C.A. **60** [1964] 5554).
[71] J. Werner, W. Schwarz, A. Schmidt (Z. Naturforsch. **36 b** [1981] 556/60).
[72] G. B. Buckton (Jahresber. Fortschr. Chem. **1861** 469/70).
[73] M. Shindo, R. Okawara (Inorg. Nucl. Chem. Letters **3** [1967] 75/7).
[74] G. B. Buckton (J. Chem. Soc. **16** [1863] 17/25).
[75] V. D. Nefedov, I. S. Kirin, V. S. Zaitsev, G. A. Semenov, B. E. Dzevitskii (Zh. Obshch. Khim. **33** [1963] 2407/10; J. Gen. Chem. [USSR] **33** [1963] 2347/9).
[76] A. N. Murin, V. D. Nefedov (Primen. Mechenykh At. Anal. Khim. Dokl. Konf., Moscow 1953 [1955], pp. 75/8; C.A. **1956** 3915/6).
[77] C. G. Moreland, G. G. Long (Inorg. Nucl. Chem. Letters **8** [1972] 347/51).
[78] C. G. Moreland, M. H. O'Brien, C. E. Douthit, G. G. Long (Inorg. Chem. **7** [1968] 834/6).
[79] C. G. Moreland, R. J. Beam (Inorg. Chem. **11** [1972] 3112/4).
[80] M. Shindo, Y. Matsumura, R. Okawara (Bull. Chem. Soc. Japan **42** [1969] 265/6).
[81] H. G. Ang, W. S. Lien (J. Fluorine Chem. **15** [1980] 453/70).
[82] H. A. Meinema, J. G. Noltes, F. Di Bianca, N. Bertazzi, E. Rivarola, R. Barbieri (J. Organometal. Chem. **107** [1976] 249/55).
[83] D. J. Parker (Diss. Univ. Wisconsin 1959; Diss. Abstr. Intern. B **20** [1959/60] 2044).
[84] J. Pebler, F. Weller, K. Dehnicke (Z. Anorg. Allgem. Chem. **492** [1982] 139/47).

2.5.1.1.2.1.2 Other Trialkylantimony Dichlorides

$(C_2H_5)_3SbCl_2$

If $Sb(C_2H_5)_3$ is dropped into a vessel filled with chlorine gas, the mixture ignites immediately and burns with a light sooty flame; most likely the title compound is formed [1], see also [6]. $Sb(C_2H_5)_3$ is described to react vigorously with HCl gas or with fuming HCl to form $(C_2H_5)_3SbCl_2$ [1], but it is known today that the product is an unidentifiable mixture [20]. A better method of preparation is to chlorinate $Sb(C_2H_5)_3$ with $SbCl_3$ in n-hexane as solvent. The mixture is refluxed for 10 min and left overnight at room temperature. The solid antimony is filtered off and the compound is purified by distillation at 123.5 °C/3 Torr. The yield is quantitative [2]. It is probably also obtained by reaction of $(C_2H_5)_3SbI_2$ with $HgCl_2$ in water, but the formula of the product is disputed [57]. A 1:1 ratio of $Sb(C_2H_5)_3$ with CH_3PCl_2 or with $C_6H_5PCl_2$ forms $(C_2H_5)_3SbCl_2$ and $(CH_3P)_5$ or $(C_6H_5P)_5$ in good yields. In the reaction with CH_3PCl_2, the reaction mixture could not be separated into the products [3]. $(C_2H_5)_3Sb{=}NSO_2R$ ($R{=}C_6H_5$, $4\text{-}CH_3C_6H_4$) dissolved in CH_2Cl_2 reacts with gaseous HCl or with $SnCl_4$ to give the title compound. In the reaction with $SnCl_4$ the yield is 60 to 70% [4]. The title compound is formed in the reaction of $(C_2H_5)_2InCl$ and $SbCl_5$ (ratio 3:2) in CH_2Cl_2 in addition to $InCl_3$. The product is separated by vacuum distillation [5].

The compound is a colorless liquid which boils at 106 to 107 °C at 0.0931 hPa (=0.07 Torr); the density is 1.674 g/cm³ at 20 °C [4] or 1.540 g/cm³ at 17 °C [1], the refractive index is $n_D^{20}=1.5725$ [4]. It is still a liquid at −12 °C. It is insoluble in water, but soluble in ethanol or ether [1].

The following vibrations (in cm^{-1}) are reported from the far IR and Raman spectra: IR (polyethylene pellet): 537 ($\nu_{as}Sb(CC)_3$), 490 ($\nu_sSb(CC)_3$), 280 (sh, $\delta SbCC$), 269 ($\nu_{as}SbCl_2$), 265 (sh, ν_sSbCl_2), 175 (sh, $\delta Sb(CC)_3$ in-plane), 145 ($\delta Sb(CC)_3$ out-of-plane), 100 ($\delta SbCl_2$); Raman: 540; 535 (dp, $\nu_{as}Sb(CC)_3$), 493 (p, $\nu_sSb(CC)_3$), 266 (p, ν_sSbCl_2), 177 (p, $\delta Sb(CC)_3$ in-plane), 144 (dp, $\delta Sb(CC)_3$ out-of-plane). The assignments are made on the basis of a C_{3v} symmetry [6]. The ^1H NMR spectrum in $CDCl_3$ shows a quartet at $\delta=1.58$ and a triplet at 2.69 ppm [3].

HCl gas escapes upon treatment with concentrated H_2SO_4 [1]. $(C_2H_5)_3SbCl_2$ reacts with AgF in CH_3OH or CH_3CN to form $(C_2H_5)_3SbF_2$ [6]. With NaOCN in refluxing CH_3CN, it reacts to form $(C_2H_5)_3Sb(NCO)_2$ [7]. From Na-8-quinolinolate and the compound, dissolved in ethanol, one obtains $(C_2H_5)_3SbCl(OC_9H_7N)$ in a yield of 51% [8]. Substituted benzoic acids $XC_6H_4CO_2H$ (X=2-NH_2, 3-NH_2, 4-NO_2, 2-CH_3, 4-CH_3) react with the title compound in benzene in the presence of triethylamine as HCl-acceptor to form the corresponding $(C_2H_5)_3$-$Sb(O_2CC_6H_4X)_2$ compounds in yields between 40 and 70%. Silver oxalate or silver terephthalate in aqueous solution form polymers with $(C_2H_5)_3SbCl_2$ in yields of about 30% [9]. $LiCH_3$ and $(C_2H_5)_3SbCl_2$ refluxed three hours in ether leads to a mixture of $CH_3Sb(C_2H_5)_4$, $(CH_3)_2Sb(C_2H_5)_3$, $(CH_3)_3Sb(C_2H_5)_2$, and $(CH_3)_4Sb(C_2H_5)$ in a ratio of 2:10:6:1 [10]. Only $(C_2H_5)_3Sb(CH_3)_2$ is obtained in a yield of 80% if the reaction is carried out at −10 °C [11]. An ethereal solution of $Mg(C_2H_5)_2$ reacts with $(C_2H_5)_3SbCl_2$ (ratio 2:1) under reflux to give $Sb(C_2H_5)_5$ and $MgCl_2$ quantitatively [18]. The compound reacts with $AlCl_3$ (1:1 molar ratio) in benzene within 20 h at 20 °C to give the adduct $(C_2H_5)_3SbCl_2 \cdot AlCl_3$ as white crystals, m.p. 50 °C, which are insoluble in n-hexane. The same reaction with a 1:2 molar ratio of $(C_2H_5)_3SbCl_2$ and $AlCl_3$ gives a dark green liquid, formulated as $(C_2H_5)_3SbCl_2 \cdot nAlCl_3$. Reaction of the title compound with $Al(C_2H_5)_3$ in a 1:1 molar ratio gives $[Sb(C_2H_5)_4]$-$[Al(C_2H_5)_2Cl]$, a 1:2.4 ratio of reactants gives $[Sb(C_2H_5)_4][(C_2H_5)_5AlCl_2]$. With $(C_2H_5)_2AlCl$ (1:1 or 1:2 ratio) $[Sb(C_2H_5)_4][C_2H_5AlCl_3]$ is produced, and with $Al(C_2H_5)_3$, $[Sb(C_2H_5)_4][AlCl_4]$ is formed [2]. Similarly, reaction with $Ga(CH_3)_3$ or $Ga(C_2H_5)_3$ in CH_2Cl_2 or benzene gives $[Sb(C_2H_5)_4][(C_2H_5)_2GaCl_2]$ and $[SbCH_3(C_2H_5)_3][(CH_3)_2GaCl_2]$, respectively [11].

In some patents "adducts" of $(C_2H_5)_3SbCl_2$ with $AlCl_3$ [12, 13] or with $Al(C_2H_5)_3$ and $TiCl_3$ [14 to 17] are described as useful substances for polymerization of olefins and polyesterification.

$(C_3H_7)_3SbCl_2$

Tripropylantimony dichloride is prepared by mixing solutions of chlorine and the stibine in CCl_4 and evaporating the solvent in vacuum; a fluid remains which decomposes upon distillation [19]. From $(C_3H_7)_3Sb{=}NSO_2R$ (R=C_6H_5, 4-$CH_3C_6H_4$) dissolved in CH_2Cl_2 and gaseous HCl, arenesulfonamide precipitates. Upon evaporation of the filtrate, the compound is obtained by distillation at reduced pressure with a boiling point of 113 to 114 °C at 0.0931 hPa (=0.07 Torr). It melts at 38 °C [4]. Together with $Al(C_3H_7)_3$ and $TiCl_3$, it is claimed to be a catalyst for the polymerization of α-olefins [16].

References on p. 25

$(i\text{-}C_3H_7)_3SbCl_2$

This compound may be prepared by direct chlorination of the corresponding stibine [6]. $Sb(C_3H_7\text{-}i)_3$ and $SbCl_3$ react in a molar ratio of 2.6:1.7 in ether at 0 °C to yield a black precipitate of elemental antimony. After evaporation of the ether, the residue is treated with methanol and acetone and concentrated. The compound is isolated in a yield of 95%. The compound is directly prepared by reaction of $i\text{-}C_3H_7MgCl$ and $SbCl_3$ in ether, 5 h at 0 °C, then 30 min reflux and workup of the mixture with water, evaporation of the ether of the organic phase, and recrystallization of the residue from methanol or sublimation of the residue at 120 °C/0.01 Torr to yield colorless crystals, m.p. 100 °C [21].

In the far IR and Raman spectra, the following vibrations (in cm^{-1}) are found: IR (polyethylene pellet): 504 ($\nu_{as}Sb(CC_2)_3$), 481 ($\nu_sSb(CC_2)_3$), 408, 295 ($\delta SbCC_2$), 268 ($\nu_{as}SbCl_2$), 180 (ϱCl_2SbCC_2), 165, 140 ($\delta Sb(CC_2)_3$); Raman: 503 (dp, $\nu_{as}Sb(CC_2)_3$), 479 (p, $\nu_sSb(CC_2)_3$), 405, 260 (p, $\delta SbCC_2$), 270 to 260 (p, ν_sSbCl_2), ca. 140 (dp, $\delta Sb(CC_2)_3$). From these data, a C_{3v} symmetry of the molecule is concluded [6]. In a ^1H NMR spectrum in benzene, resonances are found at δ=1.35(d) and 3.01 (hept) ppm. A mass spectrum was measured and discussed. Following peaks were observed (30 eV, 25 °C inlet, 150 °C source, relative intensities in brackets, relative to ^{121}Sb): m/e=320 (1), 277 (68), 232 (2), 285 (30), 242 (16), 201 (8), 250 (22), 206 (8), 207 (14), 165 (18), 164 (16), 136 (4), 43 (106) [21].

The compound decomposes, when kept for 4 h at 220 °C and 760 Torr, to $i\text{-}C_3H_7SbCl_2$ (85%) and some $Sb(C_3H_7\text{-}i)_3$. At 90 Torr and the same temperature, 60% $(i\text{-}C_3H_7)_2SbCl$, 10% $Sb(C_3H_7\text{-}i)_3$, and 25% unreacted compounds are isolated after 2 h [21]. With AgF in CH_3CN or CH_3OH, the corresponding $(i\text{-}C_3H_7)_3SbF_2$ is formed [6].

$(C_4H_9)_3SbCl_2$

Tributylantimony dichloride is obtained by reacting $Sb(C_4H_9)_3$ and $SbCl_3$ in a ratio of 0.2:0.137 in benzene, stirred overnight at room temperature, then 2 h at 50 °C. Distillation of the mixture at 0.14 Torr gives 83% of the compound with a boiling point of 120 to 125 °C [22]. Another boiling point is given as 130 to 132 °C at 1 Torr [23]. At higher pressures the compound decomposes, probably to give $(C_4H_9)_2SbCl$ [22]. This behavior was observed during the preparation of the compound from $Sb(C_4H_9)_3$ and Cl_2 in CCl_4 [19] or from $SbCl_5$ and $Hg(C_4H_9)_2$ in CCl_4 [24, 25]. The preparation of the compound from $Sb(C_4H_9)_3$ and $C_6H_5PCl_2$, $C_6H_5AsCl_2$, or $(C_6H_5)_2PCl$ is described without details in [3].

An IR spectrum in Nujol shows the following absorptions (in cm^{-1}): 2970 vs, 2940 vs, 2880 s, 2740 vw, 1468 s, 1410 m, 1388 m, 1350 m, 1295 w, 1255 m, 1170 m, 1095 m, 1048 w, 1005 w, 960 vw, 890 w, 860 w, 760 w, 712 w, 615 vw, 512 vw [22]. See [56] for a figure of the spectrum and discussion.

Tributylantimony dichloride reacts with NaOCN in CH_3CN under reflux to give $(C_4H_9)_3Sb(NCO)_2$ [7]. With $NaO_2CC_6H_4CO_2Na\text{-}2$ in CCl_4/H_2O it forms 2% of a 1:1 polymer [26]; with urea, 2 h at 100 °C, $(C_4H_9)_3Sb(NCO)_2$ is formed [27].

$(i\text{-}C_4H_9)_3SbCl_2$

Triisobutylantimony dichloride is presumably prepared by reacting $Sb(C_4H_9\text{-}i)_3$ with Cl_2 in CCl_4 solution. It melts at 91 °C recrystallized from ether or CCl_4 [28]. With NaOCN, refluxed 3 h in CH_3CN, it forms $(i\text{-}C_4H_9)_3Sb(NCO)_2$ in a yield of 75% [7, 29]. The same product is obtained by reaction of the title compound with urea for 2 h at 140 °C [27].

References on p. 25

$(C_5H_{11})_3SbCl_2$

Triamylantimony dichloride, prepared from $Sb(C_5H_{11})_3$ and Cl_2, is described as an unstable sweet smelling liquid which decomposes in water [19]. $Sb(C_5H_{11})_3$ dissolved in ether forms upon slow evaporation of the solvent in air a solid, probably $(C_5H_{11})_3SbO$. This mixture when treated with aqueous HCl leads to a liquid which decomposes at temperatures above 160 °C [30 to 32]. The substance reacts with $AgNO_3$ or Ag_2SO_4 in ethanol to form most likely $(C_5H_{11})_3Sb(NO_3)_2$ or $(C_5H_{11})_3SbSO_4$ [30].

$(C_8H_{17})_3SbCl_2$

The compound is prepared by direct chlorination of $Sb(C_8H_{17})_3$ in ether, THF, or $CHCl_3$ solution. Constants of hydrolysis in $CHCl_3$ are given as $\log K_1 = 9$ and $\log K_2 = 5$ [33]. Fluoride extraction from water to $CHCl_3$ in the presence of the compound was studied [34]. The compound reacts with urea, 2 h at 140 °C, under evolution of NH_3 and H_2O, to form the corresponding diisocyanate [27].

$(c\text{-}C_6H_{11})_3SbCl_2$

Tricyclohexylantimony dichloride is directly synthesized from $c\text{-}C_6H_{11}MgCl$ and $SbCl_3$ in ether, cooled by an ice/salt mixture. At 0 °C, the mixture is treated with dilute HCl, and the residue extracted with benzene. The organic solvents are evaporated and the remaining residue is recrystallized from benzene. The yield is 32.5% of the product, which melts with decomposition at 201 to 203 °C [35]. The compound is also obtained by reaction of $Sb(C_6H_{11})_3$, dissolved in ether [33, 36] or THF or $CHCl_3$ [33], with chlorine while cooling; 85% of white crystals precipitate, which are washed with ether; m.p. 211 to 212 °C [36].

Thermal decomposition for half an hour at 220 °C and 350 Torr yields 35.8% $(c\text{-}C_6H_{11})_2SbCl$ [35], and for 40 min at 230 °C and 360 Torr yields 45% $(c\text{-}C_6H_{11})_2SbCl$ and $c\text{-}C_6H_{11}Cl$ [36]. The compound reacts with water with formation of $(c\text{-}C_6H_{11})_3SbCl(OH)$ which may be reversibly transformed to the educt by treating with C_6H_6/HCl [35]. With $NaOCH_3$, dissolved in methanol, the title compound reacts in a ratio of 1:1 upon refluxing for one hour in benzene to form $[(c\text{-}C_6H_{11})_3SbCl]_2O$ in a yield of 70% [37]. By reaction with $(CH_3)_3SiN{=}P(C_6H_5)_3$, refluxed for five hours in toluene, one obtains $(c\text{-}C_6H_{11})_3Sb(N{=}P(C_6H_5)_3)_2$ in a yield of 72% [38]. With $t\text{-}C_4H_9OOH$ in benzene the title compound reacts to form $(c\text{-}C_6H_{11})_3Sb(OOC_4H_9\text{-}t)_2$ [39]. Constants of hydrolysis in $CHCl_3$ are given as $\log K_1 = 8$ and $\log K_2 = 4.5$ [33]. The extraction of fluoride from water to $CHCl_3$ in presence of the compound was studied [34].

$(CF_3)_3SbCl_2$

$Sb(CF_3)_3$ (0.874 g) is treated with nine successive portions of chlorine (0.189 g total) at −40 to −50 °C. Between each addition the mixture is cooled to prevent explosive decomposition. Distillation in vacuum gives the compound quantitatively as a colorless liquid which crystallizes at −34 °C. It decomposes slowly at 20 and quickly at 50 °C. From the vapor equation $\log p = 7.29 - 2024/T$ (determined between −30 and 50 °C), the extrapolated boiling point is 101 °C [40].

The compound forms adducts with water at room temperature; $(CF_3)_3SbCl_2 \cdot H_2O$ melts at 51 °C and sublimes at 18 °C at 0.9 Torr. $(CF_3)_3SbCl_2 \cdot 2H_2O$ is obtained by freeze-drying. In aqueous solution the compound acts as a strong acid, due to its hydrolysis products $2HCl + H[Sb(CF_3)_3(OH)_3]$. Pyridine · HCl and pyridine · HBr adducts of the compound are known [41] as well as a simple 1:1 pyridine adduct [40]. With an excess of NOCl at −10 °C

References on p. 25

the compound reacts to produce a yellow solid of $NO[(CF_3)_3SbCl_3]$ [41]. $(CF_3)_3SbCl_2$ is reduced by shaking with mercury to $Sb(CF_3)_3$ [40].

$[Cl(CH_3)_2SiCH_2]_3SbCl_2$ and $[(CH_3)_3SiCH_2]_3SbCl_2$

Both compounds are synthesized from $X(CH_3)_2SiCH_2Cl$ (X = Cl or CH_3) under vigorous agitation with antimony powder in an autoclave, heating for 7 h at 185 to 190 °C in the presence of catalytic amounts of $(C_4H_9)_4PI$. The conversion of antimony is about 80%. The reaction mixture is filtered, washed with benzene and acetone, and the filtrate is distilled at 78 to 80 °C, 3 Torr. Thereby, the compounds sublime. The liquid byproducts (yields 15 to 30%) are $[X(CH_3)_2SiCH_2]_2SbCl$ (X = Cl, CH_3). Both compounds are obtained in about 60% yield with melting points of 103 to 105 °C and 115 to 116 °C, respectively, recrystallized from hexane. In the mass spectra (70 eV, 25 to 50 °C direct introduction) of both compounds, the molecular ion peaks are very weak; the principal peaks are those of the ions $[M-Cl]^+$, $[M-CH_3]^+$, and the peaks of $[M-(CH_3)_3SiCH_2]^+$, and $[M-Cl(CH_3)_2SiCH_2]^+$, respectively [42].

$(C_6H_5CH_2)_3SbCl_2$

Tribenzylantimony dichloride is best prepared from $C_6H_5CH_2MgCl$ and $SbCl_3$. The Grignard mixture is poured into ice-water; thereby partial oxidation takes place. After completion of the oxidation the residue is extracted with boiling ethanol and this solution is treated with H_2O. The resulting $(C_6H_5CH_2)_3Sb(OH)_2$ is then dissolved in a mixture of ether and concentrated HCl (2:1). After some time, crystals of the compound deposit in a yield of 53%. They melt at 109 °C, recrystallized from ethanol [43]. A preparation from the same starting materials is given without details in [44]. The synthesis from $(C_6H_5CH_2)_3Sb(OH)_2$ with concentrated HCl in ether in a yield of 85% (melting point 100 to 101 °C from ethanol) [45] or in $CHCl_3$ (melting point 108 °C) [58] has also been reported. The compound can be isolated from a mixture of products obtained by grinding antimony in the presence of benzylchloride at room temperature [46].

The fundamental IR [6, 47] and Raman frequencies [47] (in solid and in solution) of the compound were measured, assigned, and discussed on the basis of a C_{3h} geometry of the molecule. For details see [6, 47]. The ^1H NMR spectrum in $CDCl_3$ at −60 °C shows a singlet at $\delta = 3.72$ (CH_2) ppm [48]; at room temperature in CCl_4 $\delta = 3.57$ (CH_2) and 7.24 (C_6H_5) were found [47]. The following NQR resonances (in MHz) were found at 300 K: ^{121}Sb: $5/2 \rightleftharpoons 3/2$ at 180.20(4), $3/2 \rightleftharpoons 1/2$ at 90.087(6); ^{123}Sb: $3/2 \rightleftharpoons 1/2$ at 54.792(3). The e^2Qq/h value for ^{121}Sb is 600.58 and for ^{123}Sb 763.34 MHz; $\eta = 0$ indicates a trigonal bipyramidal environment of the Sb atom with linear ClSbCl bonding [49]. The Mössbauer spectrum at 80 K (vs. $^{121}SnO_2$) has an isomer shift at $\delta = -6.2 \pm 0.2$ with a quadrupole coupling constant of $e^2qQ = -30 \pm 2$, and a line width of $\Gamma = 4.5 \pm 0.5$ mm/s. At 4 K (vs. $Ca^{121}SnO_3$) the values are $\delta = -5.86 \pm 0.01$, $e^2qQ = -23 \pm 0.1$, and $\Gamma = 2.89 \pm 0.03$ mm/s [50]. A molar susceptibility of $\chi_M = -249.6 \times 10^{-6}$ was measured [51, 52].

Tribenzylantimony dichloride reacts with water to form tribenzylantimony oxide [44]. Dissolved in boiling methanol and treated with aqueous NaOH, $[(C_6H_5CH_2)_3SbCl]_2O$ is obtained quantitatively upon cooling of the mixture [43]. With AgF in CH_3CN, heated for 20 h, one obtains 43% of the corresponding difluoride [48]. Organylhydroperoxides R_3COOH(R = CH_3, C_6H_5) and the title compound react in benzene, in the presence of a base like $NaNH_2$, NH_3, or an amine as HCl acceptor, to form the corresponding peroxides $(C_6H_5CH_2)_3Sb(OOCR_3)_2$ in good yields [53, 39]. Equilibrium constants for the reaction $(C_6H_5CH_2)_3SbCl_2 + (C_6H_5CH_2)_3SbX_2 \rightleftharpoons 2(C_6H_5CH_2)_3SbClX$ (X = F, Br, I) [48, 54] and also for $(C_6H_5CH_2)_3SbCl_2 + (CH_3)_3SbX_2 \rightleftharpoons (C_6H_5CH_2)_3SbX_2 + (CH_3)_3SbCl_2$ (X = Br, I) [54] in $CHCl_3$ solution were determined, see [48, 54].

$(C_{10}H_{15}O)_3SbCl_2$

"Tricamphorylantimony dichloride" is obtained by treating "sodium camphor" with $SbCl_3$ for several days in toluene. The solvent is completely removed by steam distillation, and the residue is extracted with benzene. Concentration of the extract yields crystals which melt at 244 °C with decomposition. The compound reacts with aqueous alkali with decomposition [55].

References:

[1] C. Löwig, E. Schweizer (Liebigs Ann. Chem. **75** [1850] 315/55).
[2] Y. Takashi, I. Aishima (J. Organometal. Chem. **8** [1967] 209/23).
[3] J. C. Summers, H. H. Sisler (Inorg. Chem. **9** [1970] 862/9).
[4] Z. I. Kuplennik, Zh. N. Belaya, A. M. Pinchuk (Zh. Obshch. Khim. **51** [1981] 2711/5; J. Gen. Chem. [USSR] **51** [1981] 2339/43).
[5] H. J. Widler, H. D. Hausen, J. Weidlein (Z. Naturforsch. **30 b** [1975] 645/7).
[6] L. Verdonck, G. P. Van der Kelen (Spectrochim. Acta A **31** [1975] 1707/11).
[7] W. Stamm (Trans. N. Y. Acad. Sci. [2] **28** [1966] 396/401).
[8] H. A. Meinema, E. Rivarola, J. G. Noltes (J. Organometal. Chem. **17** [1969] 71/81).
[9] A. Ouchi, M. Nakatani, Y. Takahashi, S. Kitazima, T. Sugihara, M. Matsumoto, T. Uehiro, K. Kitano, K. Kawashima, H. Honda (Sci. Papers Coll. Gen. Educ. Univ. Tokyo **25** [1975] 73/99).
[10] H. A. Meinema, J. G. Noltes (J. Organometal. Chem. **22** [1970] 653/7).

[11] N. Tempel, W. Schwarz, J. Weidlein (J. Organometal. Chem. **154** [1978] 21/32).
[12] Asahi Chemical Industry Co., Ltd. (Japan. 64-27872 [1961/64]; C.A. **63** [1965] 1896).
[13] T. Takashi, K. Fujiski, T. Aijima, T. Nakanishi, T. Shima, Asahi Chemical Industry Co., Ltd. (Japan. 70-19513 [1966/70]; C.A. **73** [1970] No. 77837).
[14] Asahi Chemical Industry Co., Ltd. (Fr. 1338970 [1962/63]; C.A. **60** [1964] 6949).
[15] Asahi Chemical Industry Co., Ltd. (Fr. 1386468 [1963/65]; C.A. **62** [1965] 16405).
[16] Y. Takashi, I. Aijima, Yu. Kobayashi, Y. Tsunoda, Asahi Chemical Industry Co., Ltd. (U.S. 3494910 [1965/70]; C.A. **72** [1970] No. 90967).
[17] M. Hideo, T. Yukichi, I. Aijima, Asahi Chemical Industry Co., Ltd. (Japan. 70-13585 [1967/70]; C.A. **73** [1970] No. 46037).
[18] Y. Takashi (J. Organometal. Chem. **8** [1967] 225/31).
[19] W. J. C. Dyke, W. J. Jones (J. Chem. Soc. **1930** 1921/7).
[20] M. Wieber (private communication).

[21] H. J. Breunig, W. Kanig (Phosphorus Sulfur **12** [1982] 149/59).
[22] R. L. McKenney, H. H. Sisler (Inorg. Chem. **6** [1967] 1178/82).
[23] A. N. Nesmeyanov, A. E. Borisov, N. G. Kizim (Izv. Akad. Nauk SSSR Ser. Khim. **1974** 1672; Bull. Acad. Sci. USSR Div. Chem. Sci. **1974** 1602).
[24] A. N. Nesmeyanov, A. E. Borisov (Izv. Akad. Nauk SSSR Ser. Khim. **1969** 974/5; Bull. Acad. Sci. USSR Div. Chem. Sci. **1969** 895).
[25] A. N. Nesmeyanov, A. E. Borisov, N. V. Novikova (Izv. Akad. Nauk SSSR Ser. Khim. **1969** 1978/82; Bull. Acad. Sci. USSR Div. Chem. Sci. **1969** 1830/3).
[26] C. E. Carraher Jr., H. S. Blaxall (Angew. Makromol. Chem. **83** [1979] 37/45).
[27] W. Stamm, Stauffer Chemical Co. (Ger. 1229529 [1963/66]; C.A. **66** [1967] No. 28896).
[28] M. E. Brinnand, W. J. C. Dyke, W. H. Jones, W. J. Jones (J. Chem. Soc. **1932** 1815/9).
[29] W. Stamm (J. Org. Chem. **30** [1965] 693/5).
[30] F. Berlé (J. Prakt. Chem. **65** [1855] 385/418).

[31] C. Cramer (Verhandl. Naturforsch. Ges. Zürich **1851** May 12th from Chem. Pharm. Centr. **26** [1855] 465/8; Jahresber. Fortschr. Chem. **1855** 590).

[32] F. Berlé (Liebigs Ann. Chem. **97** [1856] 316/22).
[33] M. Benmalek, H. Chermette, C. Martelet, D. Sandino, J. Tousset (J. Organometal. Chem. **67** [1974] 53/9).
[34] N. Benmalek, H. Chermette, C. Martelet, D. Sandino, J. Tousset (J. Inorg. Nucl. Chem. **36** [1974] 1365/8).
[35] H. Hartmann, G. Kühl (Z. Anorg. Allgem. Chem. **312** [1961] 186/94).
[36] K. Issleib, B. Hamann (Z. Anorg. Allgem. Chem. **332** [1964] 179/88).
[37] Y. Kawasaki, Y. Yamamoto, M. Wada (Bull. Chem. Soc. Japan **56** [1983] 145/8).
[38] K. Bajpai, R. C. Srivastava (Syn. Reactiv. Inorg. Metal-Org. Chem. **12** [1982] 47/54).
[39] A. Rieche, J. Dahlmann, Deutsche Akademie der Wissenschaften zu Berlin (Ger. 1155127 [1960/63]; C.A. **60** [1964] 5554).
[40] J. W. Dale, H. J. Emeléus, R. N. Haszeldine, J. H. Moss (J. Chem. Soc. **1957** 3708/13).
[41] H. J. Emeléus, J. H. Moss (Z. Anorg. Allgem. Chem. **282** [1955] 24/8).
[42] V. P. Kochergin, V. I. Shiryaev, V. F. Mironov (Zh. Obshch. Khim **48** [1978] 1428; J. Gen. Chem. [USSR] **48** [1978] 1312).
[43] L. Kolditz, M. Gitter, E. Rösel (Z. Anorg. Allgem. Chem. **316** [1962] 270/7).
[44] G. T. Morgan, F. M. G. Micklethwait (Proc. Chem. Soc. **28** [1912] 68).
[45] J. P. Tsukervanik, D. Smirnov (J. Gen. Chem. USSR **7** [1937] 1527/31; C.A. **1937** 8518).
[46] H. Grohn, H. Friederich, R. Paudert (Z. Chem. [Leipzig] **2** [1962] 24/5).
[47] L. Verdonck, G. P. Van der Kelen (Spectrochim. Acta A **29** [1973] 1675/80).
[48] C. G. Moreland, M. H. O'Brien, C. E. Douthit, G. G. Long (Inorg. Chem. **7** [1968] 834/6).
[49] T. B. Brill, G. G. Long (Inorg. Chem. **9** [1970] 1980/5).
[50] G. G. Long, J. G. Stevens, R. J. Tullbane, L. H. Bowen (J. Am. Chem. Soc. **92** [1970] 4230/5).
[51] N. K. Parab, D. M. Desai (Current Sci. [India] **26** [1957] 389).
[52] N. K. Parab, D. M. Desai (J. Indian Chem. Soc. **35** [1958] 573/5).
[53] A. Rieche, J. Dahlmann, D. List (J. Liebigs Ann. Chem. **678** [1964] 167/82).
[54] C. G. Moreland, G. G. Long (Inorg. Nucl. Chem. Letters **8** [1972] 347/51).
[55] G. T. Morgan, F. M. G. Micklethwait, G. S. Whitby (Proc. Chem. Soc. **25** [1909] 302; J. Chem. Soc. **97** [1910] 34/6).
[56] A. E. Borisov, N. V. Novikova, N. A. Chumaevskii, E. B. Shkirtil (Ukr. Fiz. Zh. **13** [1968] 75/82).
[57] W. Merck (J. Prakt. Chem. **66** [1885] 56/72).
[58] F. Challenger, A.T. Peters (J. Chem. Soc. **1929** 2610/21).

2.5.1.1.2.1.3 Trialkenylantimony Dichlorides

$(CH_2{=}CH)_3SbCl_2$

Trivinylantimony dichloride is prepared by reaction of $Sb(CH{=}CH_2)_3$ with $TlCl_3$ in ether at room temperature; 99% of TlCl is subsequently deposited and filtered. After distillation of the filtrate one obtains 68% of the compound which boils at 102 to 103 °C at 3 Torr. The relative density is $d_4^{20} = 1.67278$, and the refractive index is $n_D^{20} = 1.5875$ [1]. The compound may be also obtained quantitatively from $Sb(CH{=}CH_2)_3$ with SO_2Cl_2 in pentane. It is a colorless oily liquid which can be purified by distillation (b.p. 58 to 60 °C at 0.01 Torr), but with partial decomposition [4].

IR [2 to 4] and Raman [4] spectra were measured. See [3] for a figure of the spectrum. The observed vibrations and their assignments are given in Table 10. According to these data the molecule has a trigonal bipyramidal structure, probably with slightly twisted vinyl groups.

Table 10
IR and Raman Vibrations (in cm^{-1}) of Liquid $(CH_2{=}CH)_3SbCl_2$ [4].

IR	Raman (intensity)	assignment
1590 s-m	1594 p (21)	$\nu C{=}C$ (A′+E′)
1380 s, 1368 sh	1382 p (26); 1375 sh, dp	$\delta_s CH_2$ (A′+E′)
1233 sh, 1220 s	1234 p (40), 1226 dp (5)	δCH in-plane (A′+E′)
990 sh, 973 vs	990 p (4); 960 sh, dp	τCH_2, ωCH_2
556 s	565 dp, 561 p (9)	$\nu_{as} SbC_3$ (E′)
	522 p (100)	$\nu_s SbC_3$ (A′)
531 s	530 sh, 522?	δHCC out-of-plane (A′+E′)
270 vs, br		$\nu_{as} SbCl_2$ (A″)
	265 p (60)	$\nu_s SbCl_2$ (A′)
342 m	335 dp (1)	δSbCC (E′)
	323 p (77)	δSbCC (A′)
	323?	δCSbCl (E″)

^{1}H NMR and ^{13}C NMR spectra were measured and calculated with the LAOCOON III program. The measured ^{1}H chemical shifts are in C_6D_6: δ=5.90 (H_B), 6.565 (H_A), 6.625 (H_C) ppm, in CCl_4: δ=6.55 (H_B), 6.65 (H_A), 6.67 (H_C) ppm, in $CDCl_3$: δ=6.515 (H_B), 6.65 (H_A), 6.66 (H_C), and neat: δ=4.29 (H_A), 6.515 (H_B), 6.61 (H_C) ppm. ^{13}C NMR in C_6D_6: δ= 137.4 (C-1), 138.0 (C-2) ppm (assignments according to Formula I). See the original for the calculated chemical shifts and coupling constants [4].

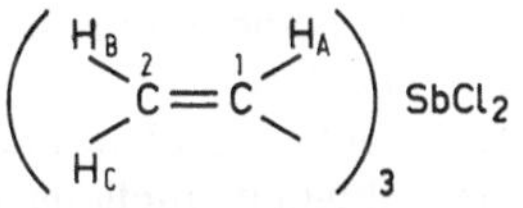

I

Reaction of the compound with $LiCH_3$ or $Na[C{\equiv}CCH_3]$ in the ratio 1:2 gives a mixture of products of the general formula $(CH_3)_nSb(CH{=}CH_2)_{5-n}$ or $(CH_2{=}CH)_nSb(C{\equiv}CCH_3)_{5-n}$, respectively, which could not be separated. Isomeric mixtures are also obtained with $NaOCH_3$. With CsF or KF, the compound reacts to form the corresponding $(CH_2{=}CH)_3SbF_2$ via $(CH_2{=}CH)_3SbClF$, identified by IR. With $SbCl_5$ in a ratio of 1:1, it reacts to form $[(CH_2{=}CH)_3SbCl][SbCl_6]$ [4].

(*cis*-$CH_3CH{=}CH)_3SbCl_2$

The compound is prepared by reaction of the corresponding tripropenylstibine with chlorine in $CHCl_3$ at 0 °C or with $TlCl_3$ in ether [5 to 7]. The yield is about 70% of the pure compound. It melts at 73.5 to 74 °C, recrystallized from ether [7], or 74 to 75 °C [9]. The IR absorptions are given and discussed in [5, 7, 9]. The data in KBr are: 1606, 1446, 1385, 1308, 1201, 1047, 940, 928, 665, 625, and 455 cm^{-1} [9].

References on p. 30

(*trans*-CH_3CH=$CH)_3SbCl_2$

The compound is prepared like the previous one in a yield of about 70% [5 to 7]. The liquid boils at 155 to 156 °C at 2 Torr, 162 °C at 4 Torr [7], or 160 to 162 °C at 4 Torr [9]. The relative density is $d_4^{20}=1.5056$, and the refractive index is $n_D^{20}=1.5820$ [7, 8]. IR vibrations are given and discussed in [5, 7, 9]. The following data are given for the liquid: 1607, 1440, 1376, 1306, 1191, 1109, 1075, 1042, 957, 724, 667, and 620 cm^{-1} [9].

$[CH_2$=$C(CH_3)]_3SbCl_2$

The compound is prepared by reacting $Sb(C(CH_3)$=$CH_2)_3$ with $TlCl_3$ in ether at room temperature or with Cl_2 in $CHCl_3$ at 0 °C. After workup of the reaction mixtures the yields are about 75%. The compound melts at 102 to 104 °C after recrystallization from ether [1].

$(CH_2$=$CHCH_2)_3SbCl_2$

Triallylantimony dichloride is obtained by treating $Sb(CH_2CH$=$CH_2)_3$ with Cl_2 in $CHCl_3$ at −55 °C. After evaporation of the solvent the residue is extracted with ether. The ether is completely removed; the product remains as a colorless liquid in a yield of 79% with a relative density of $d_4^{20}=1.4746$ and a refractive index of $n_D^{20}=1.5925$ [10]. A figure of its IR spectrum is given in [2, 3]; the νSb-C vibration is found at 510 cm^{-1} and νC=C at 1629 cm^{-1}. These vibrations are compared with those of other organometallic allyl compounds [2, 3].

The compound hydrolyzes in air. It reacts with CH_2=$CHCH_2MgCl$ with formation of $(CH_2$=$CHCH_2)_4SbCl$ in a yield of 62% [10].

(*trans*-$ClCH$=$CH)_3SbCl_2$

Tris(*trans*-2-chlorovinyl)antimony dichloride is best prepared by reacting Hg(CH=CHCl-*trans*)$_2$ with $SbCl_5$, stirred for four hours at room temperature in CCl_4. The resulting precipitate, *trans*-ClCH=CHHgCl (98%), is filtered, and the solvent is removed from the filtrate. The resulting residue is recrystallized from ethanol. It gives 62% of the compound with a melting point of 93 to 94 °C [11, 12]. The compound is also formed from $SbCl_5$, acetylene, and with [13 to 16] or without [16] some $HgCl_2$ as a catalyst when kept for 2 to 3 h at 80 to 100 °C and then at 150 to 175 °C until the absorption of acetylene ends. Fractional distillation and extraction of the lower boiling fractions in petroleum ether, evaporation of the latter, and recrystallization from ethanol yields the compound as a main product with melting point 93 to 94 °C; $d_4^{100}=1.7838$. From the mother liquor, a byproduct in a yield of 3 to 4% with a melting point of 61 to 62 °C was isolated. This was first claimed to be the (*cis*-ClCH=CH)$_3SbCl_2$ [13 to 15], but later identified as *cis-trans-trans*-isomer [12, 17, 18] (see below).

The following IR absorptions are given: 1280, 1155, 1141, 936, 802, 752, 682, and 665 cm^{-1} [19]. ^{1}H NMR measurements give resonances at δ=6.90 and 7.50 ppm. The coupling constant J(H-1,2) is 14.00 Hz [11, 12]. An NQR spectrum at 77 K has the following signals: ^{121}Sb: 1/2 ⇌ 3/2 at 84.480, 3/2 ⇌ 5/2 at 160.74 MHz, $e^2qQ/h=562.57$ MHz, η=3.3%; ^{123}Sb: 1/2 ⇌ 3/2 at 51.400, 3/2 ⇌ 5/2 at 102.36 MHz, $e^2qQ/h=716.94$ MHz, η=3.3%; ^{35}Cl: 34.962, 34.740, and 34.530 MHz [20]. The ^{121}Sb Mössbauer spectrum at 80 K vs. InSb shows an isomeric shift at δ=3.8±0.1 mm/s and a quadrupole coupling of $e^2qQ=-16.4\pm0.8$ mm/s. The structure is discussed on the basis of these values [21].

An X-ray analysis of the compound was carried out. The compound crystallizes in the monoclinic space group C2/c-C_{2h}^6 (No. 15) with Z=8; a=20.96±0.1, b=7.00±0.2, c=

References on p. 30

17.23±0.9 Å, and β=101° 50'±10' [22, 23]. The chlorovinyl groups lie flatly in an equatorial plane of the trigonal bipyramidal molecule. The angles C-C-Cl, C-C-Sb, and C-Sb-C are 120±4°. Both Cl atoms are above and below the plane with angles Cl-Sb-C of 84°. The distances C-Cl=1.70±0.03, C=C=1.31±0.04, and SbC=2.15±0.1 Å are normal. Intermolecular distances of the molecules in the unit cell are given in **Fig. 3** [23].

The compound pyrolyzes at 200 to 850 °C forming $SbCl_3$, C_2H_2, CHCl=CHCl, and other substances [14]. With Br_2 or I_2 all Sb-C bonds are cleaved. Reduction of the compound to Sb(CH=CHCl-*trans*)$_3$ in a yield of 75% is performed with $NaHSO_3$ in aqueous ethanol [24, 25]. These investigations correlate well with previous works, whereby reductive isomerization should occur [13, 14]. The compound reacts with $AgNO_3$ forming a mixture of $Ag_2C_2 \cdot AgNO_3$ and $[(ClCH{=}CH)_2Sb]_2 \cdot AgNO_3$ [14]. The compound is claimed to inhibit the corrosion of steel by 5 N H_2SO_4 [26].

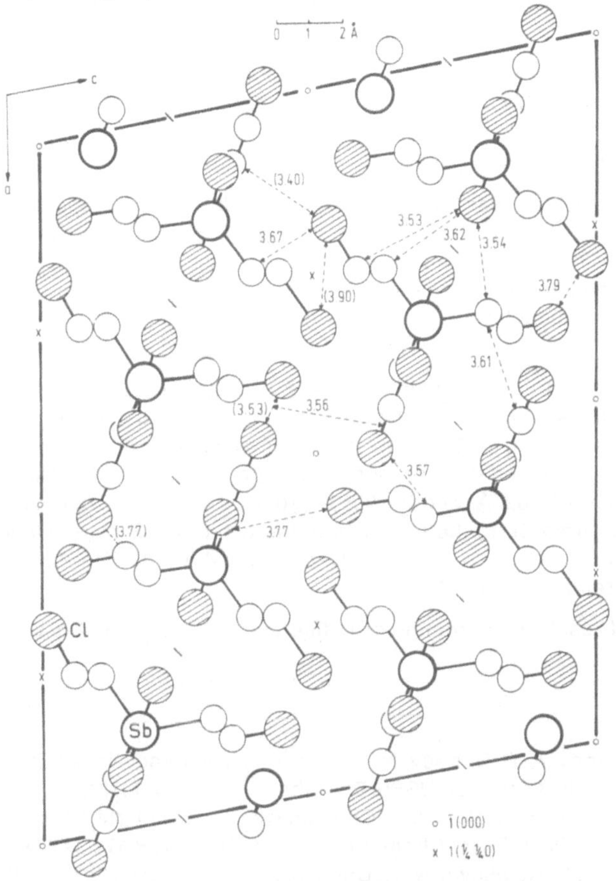

Fig. 3. Projection of the unit cell of (*trans*-ClCH=CH)$_3$SbCl$_2$ on the ac plane [23].

(*cis*-ClCH=CH)$_3$SbCl$_2$

The compound is prepared by adding $SbCl_5$ to Hg(CH=CHCl-*cis*)$_2$, both compounds dissolved in CCl_4 at room temperature. Additional heating at 40 °C for four hours, filtering of the (*cis*-ClCH=CH)HgCl precipitate, and concentrating the filtrate yield 39% of the compound, which melts at 163 to 164 °C, recrystallized from CCl_4 [11, 12]. (*cis*-ClCH=CH)$_3$SbCl$_2$

References on p. 30

is also obtained by treating the corresponding stibine with chlorine in $CHCl_3$ solution at 0 °C. Evaporation of the solvent and recrystallization from ether leads to 77% of the compound with a melting point of 163 °C [12]. The product first claimed as the title compound in [13 to 15] with a melting point of 61 °C was later identified as *cis-trans-trans*-isomer [12, 17, 18] (see below).

An IR spectrum shows the following absorptions: 1290, 1279, 1265, 1133, 1121, 920, 910, 898, 785, 700, 681, 672, 612, and 465 cm^{-1} [12, 19]. ^1H NMR resonances were found at $\delta=7.42$ (H-1) and $\delta=7.6$ (H-2) ppm with a coupling of J(H-1,2) = 7.30 Hz [11, 12]. The NQR spectrum at 77 K gives the following signals: $\nu(^{121}Sb)=80.30$ ($1/2 \rightleftharpoons 3/2$), 160.18 ($3/2 \rightleftharpoons 5/2$), 146.10 ($5/2 \rightleftharpoons 7/2$) MHz; $e^2Qq/h=534.58$ MHz; $\nu(^{123}Sb)=48.933$ ($1/2 \rightleftharpoons 3/2$), 97.27 ($3/2 \rightleftharpoons 5/2$) MHz; $e^2Qq/h=681.46$ MHz; $\nu(^{35}Cl)=34.246$, 34.248, 34.746 MHz [20]. ^{121}Sb Mössbauer spectrum at 80 K vs. InSb: $\delta=3.4\pm0.1$ mm/s and $e^2qQ=-21\pm2$ mm/s [21]. The structure of the molecule is discussed in the original references on the basis of all these spectroscopic data.

The compound, dissolved in ethanol, reacts with aqueous $NaHSO_3$ solution with reduction and formation of Sb(CH=CHCl-*cis*)$_3$ in good yields [24, 25].

(*cis*-ClCH=CH)(*trans*-ClCH=CH)$_2$SbCl$_2$

This compound, which was first claimed to be the all-*cis* compound [13 to 15], was reinvestigated in [12, 17, 18]. It is formed as a byproduct in the reaction of $SbCl_5$ with acetylene heated at 80 to 175 °C for several hours. Fractional distillation at reduced pressure, extraction of the distillate with petroleum ether, evaporation of the latter, and recrystallization of the residue from ethanol give 3 to 4% of the compound, which melts at 61 to 62 °C [13 to 15].

In the ^1H NMR spectrum (for figure see [22]) two quartets were found in a ratio of 1:2 with coupling constants J(*cis*) = 7 Hz and J(*trans*) = 14 Hz. A two-dimensional X-ray structure (R-factors 20.78 for hk0 and 17.64 for h0l projections) without localization of C and H atoms was performed. The crystal data are: monoclinic space group Pc-C_s^2 (No. 7); $a=22.20\pm0.5$, $b=16.14\pm0.05$, $c=6.97\pm0.5$ Å, and $\beta=95°36'\pm0.5°$; Z = 8, $d_m=1.78$ g/cm³, $d_c=1.19$ g/cm³. The mean distance from Sb to the Cl atoms of the two *trans*-2-chlorovinyl groups is 4.9 Å, and to the Cl atom of the *cis*-2-chlorovinyl group is 3.4 Å [18]. The data are different from those published preliminarily in [22].

The compound reacts with bromine or iodine quantitatively with Sb-C-cleavage [13, 14].

References:

[1] A. N. Nesmeyanov, A. E. Borisov, N. V. Novikova (Izv. Akad. Nauk SSSR Otd. Khim. Nauk **1961** 1578/82; Bull. Acad. Sci. USSR Div. Chem. Sci. **1961** 1473/6).

[2] A. E. Borisov, N. V. Novikova, N. A. Chumaevskii, E. B. Shkirtil (Dokl. Akad. Nauk SSSR **173** [1967] 855/8; Dokl. Phys. Chem. Proc. Acad. Sci. USSR **172/177** [1967] 248/51).

[3] A. E. Borisov, N. V. Novikova, N. A. Chumaevskii, E. B. Shkirtil (Ukr. Fiz. Zh. **13** [1968] 75/82).

[4] K. Sille, J. Weidlein, A. Haaland (Spectrochim. Acta A **38** [1982] 475/82).

[5] A. N. Nesmeyanov, A. E. Borisov, N. V. Novikova (Tetrahedron Letters **8** [1960] 23/4).

[6] A. N. Nesmeyanov, A. E. Borisov, N. V. Novikova (Izv. Akad. Nauk SSSR Otd. Khim. Nauk **1960** 147; Bull. Acad. Sci. USSR Div. Chem. Sci. **1960** 140).

[7] A. N. Nesmeyanov, A. E. Borisov, N. V. Novikova (Izv. Akad. Nauk SSSR Otd. Khim. Nauk **1961** 612/7; Bull. Acad. Sci. USSR Div. Chem. Sci. **1961** 564/8).

[8] A. N. Nesmeyanov, A. E. Borisov, N. V. Novikova (Dokl. Akad. Nauk SSSR **134** [1960] 100/1; Proc. Acad. Sci. USSR Chem. Sect. **130/135** [1960] 995/6).
[9] A. E. Borisov, N. V. Novikova, N. A. Chumaevskii (Dokl. Akad. Nauk SSSR **136** [1961] 129/32; Proc. Acad. Sci. USSR Phys. Chem. Sect. **136/141** [1961] 13/6).
[10] A. E. Borisov, N. V. Novikova, A. N. Nesmeyanov (Izv. Akad. Nauk SSSR Ser. Khim. **1963** 1506/7; Bull. Acad. Sci. USSR Div. Chem. Sci. **1963** 1368/9).
[11] A. N. Nesmeyanov, A. E. Borisov (Izv. Akad. Nauk SSSR Ser. Khim. **1969** 974/5; Bull. Acad. Sci. USSR Div. Chem. Sci. **1969** 895).
[12] A. N. Nesmeyanov, A. E. Borisov, N. V. Novikova (Izv. Akad. Nauk SSSR Ser. Khim. **1969** 1978/82; Bull. Acad. Sci. USSR Div. Chem. Sci. **1969** 1830/3).
[13] A. N. Nesmeyanov (Izv. Akad. Nauk SSSR Otd. Khim. Nauk **1945** 239/50).
[14] A. N. Nesmeyanov, A. E. Borisov (Izv. Akad. Nauk SSSR Otd. Khim. Nauk **1945** 251/60).
[15] A. E. Borisov, A. N. Nesmeyanov (Sint. Org. Soedin. **1** [1950] 150/1; C.A. **1953** 8004).
[16] A. N. Nesmeyanov, A. E. Borisov (Izv. Akad. Nauk SSSR Ser. Khim. **1971** 2103; Bull. Acad. Sci. USSR Div. Chem. Sci. **1971** 2000).
[17] A. N. Nesmeyanov, A. E. Borisov (Izv. Akad. Nauk SSSR Ser. Khim. **1968** 1922/3; Bull. Acad. Sci. USSR Div. Chem. Sci. **1968** 1838).
[18] A. N. Nesmeyanov, A. E. Borisov, E. I. Fedin (Izv. Akad. Nauk SSSR Ser. Khim. **1969** 1977/8; Bull. Acad. Sci. USSR Div. Chem. Sci. **1969** 1828/9).
[19] A. E. Borisov, V. V. Klinkova, N. A. Chumaevskii (Dokl. Akad. Nauk SSSR **200** [1971] 64/7; Soviet Phys.-Dokl. **16** [1971/72] 734/6).
[20] V. I. Svergun, A. E. Borisov, N. V. Novikova, T. B. Babushkina, E. V. Bryukhova, G. K. Semin (Izv. Akad. Nauk SSSR Ser. Khim. **1970** 484/5; Bull. Acad. Sci. USSR Div. Chem. Sci. **1970** 443/4).
[21] S. E. Gukasyan, V. P. Gor'kov, P. N. Zaikin, V. S. Shpinel (Zh. Strukt. Khim. **14** [1973] 650/5; J. Struct. Chem. [USSR] **14** [1973] 603/7).
[22] Yu. T. Struchkov, A. I. Kitaigorodskii, T. L. Khotsyanova (Zh. Fiz. Khim. **26** [1952] 530/7).
[23] Yu. T Struchkov, T. L. Khotsyanova (Dokl. Akad. Nauk SSSR **91** [1953] 565/8).
[24] A. N. Nesmeyanov, A. E. Borisov (Dokl. Akad. Nauk SSSR **60** [1948] 67/72).
[25] A. N. Nesmeyanov, A. E. Borisov (Sint. Org. Soedin. **1** [1950] 128/9).
[26] S. A. Balezin, M. A. Ignat'eva (Zh. Prikl. Khim. **29** [1956] 1647/56; J. Appl. Chem. [USSR] **29** [1956] 1777/84; C.A. **51** [1957] 3421).

2.5.1.1.2.2 R_3SbCl_2 Compounds with R = Aryl, Furyl, and Thienyl

2.5.1.1.2.2.1 Triphenylantimony Dichloride $(C_6H_5)_3SbCl_2$

Preparation. The best method to prepare the compound is to dissolve triphenylstibine in petroleum ether and to pass gaseous chlorine through this solution. Triphenylantimony dichloride precipitates and is filtered [1 to 5]. The yield is 97%, recrystallized from ethylacetate [5]. Melting points are given as 143 °C [1, 2], 141.5 °C [3], and 143 °C [4]. Another medium for the reaction is CCl_4, where 94% of the compound with a melting point of 142 to 144 °C [6] is obtained. The reaction may also be performed in ether, THF, or $CHCl_3$ [7]. Another simple method for preparation is to dissolve $Sb(C_6H_5)_3$ in toluene and to add, drop by drop, stoichiometric amounts of SO_2Cl_2. The SO_2 evolution is finished after stirring 10 min at room temperature. Addition of hexane to the solution precipitates 90% of the compound [8, 9], which melts at 144 °C, recrystallized from a $C_2H_5OH/CHCl_3$ mixture [9]. Further methods of preparation are given in Table 11. The compound forms long, colorless needles [1, 3, 36] or plates (from $CHCl_3$) [83]. Other solvents used for recrystallization are C_2H_5OH, C_2H_5OH/HCl, ether, $CHCl_3$/petroleum ether, petroleum ether/C_6H_6, and CCl_4 (from references in Table 11). The compound can be further purified from traces of elemental Sb on Wofatit L-150 or Wofatit SBS-400 with acetone as eluent and subsequent recrystallization from acetone [107]. The separation of $(C_6H_5)_3SbCl_2$ from other organoantimony compounds during thin layer chromatography on alumina was studied with different solvents, e.g., C_6H_{14}, C_7H_{16}, C_6H_6, $CHCl_3$, C_6H_6/C_2H_5OH, $CHCl_3/(CH_3)_2CO$, or $C_2H_5O_2CCH_3/C_2H_5OH$ [145].

Table 11
Preparation and Formation of $(C_6H_5)_3SbCl_2$.

reactants	reaction conditions	products (yield in %)[a]	Ref.
$SbCl_3 + C_6H_5Br + Na$	—	$Sb(C_6H_5)_3$ (main product), I (byproduct)	[1, 10]
$SbCl_3 + Hg(C_6H_5)_2$	in xylene, 130 °C	I, $(C_6H_5)_2SbCl_3 \cdot H_2O$	[11]
$SbCl_3 \cdot 2[C_6H_5N_2]Cl + Zn$	in $CH_3CO_2C_2H_5$ at 60 °C, or other solvents	C_6H_5SbO (46), $(C_6H_5)_2SbO(O_2CCH_3)$ (16), I (18)	[12, 13, 83]
$SbCl_3 \cdot 2[C_6H_5N_2]Cl + M$ M = Fe, Sb, Cu, Mg, Al, Al/Hg, Zn/Cu, Zn/Hg	in $CH_3CO_2C_2H_5$	C_6H_5SbO (6 to 28); $(C_6H_5)_2SbO(O_2CCH_3)$ (3 to 14), I (2 to 15)	[13]
$SbCl_3 \cdot 2[C_6H_5N_2]Cl$	in $CH_3CO_2C_2H_5$/conc. HCl + Zn powder + $CaCl_2$	I as main product (46.2)	[13]
$SbCl_3 \cdot 2[C_6H_5N_2]Cl + Sb$	in $(CH_3)_2CO + CaCO_3$	I as byproduct (10)	[14]
$ZnCl_2 \cdot 2[C_6H_5N_2]Cl + Sb$	in $(CH_3)_2CO + CaCO_3$	I as byproduct (20)	[14]
$[C_6H_5N_2]Cl + Sb$	in $(CH_3)_2CO + CaCO_3$	I as byproduct (ca. 10)	[15]
$SbCl_5 + Hg(C_6H_5)_2$	in CCl_4, 1 h at 50 °C	$C_6H_5HgCl\downarrow$ + I (good)	[16, 17]
$(C_6H_5)_2SbCl_3 + Hg(C_6H_5)_2$	in C_6H_6, 1 h at 20 °C	$C_6H_5HgCl\downarrow$ + I (54)	[18]
$(C_6H_5)_2SbCl_3 \cdot (C_6H_5)_2ICl$ + Zn	ratio 1:4 in $(CH_3)_2CO$, 2 h at 20 °C	I (50)	[19]
$Bi(C_6H_5)_3 + SbCl_3$	ratio 1:1 in ether	$(C_6H_5)_2BiCl$ + I	[20, 178]
$Sb(C_6H_5)_3 + PCl_3$	in petroleum ether	I + $(C_6H_5)PCl_2$ (trace)	[20]
$Sb(C_6H_5)_3 + AsCl_3$	in ether, 3 h reflux	I + As + $(C_6H_5)AsCl_2$ (trace)	[20]
$Sb(C_6H_5)_3 + BiCl_3$	ratio 1:1 in ether, overnight at 20 °C	I + inorganic Bi compound	[20, 178]
$Sb(C_6H_5)_3 + SiCl_4$	in boiling C_6H_6	I + other phenylantimony chlorides	[20]
$Sb(C_6H_5)_3 + TiCl_4$	in ether	I	[20]

References on p. 45

Table 11 [continued]

reactants	reaction conditions	products (yield in %)[a)]	Ref.
$Sb(C_6H_5)_3+TlCl_3$	in toluene/ether	I (ca. 100) + $TlCl$	[20, 21]
$Sb(C_6H_5)_3+CuCl_2$	in $(CH_3)_2CO$ or CH_3OH	I + CuCl (in $(CH_3)_2CO$) or Cu^I complexes with $Sb(C_6H_5)_3$ (in CH_3OH)	[1, 22 to 24]
$Sb(C_6H_5)_3+TeCl_4$	in cold C_6H_6	I + Te	[25]
$Sb(C_6H_5)_3+MnCl_3$	in ether	I	[26]
$Sb(C_6H_5)_3+WCl_6$	in C_6H_6 for 24 h	I + WCl_3	[27]
$Sb(C_6H_5)_3+SnCl_4$	ratio 2:1 in C_5H_{12}	I + $SnCl_2$	[28]
$Sb(C_6H_5)_3+SbCl_5$	ratio 1:1 in toluene at −20 °C	I (75)	[29]
$Sb(C_6H_5)_3+AlCl_3$	ratio 1:3 in $CHCl_3$, 5 h at 20 °C	I (56)	[184]
$Sb(C_6H_5)_3+FeCl_3$	ratio 1:3 in $CHCl_3$, 5 to 6 h at ca. 100 °C	I (95)	[184]
$Sb(C_6H_5)_3+S_2Cl_2$	in ether, 0.5 h at 20 °C	I (74) + S_8	[30]
$Sb(C_6H_5)_3+SCl_2$	in ether (20 °C) or C_6H_6 (10 °C)	I (85) + S_8	[30, 31]
$Sb(C_6H_5)_3+SOCl_2$	in C_6H_6 at 10 °C	I (90) + SO_2 + S_8	[31, 32]
$Sb(C_6H_5)_3+[C_6H_5N_2]Cl$	ratio 1:1 in CH_3CO_2H	I (42)	[33]
$Sb(C_6H_5)_3+RSO_2NCl_2$ R = C_6H_5, 4-ClC_6H_4, 4-BrC_6H_4, 4-$CH_3C_6H_4$	ratio 2:1 in CCl_4, 15 min	I (ca. 100) + $(C_6H_5)_3Sb{=}NSO_2R$	[34]
$Sb(C_6H_5)_3+NH_2Cl$	in ether	$[(CH_3)_3SbCl]NH$ (55 to 71), I (25 to 43)	[35]
$Sb(C_6H_5)_3$ + hydrated chloramine T	in aqueous dioxane, 1 h at 20 °C	I (79) + $CH_3C_6H_4SO_2NH_2$	[36]
$(C_6H_5)_3Sb(Cl)N(SiCl_3)Si(CH_3)_3$	thermal decomposition	I and others	[37]
$Sb(C_6H_5)_5+Cl_2$	ratio 1:2 in warm C_6H_6	I (90) + C_6H_5Cl	[38]
$Sb(C_6H_5)_5+CH_3OH$ or C_2H_5OH	30 to 40 h reflux, workup with aqueous HCl	I (80 to 90)	[39]
$(C_6H_5)_3SbF_2+BCl_3$	in CH_2Cl_2, 30 min at 20 °C, vacuum distillation	I (98) + BF_3	[40]
$Sb(C_6H_5)_3+KMnO_4$	in aqueous alkali or acid, workup with aqueous HCl	I	[41]
$(C_6H_5)_3Sb(OH)_2$ + ethanolic HCl	–	I	[42]

References on p. 45

Table 11 [continued]

reactants	reaction conditions	products (yield in %)[a]	Ref.
$(C_6H_5)_3SbO$ + conc. HCl	in $(CH_3)_2CO$	I	[43]
$(C_6H_5)_3SbO + COCl_2$	in CH_3CN at 0 °C	I + CO_2	[44]
$(C_6H_5)_3SbO + SOCl_2$	in CH_3CN at 60 °C	I	[44]
$(C_6H_5)_3Sb(O_2CCH_3)_2$ + HCl	in C_2H_5OH, CH_3OH, or H_2O	I	[45 to 47]
$[(C_6H_5)_3SbCl]_2O$ + conc. HCl	—	I (ca. 100)	[35, 48]
$[(C_6H_5)_3SbNO_3]_2O$ + conc. HCl	in hot C_2H_5OH	I	[49]
$(C_6H_5)_3Sb(Cl)OC_4H_9$-t + HCl (0.01 N)	in CH_2Cl_2 at 20 °C, immediate reaction	I (ca. 100) + t-C_4H_9OH	[50]
$(C_6H_5)_3Sb(Cl)OC_4H_9$-t + CH_3COCl	in CH_2Cl_2, 20 min reflux	I (ca. 100) + $CH_3CO_2C_4H_9$-t	[50]
$(C_6H_5)_3Sb(OH)OSi(C_6H_5)_3$ or $[(C_6H_5)_3SbOSi(C_6H_5)_3]_2O$ + HCl	in hot C_2H_5OH	$(C_6H_5)_3SiOH$ + I (97)	[51]
$(C_6H_5)_3Sb(OSi(C_6H_5)_3)_2$ + HCl	in C_2H_5OH	$(C_6H_5)_3SiOH$ + I (ca. 100)	[51, 52]
$(C_6H_5)_3SbS_2O_4$ + HCl	in C_2H_5OH	I	[53]
$(C_6H_5)_3Sb{=}NSO_2R$ + HCl gas (R = C_6H_5, 4-$CH_3C_6H_4$)	in CH_2Cl_2	$C_6H_5CONSO_2R$ + I (90) from filtrate	[34, 54]
$(C_6H_5)_3Sb{=}NSO_2C_6H_5 + Cl_2$	in CH_2Cl_2	$C_6H_5SO_2NCl_2$ + I (90)	[34]
$(C_6H_5)_3Sb{=}NSO_2C_6H_5$ + C_6H_5COCl	ratio 1:2 in $(CH_2Cl)_2$, 8 h reflux	$C_6H_5SO_2N(OCC_6H_5)_2$ (55) + I (85)	[54]
$(C_6H_5)_3Sb{=}NSO_2C_6H_5$ + $(CH_3)_3SiCl$	ratio 1:3 in $(CH_2Cl)_2$, 30 min at 40 to 50 °C	$C_6H_5SO_2N(Si(CH_3)_3)_2$ (55) + I (97)	[54]
$(C_6H_5)_3Sb{=}NSO_2C_6H_5$ + $GeCl_4$	in $(CH_2Cl)_2$, 30 min at 40 to 45 °C	$C_6H_5SO_2N{=}GeCl_2$ (80) + I (94)	[54]
$(C_6H_5)_3Sb{=}NSO_2C_6H_5$ + $SnCl_4$	in $(CH_2Cl)_2$, 30 min at 40 to 45 °C	$C_6H_5SO_2N{=}SnCl_2$ (85) + I (98)	[54]
$(C_6H_5)_3Sb(N{=}P(C_6H_5)_3)_2$ + $C_6H_5TeCl_3$	ratio 1:2 in C_6H_6, 3 h reflux	$C_6H_5TeCl_2(NP(C_6H_5)_3)\downarrow$ + I (85) from conc. filtrate	[55]

[a] $(C_6H_5)_3SbCl_2$ is abbreviated I.

References on p. 45

Physical Properties. IR spectra of triphenylantimony dichloride were recorded and discussed in several publications [26, 56 to 62]. A Raman spectrum is described in [62]. Complete IR (in Nujol or KBr) and solid state Raman data with assignment of the vibrations (Whiffen's nomenclature) are given in [63]. The SbCl stretching vibrations (in cm^{-1}) are observed at 279 vs (ν_{as}SbCl) in the IR spectrum and at 269 vvs (ν_sSbCl) in the Raman spectrum. Bands at 249 s (IR) and 248 w (Raman) are assigned to ν_sSbC and at 292 s (IR) and 294 vw (Raman) to ν_{as}SbC [63].

UV spectra, measured in 95% C_2H_5OH, are compared with those of other elementorganic compounds [64 to 66]; $\lambda_{max}(\varepsilon \cdot 10^{-4}$ in $L \cdot mol^{-1} \cdot cm^{-1})$ values of 218(2.90), 259.5(0.130), 263.5(0.146), 269.5(0.106) nm, and a point of inflection at 252(0.107) nm were found [66].

^{1}H NMR spectra show resonances at δ=7.38 to 7.76 and 8.10 to 8.46 ppm in $CDCl_3$ [35], or δ=7.55(H-3,4) and 8.25(H-2) ppm in $CDCl_3$ or $(CD_3)_2CO$ [9]. The ^{13}C NMR values vs. C_6H_6 are δ=2.2(C-3,5), 4.3(C-4), 6.6(C-2,6), and 12.5(C-1) ppm. The corresponding values in $CDCl_3$ are δ=129.44(C-3,5), 131.63(C-4), 133.92(C-2,6), and 139.72(C-1) ppm with J(C-2,H)=167.0, J(C-3,H)=164.6, and J(C-4,H)=161.9 Hz [68].

^{121}Sb Mössbauer spectra gave the following data (in mm/s):

T	δ	referred to	e^2qQ	Γ	Ref.
80 K	2.8±0.1	InSb	−21.7±0.9	−	[69]
4 K	2.42	InSb	−20.5	−	[70]
80 K	−6.9±0.2	$^{121}SnO_2$	−	3.7	[71]
80 K	−6.1±0.2	$^{121}SnO_2$	−20±1	2.6±0.2	[72]
4 K	−6.02±0.02	$Ca^{121}SnO_3$	−20.6±0.3	2.55±0.04	[73]

From these data the structure of the molecule has been discussed. Quantum chemical calculations in connection with the spectra are given in [70, 74, 75]; the quadrupole coupling constant was calculated to −19.9 mm/s [76] and compared with that of $[(C_6H_5)_3SnCl_2]^-$ in [77].

NQR spectra were measured and discussed. The value for ν^{35}Cl is given as 16.0 MHz at 300 K [78]. The other values (ν and e^2Qq/h in MHz) at 300 [78] and 77 K [79] are:

T	ν^{121}Sb		ν^{123}Sb			e^2Qq/h		η
in K	5/2⇌3/2	3/2⇌1/2	7/2⇌5/2	5/2⇌3/2	3/2⇌1/2	^{121}Sb	^{123}Sb	
300	177.29	90.019	162.17	107.17	56.75	592.35	755.16	0.110
77	178.26	90.36		107.8	56.24	595.48	758.97	0.103

From these and Mössbauer data, an orbital population analysis of the Sb atom was performed [70]. Core level binding energies are detected by XPS spectroscopy. The following values (against phenyl C 1s=285.00 eV as the standard) were found: Sb 4d(5/2)=35.23 and Cl 2p(3/2)=198.81 eV [80].

It was concluded from a crystal structure investigation that associated cations $[Sb_2(C_6H_5)_6]^{4+}$ are surrounded tetrahedrally by Cl^- ions [81]. This was found to be incorrect upon reinvestigation [82, 83]. Analysis of the Patterson projections gave the coordinates of the Sb and Cl atoms and the approximate coordinates of the C atoms, which are consistent with a trigonal bipyramidal structure of the molecule [82]. The exact structure was estab-

References on p. 45

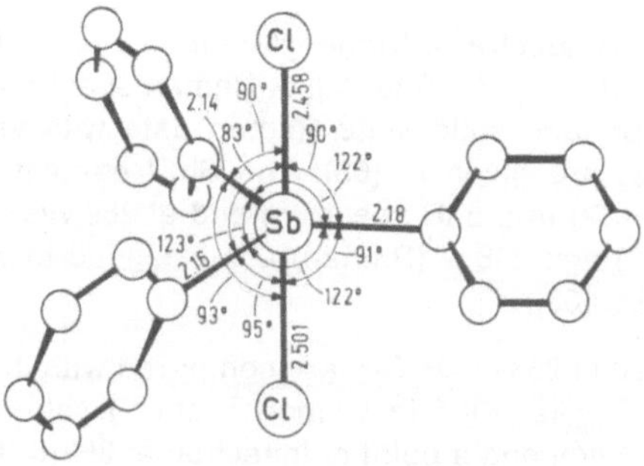

Fig. 4. Molecular structure of $(C_6H_5)_3SbCl_2$ [83].

lished and refined to R=0.17 [83]. Crystals were obtained from $CHCl_3$ solution. The following parameters were found: rhombohedral symmetry, space group $P2_12_12_1$-D_2^4 (No. 19) with $a=13.17\pm0.03$, $b=11.08\pm0.03$, and $c=12.39\pm0.03$ Å; Z=4, $d_c=1.56$ and $d_m=1.55$ g/cm³. The main angles and distances of the trigonal bipyramidal molecule with sloped (45°) phenyl rings in the equatorial plane and the two chlorine atoms in axial positions are shown in **Fig. 4** [83]. Concerning intermolecular interaction in the crystal, depending on the space group, see [84].

Dipole moments of 1.35 [88] and 1.19 D in C_6H_6 [89] and a molar susceptibility of $\chi_M=-215.1\times10^{-6}$ [90, 91] were determined.

A mass spectrum of the compound was measured (70 eV, source temperature 250 °C, direct inlet at 25 to 150 °C), discussed and compared with $(C_6H_5)_3SbBr_2$ and $(CH_3)_3SbX_2$ (X=Cl, Br). The following fragments are given: $[Sb(C_6H_5)_3Cl_n]^+$ (n=0, 1), $[Sb(C_6H_5)_2Cl_n]^+$ (n=0 to 2), $[SbC_6H_5Cl_n]^+$ (n=0 to 2), $[Sb(C_6H_4)_n]^+$ (n=1, 2), $[SbCl_n]^+$ (n=0 to 2), $[C_{12}H_n]^+$ (n=8 to 10), $[C_6H_n]^+$ (n=5, 6), $[C_4H_3]^+$, $[HCl]^+$, and $[Cl]^+$ [85].

No electrolytic conductivity is observed in CH_3CN [5, 86] or CH_3OH. Small values of specific conductivity result from partial hydrolysis [5]. The polarographic half-wave potential in 1N HCl against saturated calomel at 25 °C is −0.91 V, but in 1N NH_4OH unsatisfactory results were obtained [92]. Two potentials of −2.0 and −3.3 V were measured in glyme with $[N(C_4H_9)_4]ClO_4$ as supporting electrolyte against 10^{-3} M $Ag/AgClO_4$ for the two step reaction $(C_6H_5)_3SbCl_2+2e\rightarrow(C_6H_5)_3Sb+2Cl^-$, $(C_6H_5)_3Sb+2e\rightarrow(C_6H_5)_2Sb^-+C_6H_5^-$ [93]. Polarograms in C_2H_5OH/H_2O solutions at different concentrations and in the presence of HCl, NaOH, or buffers are shown in [94].

The compound is slightly soluble in ether, cold C_2H_5OH [1], H_2O [5], soluble in boiling C_2H_5OH, C_6H_6, and CS_2. It is rather stable towards hydrolysis in H_2O [1]. It dissolves as a monomer in $CHBr_3$ as detected by cryoscopy [5]. Solubilities (at 25 °C) are 1 in C_6H_6, 0.3 in CCl_4, and 1.8 mol/L in $CHCl_3$. Distribution coefficients between these organic phases and water are 10^5, 3×10^4, and 2×10^5, respectively [87].

Reactions. Solid $(C_6H_5)_3SbCl_2$ was irradiated with neutrons to study the n,γ- [107, 147, 186] and n,2n-reactions [146]. The fraction of radioactive isotopes retained in the form of the initial compound and the fraction of nuclei stabilized in the form of inorganic ions was determined [107, 142, 146, 147]. During the β-decay process of $(C_6H_5)_3{}^{125}SbCl_2$, the compounds $Te(C_6H_5)_2$, $(C_6H_5)_2TeCl_2$, and $(C_6H_5)_3TeCl$ are formed and can be separated by paper chromatography [148].

No isotope exchange could be observed in the systems $(C_6H_5)_3SbCl_2/{}^{125}SbCl_3$ and $(C_6H_5)_3SbCl_3/{}^{125}SbCl_5$ in ethereal or alcoholic medium, even at 100 °C in sealed ampules

References on p. 45

[147]. The kinetics of the isotope exchange in the system $(C_6H_5)_3SbCl_2/^{125}Sb(C_6H_5)_3$ were studied in n-propanol at 50, 60, and 70 °C [149, 150].

The redistribution reactions $(C_6H_5)_3SbCl_2+(C_6H_5)_3SbF_2 \rightleftharpoons 2(C_6H_5)_3SbClF$ and $(C_6H_5)_3SbCl_2+(CH_3)_3SbF_2 \rightleftharpoons (C_6H_5)_3SbF_2+(CH_3)_3SbCl_2$ were investigated in $CHCl_3$ by ^1H and ^{19}F NMR. The equilibrium constants at 32 °C are $K=10\pm0.1$ (9.0 ± 1.2 at 35 °C [152]) and 600 ± 200, respectively [151].

$(C_6H_5)_3SbCl_2$ gives color reactions with acid dyes like eriochrome black T, chrome dark blue, and xylenol orange forming extractable products. The analytical detection limit is 10^{-3} to 10^{-4} M [143].

Further reactions of $(C_6H_5)_3SbCl_2$ are summarized in Tables 12 to 14.

Table 12
Reactions of $(C_6H_5)_3SbCl_2$.

reactants	reaction conditions	products (yield in %)	Ref.
–	30 min at 250 to 270 °C and 150 Torr, subsequent distillation	$(C_6H_5)_2SbCl$ (35)	[46, 95]
H_2O	in aqueous C_2H_5OH	$(C_6H_5)_3Sb(OH)Cl$	[48, 178]
–	in H_2O some days reflux	$(C_6H_5)_3SbO$	[96]
$NaOH/H_2O$	in boiling CH_3OH, C_2H_5OH, or C_6H_6	$[(C_6H_5)_3SbCl]_2O$ (97)	[5, 18, 41]
	hydrolysis constants in C_6H_6: $\log K_1=9$, $\log K_2=7$	$(C_6H_5)_3SbCl(OH)$, $(C_6H_5)_3Sb(OH)_2$	[87]
KOH	in C_2H_5OH	$(C_6H_5)_3Sb(OH)_2$	[1, 2, 184]
NH_3	in C_2H_5OH at 20 °C	NH_4Cl precipitates	[1]
KOH/C_2H_5OH	in $(CH_3)_2CO$, 1 h at 30 °C	$(C_6H_5)_3SbO$	[97]
H_2S gas	in C_2H_5OH/NH_3, 1 h at 20 °C	$Sb(C_6H_5)_3+HCl+S$	[1]
conc. H_2SO_4	–	$(C_6H_5)_3SbSO_4$	[98]
HNO_3	at 40 °C	$(O_2NC_6H_4)_3SbCl_2$ (?)	[41]
KF	in aqueous C_2H_5OH, 1 h reflux	$(C_6H_5)_3SbF_2$ (85)	[9, 99]
	in aqueous $(CH_3)_2CO$, chromatography and anion exchange	$(C_6H_5)_3SbF_2$ and other compounds	[107]
HN_3+NaN_3 in excess	in C_6H_6, 12 h at 20 °C	$(C_6H_5)_3Sb(N_3)_2$	[100]
NaN_3 in excess	in C_6H_6	$(C_6H_5)_3Sb(N_3)_2$	[101]
NaSCN or AgSCN	in CH_3OH or $(CH_3)_2CO$	$(C_6H_5)_3Sb(SCN)_2$	[101]

References on p. 45

Table 12 [continued]

reactants	reaction conditions	products (yield in %)	Ref.
$Pb(SCN)_2$	8 h in C_6H_6	$(C_6H_5)_3Sb(SCN)_2$ + $(C_6H_5)_3Sb(SCN)OH$	[102]
AgOCN	in ether, 1 d at 20 °C	$(C_6H_5)_3Sb(NCO)_2$ (70)	[100, 101, 103]
$AgNO_3$	in ether or wet CH_3CN	$[(C_6H_5)_3SbNO_3]_2O$	[48, 104]
excess N_2O_4	without solvent at 0 °C	$(C_6H_5)_3Sb(NO_3)_2$	[104]
Ag_2SO_4	boiling H_2O	probably $[(C_6H_5)_3Sb(HSO_4)]_2O$	[48]
$AgClO_4$	in C_2H_5OH	$[(C_6H_5)_3SbClO_4]_2O$ caution, explosive!	[98]
$AgClO_4$ + D, D = $(CH_3)_2SO$, $(C_6H_5)_2SO$, $(C_6H_5)_3PO$, or $(C_6H_5)_3AsO$	2 h in C_6H_6 and extraction of precipitate with C_2H_5OH	$[(C_6H_5)_3SbD_2](ClO_4)_2$	[105]
$LiBH_4$	ratio 1:5 in ether at −120 to −65 °C	H_2, B_2H_6, LiCl, and $Sb(C_6H_5)_3$	[106]
$LiAlH_4$	ratio 1:5 in ether at −90 °C	H_2, Al_2H_6, LiCl, and $Sb(C_6H_5)_3$	[106]
LiC_6H_5	in ether, 1 h shaking, workup with H_2O	$Sb(C_6H_5)_5 \cdot 0.5$ c-C_6H_{12} (90), recrystallized from c-C_6H_{12}	[38]
1,4-$Li_2C_6H_4$	in ether at −25 °C	$[(C_6H_5)_3SbC_6H_4]_n$ (n = 2 to 3)	[108]
C_6H_5MgBr	ratio 1:3 in ether, some C_6H_6, 3 d at 20 °C, workup with ice/HBr	$(C_6H_5)_4SbBr$ (53)	[6, 109]
C_6H_5MgBr in ether	in $CH_3OCH_2CH_2OCH_3$, 8 h at 20 °C	$Sb(C_6H_5)_5$ (62)	[110]
$Pb_2(C_6H_5)_6$	ratio 1:1, 6 h reflux in $CHCl_3$	$Pb(C_6H_5)_4$ (80), $(C_6H_5)_2PbCl_2$ (60), $Sb(C_6H_5)_3$ (45)	[111]
$NaOCH_3$	in boiling CH_3OH	$(C_6H_5)_3Sb(Cl)OCH_3$ (86)	[5, 112]
C_2H_5OH	in $C_2H_5OH + NH_3$	$(C_6H_5)_3Sb(OC_2H_5)_2$	[112]
$LiO(CH_2)_2OH$	in $CH_3CO_2C_2H_5/C_6H_{14}$	$(C_6H_5)_3Sb(OCH_2CH_2OH)_2$	[113]
Na-acac	in C_6H_6	$(C_6H_5)_3Sb(Cl)acac$	[114]

References on p. 45

Table 12 [continued]

reactants	reaction conditions	products (yield in %)	Ref.
HOC_6H_4X, X = H, 2-, 3-, 4-Cl, 4-Br, 2-,3-, 4-CH_3, 4-OCH_3, 4-C_4H_9-t	ratio 1:2, in C_6H_6 + $N(C_2H_5)_3$	$(C_6H_5)_3Sb(OC_6H_4X)_2$ (40 to 70)	[115]
HOC_6H_4X, X = 2-,3-,4-NO_2	ratio 1:2, in C_6H_6 + $N(C_2H_5)_3$	$[(C_6H_5)_3SbOC_6H_4X]_2O$	[115]
Na-8-quinolinolate	ratio 1:1 in C_2H_5OH	$(C_6H_5)_3SbCl(OC_9H_6N)$	[116]
$NaOC_6H_4CR{=}NR'$-2, R = H, R' = CH_3, C_2H_5, C_3H_7, C_4H_9, C_6H_5, 4-$CH_3C_6H_4$; R = CH_3, R' = CH_3, C_4H_9, i-C_4H_9	ratio 1:1 in C_6H_6, 2 h reflux	$(C_6H_5)_3SbCl(OC_6H_4CR{=}NR'$-2) (60)	[117]
2-HOC_6H_4OH	in CH_2Cl_2 + NH_3 gas at −5 °C	$(C_6H_5)_3Sb(-OC_6H_4O$-2-) · 0.5 H_2O	[118]
$(NaOC_6H_4O(CH_2)_2$-2$)_2O$	–	$(C_6H_5)_3Sb(OC_6H_4O(CH_2)_2-2)_2O$	[119]
$AgO_2CC_6H_5$	in $CHCl_3$, shaking for 12 h	$(C_6H_5)_3Sb(O_2CC_6H_5)_2$	[103]
$HO_2CC_6H_4X$, X = H, 2-,3-,4-Cl, 2-,3-,4-NO_2, 2-,4-OH, 2-,3-,4-NH_2, 2-,3-,4-CH_3, 4-OCH_3	ratio 1:2 in C_6H_6 with $N(C_2H_5)_3$	$(C_6H_5)_3Sb(O_2CC_6H_4X)_2$ (40 to 70)	[115]
$(HO_2CCH_2)_2NH$	in aqueous THF + NaOH	$(C_6H_5)_3Sb(O_2CCH_2)_2NH$	[120]
$HON{=}CR_2$, R = CH_3, C_2H_5, C_3H_7	ratio 1:2 in C_6H_6 with $N(C_2H_5)_3$	$(C_6H_5)_3Sb(N{=}CR_2)_2$	[121]
$NaON{=}CR_2$, R = CH_3, C_2H_5, C_3H_7	ratio 1:2 in C_6H_6	$(C_6H_5)_3Sb(N{=}CR_2)_2$	[121]
$HON{=}CRR'$, R = H, R' = 4-$CH_3OC_6H_4$, 2-furyl; R = CH_3, R' = C_6H_5, 4-$NO_2C_6H_4$; R = R' = C_6H_5; R, R' = $(CH_2)_5$	ration 1:2 in C_6H_6 with $N(C_2H_5)_3$, 3 h reflux	$(C_6H_5)_3Sb(N{=}CRR')_2$	[122]
2-$NaOC_6H_4$-$CH{=}NC_6H_4XNa$-2' (X = O,S), $C_6H_5(NaO)C{=}CH$-$C(CH_3){=}NC_6H_4ONa$-2, $CH_3(NaO)C{=}CH$-$C(CH_3){=}NC_6H_4ONa$-2	in CH_3OH at 20 °C	corresponding 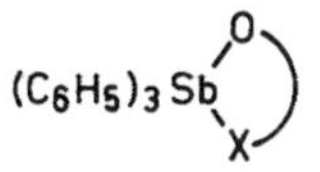 X = O, S (63 to 68)	[123, 124]

References on p. 45

Table 12 [continued]

reactants	reaction conditions	products (yield in %)	Ref.
$CH_3(NaO)C{=}CHC(CH_3){=}NC_6H_4SNa$-2	in CH_3OH at 20 °C	as above (X = S), not isolated due to spontaneous decomposition	[124]
c-$C_6H_{11}OOH$	in C_6H_6 with NH_3 or amine	$(C_6H_5)_3Sb(OOC_6H_{11}$-c$)_2$ (74)	[125, 126]
ROOH, R = H, t-C_4H_9, $C_6H_5(CH_3)_2C$, 1,2,3,4-tetrahydronaphthalinyl, 3,4-dihydro-1H-benzoisopyranyl, cyclohexenyl	in C_6H_6 with $NaNH_2$ at room temperature	$(C_6H_5)_3Sb(OOR)_2$	[127]
$(C_6H_5)_3SiOH$	ratio 1:2 in ether + NH_3	$(C_6H_5)_3Sb(OSi(C_6H_5)_3)_2$ (85)	[51]
$(CH_3)_3SiONa$	ratio 1:1 in C_6H_6, 1 h at 25 °C	$(C_6H_5)_3SbCl(OSi(CH_3)_3)$ (27)	[128]
	ratio 1:2 in C_6H_6, 1 h at 80 °C	$(C_6H_5)_3Sb(OSi(CH_3)_3)_2$ (56)	[128]
$(C_6H_5)_3SiOOH$	in ether + $N(C_2H_5)_3$	$(C_6H_5)_3Sb(OOSi(C_6H_5)_3)_2$ (74)	[51]
RSH, R = C_3H_7, C_4H_9	ratio 1:2 in C_6H_6 + NH_3	$Sb(C_6H_5)_3$ + $(RS)_2$ (100)	[129]
$AgSCF_3$	in CH_3CN, 1 h at 50 °C	$(C_6H_5)_3SbF_2$ (77)	[40]
$NaSC(S)NR_2 \cdot nH_2O$, R = CH_3, C_2H_5, C_6H_5	2 h in $CHCl_3$ or C_6H_6/CH_3CN	$Sb(C_6H_5)_3$ (60) + $[R_2NC(S)]_2S_2$ (90)	[130]
$(KS)_2C{=}N{-}CN$	in $HCON(CH_3)_2$	probably $(C_6H_5)_3Sb(S_2CNCN)$	[131]
$(KS)_2C{=}C(CN)_2$	in $HCON(CH_3)_2$	probably $(C_6H_5)_3Sb(S_2C{=}C(CN)_2)$	[132]
$(R_3Sn)_2S$, R = C_4H_9, C_6H_5	in $CHCl_3$ at −5 °C	$(C_6H_5)_3SbS$ (80 to 90) + R_3SnCl	[133]
$(C_6H_5)_3SbO$	in C_6H_6, 30 min reflux	$[(C_6H_5)_3SbCl]_2O$ (96)	[134]
$(CH_3)_3SiX$, X = $NCH_3{-}C(C_6H_5){=}NCH_3$	in CH_2Cl_2	$(CH_3)_3SiCl$ + $(C_6H_5)_3Sb(Cl)X$	[135]

References on p. 45

Table 13
Formation of Polymers from $(C_6H_5)_3SbCl_2$ and Polyfunctional Organic Reactants.

reactant (% yield of polymer)	reaction conditions; properties of products	Ref.
antimony polyesters:		
$HO_2C(CH_2)_nCO_2H$, n = 0 to 4 (ca. 30)	in $H_2O + Ag_2O$ or in C_6H_6 with $N(C_2H_5)_3$	[115]
phthalic acid (ca. 30) isophthalic acid (ca. 30) terephthalic acid (ca. 30)	in $H_2O + Ag_2O$	[115]
tetramethylterephthalic acid (49 to 56) dimethylterephthalic acid (ca. 3) terephthalic acid (6 to 46) bromoterephthalic acid (ca. 7) nitroterephthalic acid (19) dichloroterephthalic acid (ca. 18) $CH_2{=}C(CO_2H)CH_2CO_2H$ (0) 4-oxo-4H-pyran-2,6-dicarboxylic acid (7) mercaptosuccinic acid (31) oxalic acid (27 to 39) maleic acid (ca. 3) fumaric acid (28 to 93)	$(C_6H_5)_3SbCl_2$ in CCl_4 added to aqueous solution of diacid (neutralized with NaOH), stirred for 30 min at 25 °C; dec. at about 300 °C	[136, 137]
2,5-dimethylterephthalic acid	probably as above; biologically active	[140]
polyacrylic acid	$(C_6H_5)_3SbCl_2$ in $CHCl_3$ or CH_3NO_2 or C_6H_6, acid in H_2O/NaOH; Sb inclusion from 42 to 98% depending on organic phase	[144]
$(C_5H_5CO_2H)_2Fe$ (4 to 40)	in CCl_4/H_2O/NaOH as above, stirred for 30 s at 25 °C; biologically active	[136, 140]
$[(C_5H_5CO_2H)_2Co^+]PF_6^-$ (80 to 90)	in CCl_4/H_2O/NaOH	[141]
antimony polyoximes:		
2,5-cyclohexadiene-1,4-dione dioxime (25) 1,4-benzenedicarboxaldehyde dioxime (24)	$(C_6H_5)_3SbCl_2$ in $CHCl_3$ added to stirred solution of dioxime in H_2O, neutralized with NaOH (pH 11 to 13), stirred for 15 min at 25 °C; dec. at 200 to 300 °C	[138]
antimony polyamines:		
1,6-diaminohexane (36) 1,12-diaminododecane 2,6-diamino-8-purinol (39) 2,6-diaminoanthraquinone (40) 2,6-diamino-5-nitropyrimidine (14)	$(C_6H_5)_3SbCl_2$ in CCl_4 added to stirred solution of diamine and NaOH in H_2O for 30 s at 25 °C; biological activity against bacteria and cancer cell lines	[139]

References on p. 45

Table 13 [continued]

reactant (% yield of polymer)	reaction conditions; properties of products	Ref.
2,4-diamino-5-[(3,4-dimethylphenyl)-methyl]pyrimidine (21)	conditions on p. 41	[139]
4,4'-diaminodiphenylsulfon (18)		
adenine (9)		
4-phenylenediamine (12)		
2-methoxy-4-phenylenediamine (76)		
2,3,5,6-tetramethyl-4-phenylenediamine (8)		
2,5-dichloro-4-phenylenediamine (39)		
2-nitro-4-phenylenediamine (23)		
4,4'-methylenedianiline (54)		
4,4'-diaminobenzanilide (43)		
$[SC(S)NH(CH_2)_2NHC(S)S]^{2-}Zn^{2+}$ (36)		
4,6-diamino-2(1H)-pyrimidinethione 2,6-dichloro-4-phenylenediamine	probably as above; biologically active	[140]

Table 14
Complex Formation of $(C_6H_5)_3SbCl_2$.

reactant (ratio)	reaction conditions	product (properties)	Ref.
BCl_3 (1:2)	in C_6H_6 or CH_3CN at 20 °C	$(C_6H_5)_3SbCl_2 \cdot 2BCl_3$ (white powder)	[153]
$SbBr_3$ (1:1)	in $CHCl_3$	$(C_6H_5)_3SbCl_2 \cdot SbBr_3$ (m.p. 102 °C)	[29]
$SbCl_3$ (1:1)	in CCl_4 or C_6H_{14}	$(C_6H_5)_3SbCl_2 \cdot SbCl_3$ (see p. 43)	[29]
$SbCl_5$ (1:1)	in CCl_4	$(C_6H_5)_3SbCl_2 \cdot SbCl_5$ (see p. 44)	[29]
$SbCl_5$ (1:2)	in $CHCl_3$	$(C_6H_5)_3SbCl_2 \cdot 2SbCl_5$ (pale yellow oil)	[29]
$TiCl_4$ (2:1)	in C_6H_{14}, 15 h at 60 °C	$(C_6H_5)_3SbCl_2 \cdot (TiCl_4)_{0.55}$	[154]
(1:1)		$(C_6H_5)_3SbCl_2 \cdot (TiCl_4)_{0.91}$	
(1:2)		$(C_6H_5)_3SbCl_2 \cdot (TiCl_4)_{1.24}$	
$TiCl_4$ (1:1)	in C_6H_6 at −5 °C, warming up to 20 °C	$(C_6H_5)_3SbCl_2 \cdot TiCl_3$ (brown powder, $\mu_M = 1.75$ B.M.)	[155]
$TiCl_4$ (1:1)	in C_6H_6, 15 h at 70 °C	$(C_6H_5)_3SbCl_2 \cdot TiCl_4$ (yellow powder; diamagnetic)	[155]
$ZrCl_4$ (1:1)	probably in C_6H_6	$(C_6H_5)_3SbCl_2 \cdot ZrCl_4$ (dirty white)	[27]
$NbCl_5$ (1:1)	probably in C_6H_6	$(C_6H_5)_3SbCl_2 \cdot NbCl_5$ (yellow)	[27]
WCl_6 (1:1)	probably in C_6H_6	$(C_6H_5)_3SbCl_2 \cdot WCl_6$ (green-brown)	[27]
VCl_3 (1:1)	in C_6H_6 or CH_3CN, 15 h reflux	$(C_6H_5)_3SbCl_2 \cdot VCl_3$ (gray powder, m.p. >260 °C)	[153]
$FeCl_3$ (1:2)	in C_6H_6, 15 h reflux	$(C_6H_5)_3SbCl_2 \cdot 2FeCl_3$ (brown solid)	[156]

References on p. 45

Uses. Triphenylantimony dichloride can be used for extracting fluoride ions from water into organic solvents like CCl_4 [157 to 160], C_6H_6 [159, 160], or $CHCl_3$ [161 to 163]. Distribution coefficients were measured (see originals). The compound may act as a catalyst for the reaction of epoxides with CO_2 forming cyclic carbonates in quantitative yield [164 to 168] or as a catalyst for the photoinitiated polymerization of vinyl monomers via a radical mechanism [169]. The use as an additive to epoxy resins to prevent flammability and depolymerizations was studied in several works [170 to 174]. Halogen-containing polyesters are also protected against flammability by the compound [176, 177]. Polymers and polycondensates mixed with chlorinated paraffin waxes and $(C_6H_5)_3SbCl_2$ are transparent and difficult to inflame [175]. It is useful for the preparation of photocrosslinkable organometallic polyesters with different ketones, phenoles, and α,ω-alkanedichlorides [179]. The compound retards the oxidation of furfurol in alkali and accelerates the oxidation of styrene [180]. Chlorine selective, solid membrane electrodes may be made from the compound and paraffin wax on graphite [181]. Triphenylantimony dichloride prevents growing of microorganisms including bacteria and fungi in paints, plastics, textile, and paper products [182]. Concerning its fungitoxicity using acetone as carrier and a comparison with other organoantimony compounds, see [183].

$(C_6H_5)_3SbCl_2 \cdot SbCl_3$

Addition of a solution of $Sb(C_6H_5)_3$ in CCl_4 to $SbCl_5$ (ratio 1:1) in the same solvent leads to immediate precipitation of the compound. The adduct also precipitates immediately upon mixing of solutions of $(C_6H_5)_3SbCl_2$ and $SbCl_3$ in n-hexane. After recrystallization from CCl_4 the compound melts at 126 °C.

The IR spectrum below 500 cm^{-1} is essentially a superposition of the spectra of $(C_6H_5)_3SbCl_2$ and $SbCl_3$: 455 sh, 452 s (C_6H_5 y mode), 375 s, 371 sh (ν_sSbCl_3), 340 s, 322 s ($\nu_{as}SbCl_3$), 295 s, 284 sh, 245 s (C_6H_5 t mode), 233 w (C_6H_5 u mode) cm^{-1}.

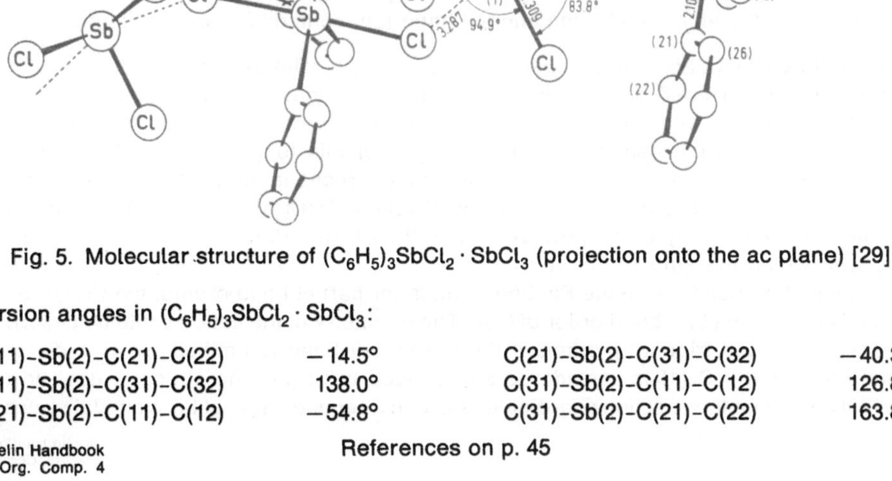

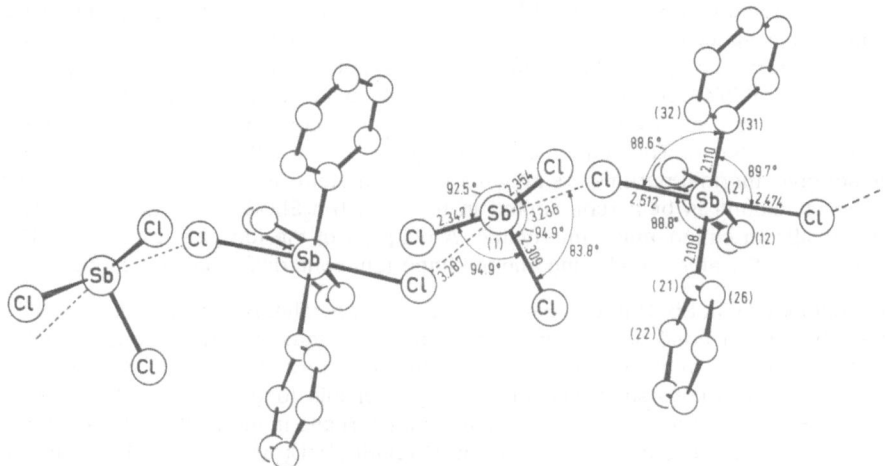

Fig. 5. Molecular structure of $(C_6H_5)_3SbCl_2 \cdot SbCl_3$ (projection onto the ac plane) [29].

Torsion angles in $(C_6H_5)_3SbCl_2 \cdot SbCl_3$:

C(11)-Sb(2)-C(21)-C(22)	−14.5°	C(21)-Sb(2)-C(31)-C(32)	−40.3°
C(11)-Sb(2)-C(31)-C(32)	138.0°	C(31)-Sb(2)-C(11)-C(12)	126.8°
C(21)-Sb(2)-C(11)-C(12)	−54.8°	C(31)-Sb(2)-C(21)-C(22)	163.8°

References on p. 45

An X-ray crystal structure of the monoclinic crystals was performed; crystal data: space group $P2_1/c-C^5_{2h}$, a=9.118(4), b=10.888(4), c=22.435(8) Å, and β=97.84(2)°; Z=4; d_c= 1.96 g/cm³. The structure was refined to R=0.038. It is shown in **Fig. 5**, p. 43.

The major feature of the structure is the formation of weak bonds between the two axial chlorine atoms of a $(C_6H_5)_3SbCl_2$ molecule and adjacent $SbCl_3$ molecules to give polymeric chains parallel to the a axis. The trigonal bipyramidal geometry of $(C_6H_5)_3SbCl_2$ in the adduct is little different from that in the pure compound, the major difference being in the orientation of the phenyl groups. In the parent compound, these are related by a pseudo threefold axis; but in the adduct, the plane of the C(21)-C(26) group is almost coincident with the Sb(2)-C(11)-C(21)-C(31) equatorial plane. The orientation of the phenyl groups is more accurately defined by the torsion angles. Such minor changes are perhaps to be expected as a result of decreased nonbonded interactions in the more spacious lattice of the adduct. Coordination about the Sb(1) atom is square pyramidal (or octahedral if the antimony lone pair is considered to be stereochemically active), showing three short Sb-Cl distances (mean 2.337 Å) and two longer contacts (mean 3.262 Å) [29].

$(C_6H_5)_3SbCl_2 \cdot SbCl_5$

A solution of $SbCl_5$ in CCl_4 is added dropwise to a solution of $(C_6H_5)_3SbCl_2$ (ratio 1:1) in the same solvent. Extremely moisture-sensitive crystals are slowly deposited. Recrystallization from $CHCl_3$ gave a melting point of 115 to 116 °C. With donor solvents, only uncomplexed $(C_6H_5)_3SbCl_2$ could be recovered.

IR spectra of $(C_6H_5)_3SbCl_2 \cdot SbCl_5$ below 600 cm^{-1} show bands at 451 s, 443 sh, 347 vs, 330 sh, 295 s, and 286 sh cm^{-1}. Those at 286 and 451 cm^{-1} are characteristic phenyl group modes, while the sharp intense SbCl stretching band at 347 cm^{-1} could easily be assigned to $[SbCl_6]^-$.

An X-ray structure was performed from a single crystal of rather poor quality. Crystal data: monoclinic, space group $C2/c-C^6_{2h}$ (No. 15), a=25.033(8), b=10.147(4), c=26.043(8) Å, and β=133.46(3)°; Z=8; d_c=2.00 g/cm³. Structure determination was established from data collected for 2746 observed reflections. During data collection, the intensity of the standard reflections decreased substantially, probably as a consequence of crystal deterioration in the X-ray beam. The refinement converged at R=0.119 with isotropic, and 0.084 with anisotropic, thermal parameters and the hydrogen atoms fixed at their calculated positions. The compound is best represented in terms of $[Ph_3SbCl]^+$ and $[SbCl_6]^-$ ions, but there is significant cation-anion interaction. A diagram of the ion pair is shown in **Fig. 6**, which also gives the atom numbering and the important bond distances and angles.

The antimony atom Sb(1) in the anion is in distorted octahedral coordination to chlorine with the Sb(1)-Cl(1) bond significantly longer (2.414 Å) than the other bonds which are equal (mean 2.344 Å) within the limits of the determinations. Bond angle deviations from the ideal values are consistent with the presence of this longer bond with a reduction of the Cl(1)-Sb(1)-Cl(2) angle to 87.4°, leading to a decrease in the Cl(2)-Sb(1)-Cl(4) angle to 176.5° and a compensating increase in the Cl(2)-Sb(1)-Cl(5) angle to 91.7°. The anion and cation are bridged by Cl(1); however, while the Cl(1)-Sb(2) distance is long (3.231 Å), it falls well within the sum of the van der Waals radii (4.00 Å), implying the presence of a weak secondary bond. From the Pauling formula for partial bond orders, the Cl(1)···Sb(2) distance is equivalent to a bond order of 0.06. The geometry of the cation could be described either as a distorted trigonal pyramid if the secondary bond is omitted, or as a distorted trigonal bipyramid if Cl(1) is considered still bonded to Sb(2). The structure does in fact represent a point on the reaction coordinate between a regular trigonal bipyramidal structure

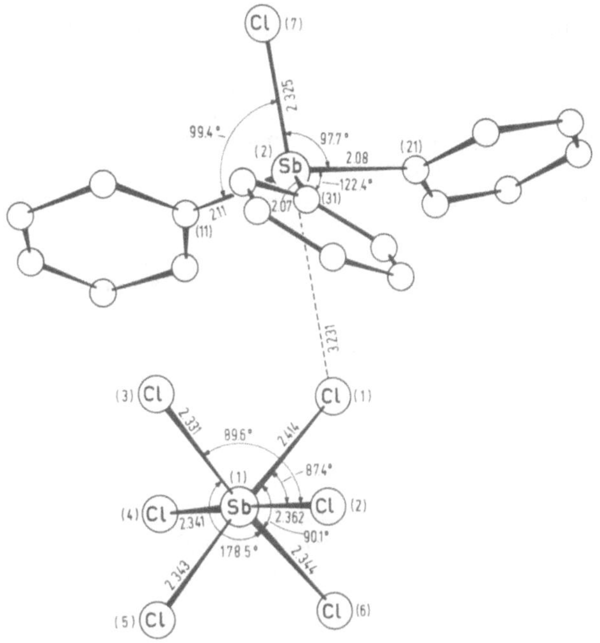

Fig. 6. Molecular structure of $[(C_6H_5)_3SbCl]^+[SbCl_6]^-$ [29].

Torsion angles in $[(C_6H_5)_3SbCl]^+[SbCl_6]^-$:

C(11)-Sb(2)-C(21)-C(22)	−27.6°	C(21)-Sb(2)-C(31)-C(32)	51.1°
C(11)-Sb(2)-C(31)-C(32)	−158.5°	C(31)-Sb(2)-C(11)-C(12)	170.2°
C(21)-Sb(2)-C(11)-C(12)	−36.7°	C(31)-Sb(2)-C(21)-C(22)	124.4°

and the tetrahedral geometry observed for the cation in $[Sb(CH_3)_4]^+[SbCl_6]^-$. In view of the great disparity in the Sb(2)-Cl(1) and Sb(2)-Cl(7) distances, formulation as a stibonium salt is the more realistic alternative. The torsion angles for the phenyl groups indicate that the C(11)-C(16) phenyl group is tilted in the opposite sense to the C(21) and C(31) groups. This is in contrast to the situation in both free $(C_6H_5)_3SbCl_2$ and the $SbCl_3$ adduct [29].

References:

[1] A. Michaelis, A. Reese (Liebigs Ann. Chem. **233** [1886] 39/60).
[2] A. Michaelis, L. Weitz (Ber. Deut. Chem. Ges. **20** [1887] 52).
[3] P. Pfeiffer, I. Heller (Ber. Deut. Chem. Ges. **37** [1904] 4620/3).
[4] W. J. Lile, R. J. Menzies (J. Chem. Soc. **1950** 617/21).
[5] L. Kolditz, M. Gitter, E. Rösel (Z. Anorg. Allgem. Chem. **316** [1962] 270/7).
[6] H. H. Willard, L. R. Perkins, F. F. Blicke (J. Am. Chem. Soc. **70** [1948] 737/8).
[7] K. S. Mingaleva, G. M. Bogolyubov, Yu. N. Shlyk, A. A. Petrov (Zh. Obshch. Khim. **39** [1969] 2679/93; J. Gen. Chem. [USSR] **39** [1969] 2616/9).
[8] A. J. Banister, L. F. Moore (J. Chem. Soc. A **1968** 1137/8).
[9] B. Raynier, B. Waegell, R. Commandeur, H. Mathais (Nouv. J. Chim. **3** [1979] 393/401).
[10] A. Michaelis, A. Reese (Ber. Deut. Chem. Ges. **15** [1882] 2876/7).
[11] J. Hasenbäumer (Ber. Deut. Chem. Ges. **31** [1898] 2910/4).

[12] A. N. Nesmeyanov, K. A. Kocheshkov (Izv. Akad. Nauk SSSR Otd. Khim. Nauk **1944** 416/31).
[13] A. N. Nesmeyanov, N. K. Gipp, L. G. Makarova, K. K. Mozgova (Izv. Akad. Nauk SSSR Otd. Khim. Nauk **1953** 298/302; Bull. Acad. Sci. USSR Div. Chem. Sci. **1953** 271/4).
[14] F. B. Makin, W. A. Waters (J. Chem. Soc. **1938** 843/8).
[15] W. A. Waters (J. Chem. Soc. **1937** 2007/4).
[16] A. N. Nesmeyanov, A. E. Borisov (Izv. Akad. Nauk SSSR Ser. Khim. **1969** 974/5; Bull. Acad. Sci. USSR Div. Chem. Sci. **1969** 895).
[17] A. N. Nesmeyanov, A. E. Borisov, N. V. Novikova (Izv. Akad. Nauk SSSR Ser. Khim. **1969** 1978/82; Bull. Acad. Sci. USSR Div. Chem. Sci. **1969** 1830/3).
[18] A. N. Nesmeyanov, A. E. Borisov (Izv. Akad. Nauk SSSR Ser. Khim. **1969** 939/40; Bull. Acad. Sci. USSR Div. Chem. Sci. **1969** 853/5).
[19] O. A. Reutov, A. N. Lovtsova (Vestn. Mosk. Univ. Ser. Mat. Mekhan. Astron. Fiz. Khim. **13** [1958] 191/6).
[20] F. Challenger, F. Pritchard, J. R. A. Jinks (J. Chem. Soc. **125** [1924] 864/75).

[21] A. E. Goddard (J. Chem. Soc. **121** [1922] 36/40).
[22] G. Ondrejovic, D. Makanova, D. Valigura, J. Gazo (Z. Chem. [Leipzig] **13** [1973] 193/4).
[23] D. Valigura, G. Ondrejovic, D. Makanova, J. Gazo (Chem. Zvesti **28** [1974] 599/603).
[24] D. Makanova, G. Ondrejovic (Proc. 9th Conf. Coord. Chem., Bratislava, Czech., 1983, pp. 277/81).
[25] F. J. Berry, N. Gündüz, M. Roshani, B. C. Smith (Commun. Fac. Sci. Univ. Ankara B **22** [1975] 21).
[26] W. Levason, C. A. McAuliffe (J. Inorg. Nucl. Chem. **37** [1975] 340/2).
[27] H. K. Sharma, S. N. Dubey, D. M. Puri (Proc. Indian Acad. Sci. Chem. Sci. **90** [1981] 555/8).
[28] D. Cunningham, M. J. Frazer, J. D. Donaldson (J. Chem. Soc. A **1971** 2049).
[29] M. Hall, D. B. Sowerby (J. Chem. Soc. Dalton Trans. **1983** 1095/9).
[30] C. Glidewell (J. Organometal. Chem. **116** [1976] 199/209).

[31] E. H. Kustan, B. C. Smith, M. E. Sobeir, A. N. Swami, M. Woods (J. Chem. Soc. Dalton Trans. **1972** 1326/90).
[32] B. C. Smith, M. E. Sobeir (Chem. Ind. **1969** 621).
[33] A. B. Bruker, N. M. Nikiforova (Zh. Obshch. Khim. **18** [1948] 1133/6).
[34] A. M. Pinchuk, Z. I. Kuplennik, Zh. N. Belaya (Zh. Obshch. Khim. **46** [1976] 2242/6; J. Gen. Chem. [USSR] **46** [1976] 2155/8).
[35] R. L. McKenney, H. H. Sisler (Inorg. Chem. **6** [1967] 1178/82).
[36] D. K. Padma, R. A. Shaw, A. R. Vasudeva Murthy, M. Woods (Phosphorus **4** [1974] 25/8).
[37] L. P. Filonenko, G. K. Bespal'ko, A. M. Pinchuk (Zh. Obshch. Khim. **49** [1979] 2634/5; J. Gen. Chem. [USSR] **49** [1979] 2338/9).
[38] G. Wittig, K. Clauss (Liebigs Ann. Chem. **577** [1952] 26/39).
[39] G. A. Razuvaev, N. A. Osanova, N. P. Shulaev, B. M. Tsigin (Zh. Obshch. Khim. **30** [1960] 3234/7; J. Gen. Chem. [USSR] **30** [1960] 3203/5).
[40] L. M. Yagupol'skii, N. V. Kondratenko, V. I. Popov (Zh. Obshch. Khim. **46** [1976] 620/3; J. Gen. Chem. [USSR] **46** [1976] 618/21).

[41] P. May (Proc. Chem. Soc. **26** [1910] 218; J. Chem. Soc. **97** [1910] 1956/60).
[42] J. F. Carson, F. F. Wong (J. Org. Chem. **26** [1961] 1467/70).
[43] G. H. Briles, W. E. McEwen (Tetrahedron Letters **1966** 5299/302).
[44] R. Appel, W. Heinzelmann; Badische Anilin- & Soda-Fabrik A.-G. (Ger. 1192205 [1962/65]; C. A. **63** [1965] 8405).

[45] A. N. Nesmeyanov, O. A. Reutov, O. A. Ptitsyna, P. A. Tsurkan (Izv. Akad. Nauk SSSR Otd. Khim. Nauk **1958** 1435/44; Bull. Acad. Sci. USSR Div. Chem. Sci. **1958** 1384/92).
[46] H. Schmidt (Liebigs Ann. Chem. **429** [1922] 123/52).
[47] V. I. Lodochnikova, E. M. Panov, K. A. Kocheshkov (Zh. Obshch. Khim. **34** [1964] 946/9; J. Gen. Chem. [USSR] **34** [1964] 940/3).
[48] G. T. Morgan, F. M. G. Micklethwait, G. S. Whitby (Proc. Chem. Soc. **25** [1909] 302; J. Chem. Soc. **97** [1910] 34/6).
[49] D. R. Lyon, F. G. Mann, G. H. Cookson (J. Chem. Soc. **1947** 662/70).
[50] J. Dahlmann, L. Austenat (Liebigs Ann. Chem. **729** [1969] 1/7).

[51] G. A. Razuvaev, T. G. Brilkina, E. V. Krasil'nikova, T. I. Zinov'eva, A. I. Filimonov (J. Organometal. Chem. **40** [1972] 151/7).
[52] G. A. Razuvaev, N. A. Osanova, T. G. Brilkina, T. I. Zinov'eva, V. V. Sharutin (J. Organometal. Chem. **99** [1975] 93/106).
[53] S. I. A. El Sheikh, B. C. Smith (Chem. Commun. **1968** 1474).
[54] Z. I. Kuplennik, A. M. Pinchuk (Zh. Obshch. Khim. **49** [1979] 155/60; J. Gen. Chem. [USSR] **49** [1979] 135/9).
[55] K. Bajpai, R. C. Srivastava (Syn. Reactiv. Inorg. Metal-Org. Chem. **12** [1982] 47/54).
[56] K. A. Jensen, P. H. Nielsen (Acta Chem. Scand. **17** [1963] 1875/85).
[57] G. O. Doak, G. G. Long, L. D. Freedman (J. Organometal. Chem. **4** [1965] 82/91).
[58] A. E. Borisov, N. V. Novikova, N. A. Chumaevskii, E. B. Shkirtil (Dokl. Akad. Nauk SSSR **173** [1967] 855/8; Dokl. Phys. Chem. Proc. Acad. Sci. USSR **172/177** [1967] 248/51).
[59] A. E. Borisov, N. V. Novikova, N. A. Chumaevskii, E. B. Shkirtil (Ukr. Fiz. Zh. **13** [1968] 75/82).
[60] K. M. Mackay, D. B. Sowerby, W. C. Young (Spectrochim. Acta A **24** [1968] 611/31).

[61] R. G. Goel, E. Maslowsky Jr., C. V. Senoff (Inorg. Chem. **10** [1971] 2572/7).
[62] R. G. Goel, E. Maslowsky Jr., C. V. Senoff (Inorg. Nucl. Chem. Letters **6** [1970] 833/5).
[63] B. A. Nevett, A. Perry (Spectrochim. Acta A **33** [1977] 755/60).
[64] C. N. R. Rao, J. Ramachandran, M. S. C. Iah, S. Somasekhara, T. V. Rajakumar (Nature **183** [1959] 1475/6).
[65] C. N. R. Rao, J. Ramachandran, A. Balasubramanian (Can. J. Chem. **39** [1961] 171/9).
[66] H. H. Jaffé (J. Chem. Phys. **22** [1954] 1430/3).
[67] A. Ouchi, T. Uehiro, Y. Yoshino (J. Inorg. Nucl. Chem. **37** [1975] 2347/9).
[68] J. Havranek, A. Lycka (Sb. Ved. Pr. Vys. Sk. Chemickotechnol. Pardubice **43** [1980] 123/7).
[69] S. E. Gukasyan, V. P. Gor'kov, P. N. Zaikin, V. S. Shpinel (Zh. Strukt. Khim. **14** [1973] 650/5; J. Struct. Chem. [USSR] **14** [1973] 603/7).
[70] L. H. Bowen, G. G. Long (Inorg. Chem. **15** [1976] 1039/44).

[71] S. E. Gukasyan, V. S. Shpinel (Phys. Status Solidi **29** [1968] 49/52).
[72] G. G. Long, J. G. Stevens, R. J. Tullbane, L. H. Bowen (J. Am. Chem. Soc. **92** [1970] 4230/5).
[73] J. G. Stevens, S. L. Ruby (Phys. Letters A **32** [1970] 91/2).
[74] D. Baltrunas, S. P. Ionov, A. Yu. Aleksandrov, E. F. Makarov (Chem. Phys. Letters **20** [1973] 55/8).
[75] J. N. R. Ruddick, J. R. Sams, J. C. Scott (Inorg. Chem. **13** [1974] 1503/7).
[76] G. M. Bancroft, V. G. K. Das, T. K. Sham, M. G. Clark (J. Chem. Soc. Dalton Trans. **1976** 643/54).
[77] G. M. Bancroft, V. G. K. Das, K. D. Butler (J. Chem. Soc. Dalton Trans. **1974** 2355/8).
[78] T. B. Brill, G. G. Long (Inorg. Chem. **9** [1970] 1980/5).

[79] V. I. Svergun, A. E. Borisov, N. V. Novikova, T. A. Babushkina, E. V. Bryukhova, G. K. Semin (Izv. Akad. Nauk SSSR Ser. Khim. **1970** 484/5; Bull. Acad. Sci. USSR Div. Chem. Sci. **1970** 443/4).
[80] S. Hoste, D. F. Van de Vondel, G. P. Van der Kelen (J. Electron. Spectrosc. Relat. Phenomena **17** [1979] 191/5).

[81] E. V. Stroganov (Vestn. Leningr. Univ. Fiz. Khim. **1959** No. 4, pp. 103/6).
[82] T. N. Polynova, M. A. Porai-Koshits (Zh. Strukt. Khim. **1** [1960] 159/61; J. Struct. Chem. [USSR] **1** [1960] 146/8).
[83] T. N. Polynova, M. A. Porai-Koshits (Zh. Strukt. Khim. **7** [1966] 742/51; J. Struct. Chem. [USSR] **7** [1966] 691/9).
[84] P. M. Zorkii, S. G. Lazareva (Zh. Strukt. Khim. **9** [1968] 95/100; J. Struct. Chem. [USSR] **9** [1968] 78/82).
[85] H. Preiss (Z. Anorg. Allgem. Chem. **389** [1972] 280/92).
[86] A. D. Beveridge, G. S. Harris, F. Inglis (J. Chem. Soc. A **1966** 520/8).
[87] M. Benmalek, H. Chermette, C. Martelet, D. Sandino, J. Tousset (J. Organometal. Chem. **67** [1974] 53/9).
[88] C. P. Smyth (J. Org. Chem. **6** [1941] 421).
[89] P. F. Oesper, C. P. Smyth (J. Am. Chem. Soc. **64** [1942] 173/5).
[90] N. K. Parab, D. M. Desai (Current Sci. [India] **26** [1957] 389).

[91] N. K. Parab, D. M. Desai (J. Indian Chem. Soc. **35** [1958] 569/75).
[92] M. K. Saikina (Uch. Zap. Kaz. Gos. Univ. **116** [1956] 129/86; C. A. **1957** 7191).
[93] R. E. Dessy, T. Chivers, W. Kitching (J. Am. Chem. Soc. **88** [1966] 467/70).
[94] V. F. Toropova, M. K. Saikina (Sb. Statei Obshch. Khim. Akad. Nauk SSSR **1** [1953] 210/5).
[95] K. Issleib, B. Hamann (Z. Anorg. Allgem. Chem. **343** [1966] 196/203).
[96] J. Bernstein, M. Halmann, S. Pinchas, D. Samuel (J. Chem. Soc. **1964** 821/4).
[97] D. L. Venezky, C. W. Sink, B. A. Nevett, W. F. Fortescue (J. Organometal. Chem. **35** [1972] 131/42).
[98] G. O. Doak, G. G. Long, L. D. Freedman (J. Organometal. Chem. **4** [1965] 82/91).
[99] V. P. Glushkova, T. V. Talalaeva, Z. P. Razmanova, G. S. Zhdanov, K. A. Kocheshkov (Sb. Statei Obshch. Khim. Akad. Nauk SSSR **2** [1953] 992/6).

[100] R. G. Goel, D. R. Ridley (Inorg. Chem. **13** [1974] 1252/5).
[101] R. G. Goel, D. R. Ridley (Inorg. Nucl. Chem. Letters **7** [1971] 21/3).
[102] F. Challenger, A. L. Smith, F. J. Paton (J. Chem. Soc. **123** [1923] 1046/54).
[103] F. Challenger, V. K. Wilson (J. Chem. Soc. **1927** 209/13).
[104] G. C. Tranter, C. C. Addison, D. B. Sowerby (J. Organometal. Chem. **12** [1968] 369/76).
[105] R. G. Goel, H. S. Prasad (J. Organometal. Chem. **59** [1973] 253/7).
[106] E. Wiberg, K. Mödritzer (Z. Naturforsch. **11b** [1956] 753/5).
[107] G. Grossmann, A. Winzer (Isotopentechnik **2** [1962] 193/8).
[108] N. A. Adrova, M. M. Koton, L. K. Prokhorova (Vysokomolekul. Soedin., Geterotsepnye Vysokomolecul. Soedin. **1964** 9/10; C. A. **61** [1964] 5784).
[109] W. E. McEwen, G. H. Briles, B. E. Giddings (J. Am. Chem. Soc. **91** [1969] 7079/84).
[110] L. I. Zakharkin, O. Yu. Okhlobystin, K. A. Bilevich (Tetrahedron **21** [1965] 881/6).

[111] S. N. Bhattacharya, A. K. Saxena (Indian J. Chem. A **17** [1979] 307/9).
[112] J. Dahlmann, A. Rieche (Chem. Ber. **100** [1967] 1544/9).
[113] S. B. Maerov (J. Polym. Sci. Polym. Chem. Ed. **17** [1979] 4033/40).
[114] H. A. Meinema, A. Mackor, J. G. Noltes (J. Organometal. Chem. **37** [1972] 285/95).
[115] A. Ouchi, M. Nakatani, Y. Takahashi, S. Kitazima, T. Sugihara, M. Matsumoto,

T. Uehiro, K. Kitano, K. Kawashima, H. Honda (Sci. Papers Coll. Gen. Educ. Univ. Tokyo **25** [1975] 73/99; C.A. **86** [1977] No. 5561).
[116] H. A. Meinema, E. Rivarola, J. G. Noltes (J. Organometal. Chem. **17** [1969] 71/81).
[117] V. K. Jain, R. Bohra, R. C. Mehrotra (Australian J. Chem. **33** [1980] 2749/52).
[118] M. Hall, D. B. Sowerby (J. Am. Chem. Soc. **102** [1980] 628/32).
[119] Yu. A. Sokolova, O. A. D'yachenko, L. O. Atovmyan, N. I. Liptuga, M. O. Lozinskii (Izv. Akad. Nauk SSSR Ser. Khim. **1980** 1446/8).
[120] H. G. Langer, Dow Chemical Co. (U.S. 3442922 [1970]; C. A. **72** [1970] No. 12880).

[121] V. K. Jain, R. Bohra, R. C. Mehrotra (J. Indian Chem. Soc. **57** [1980] 408/10).
[122] K. Bajpai, R. C. Srivastava (Syn. Reactiv. Inorg. Metal-Org. Chem. **11** [1981] 7/13).
[123] F. Di Bianca, E. Rivarola (Atti Accad. Sci. Lettere Arti Palermo Parte I **31** [1972] 167/72).
[124] F. Di Bianca, E. Rivarola, A. L. Spek, H. A. Meinema, J. G. Noltes (J. Organometal. Chem. **63** [1973] 293/300).
[125] A. Rieche, J. Dahlmann, D. List (Angew. Chem. **73** [1961] 494).
[126] A. Rieche, J. Dahlmann, D. List (Liebigs Ann. Chem. **678** [1964] 167/82).
[127] A. Rieche, J. Dahlmann (Ger. 1155127 [1960/63]; C.A. **60** [1964] 5554).
[128] H. Schmidbaur, H. S. Arnold, E. Beinhofer (Chem. Ber. **97** [1964] 449/58).
[129] S. Chatterjee (J. Inst. Chem. [India] **49** [1977] 263/4).
[130] E. J. Kupchik, P. J. Calabretta (Inorg. Chem. **4** [1965] 973/8).

[131] W. L. Mosby, American Cyanamid Co. (U.S. 3365478 [1965/68]; C.A. **68** [1968] No. 95978).
[132] W. L. Mosby, American Cyanamid Co. (U.S. 3429905 [1965/69]; C.A. **70** [1969] No. 115333).
[133] S. N. Bhattacharya, P. Raj, A. K. Saxena (Indian J. Chem. A **16** [1978] 1071/4).
[134] W. E. McEwen, G. H. Briles, D. N. Schulz (Phosphorus **2** [1972] 147/53).
[135] K. Hartke, H.-M. Wolff (Chem. Ber. **113** [1980] 1394/405).
[136] C. E. Carraher Jr., H. S. Blaxall (Angew. Makromol. Chem. **83** [1979] 37/45).
[137] C. E. Carraher Jr., H. S. Blaxall (Polym. Prepr. Am. Chem. Soc. Div. Polym. Chem. **16** [1975] 261/3).
[138] C. E. Carraher Jr., L. J. Hedlund (J. Macromol. Sci. Chem. A **14** [1980] 713/28).
[139] C. E. Carraher Jr., M. D. Naas, D. J. Giron, D. R. Cerutis (J. Macromol. Sci. Chem. A **19** [1983] 1101/20).
[140] C. E. Carraher Jr., D. J. Giron, D. R. Cerutis, W. R. Burt, R. S. Venkatachalam, T. J. Gehrke, S. Tsuji, H. S. Blaxall (ACS Symp. Ser. No. 186 [1982] 13/25).

[141] J. E. Sheats, C. H. Carraher Jr., H. S. Blaxall (Polym. Prepr. Am. Chem. Soc. Div. Polym. Chem. **16** [1975] 655/8).
[142] W. Herr (Z. Anorg. Allgem. Chem. **258** [1959] 94/8).
[143] S. N. Maslennikova (Tr. Khim. Khim. Tekhnol. **1974** 50/1; C.A. **83** [1975] No. 157558).
[144] C. E. Carraher Jr., M. J. Moran (Organomet. Polym. **1977** 107/14).
[145] V. E. Zhuravlev, N. I. Trofimova, L. Kashina (Tr. Estestvennonauchn. Inst. Permsk. Gos. Univ. **13** No. 1 [1972] 179/82; C.A. **80** [1974] No. 33626).
[146] V. D. Nefedov, I. M. Rozman, Yu. A. Ryukhin, E. A. Makoveev (Radiokhimiya **5** [1963] 643/6; Soviet Radiochem. **5** [1963] 604/6).
[147] A. N. Murin, V. D. Nefedov (Primen. Mechenykh At. Analit. Khim. Dokl. Konf., Moscow 1953 [1955], pp. 75/8; C.A. **1956** 3915).
[148] V. D. Nefedov, I. S. Kirin, V. M. Zaitsev (Radiokhimiya **4** [1962] 351/5; Soviet Radiochem. **4** [1962] 311/4).
[149] N. I. Trofimova, V. E. Zhuravlev, E. N. Sinotova, A. I. Sarbash (Tr. Estestvennonauchn. Inst. Permsk. Gos. Univ. **13** No. 3 [1975] 177/86; C.A. **86** [1977] No. 105407).

[150] N. I. Trofimova, V. E. Zhuravlev, E. N. Sinotova, N. E. Shchepina, M. V. Moshkovskaya (Tr. Estestvennonauchn. Inst. Permsk. Gos. Univ. **13** No. 3 [1975] 187/93; C.A. **86** [1977] No. 88696).

[151] C. G. Moreland, G. G. Long (Inorg. Nucl. Chem. Letters **8** [1972] 347/51).
[152] C. G. Moreland, M. H. O'Brien, C. E. Douthit, G. G. Long (Inorg. Chem. **7** [1968] 834/6).
[153] H. K. Sharma, S. N. Dubey, D. M. Puri (J. Indian Chem. Soc. **59** [1982] 1031/3).
[154] Y. Takashi (Bull. Chem. Soc. Japan **40** [1967] 1194/201).
[155] H. K. Sharma, S. N. Dubey, D. M. Puri (Indian J. Chem. A **20** [1981] 620/1).
[156] H. K. Sharma, S. Singh, S. N. Dubey, D. M. Puri (Indian J. Chem. A **21** [1982] 619/21).
[157] H. Chermette, C. Martelet, D. Sandino, M. Benmalek, J. Tousset (Anal. Chim. Acta **59** [1972] 373/80).
[158] J. G. Stevens, S. L. Ruby (Phys. Letters A **32** [1970] 91/2).
[159] H. Chermette, C. Martelet, D. Sandino, J. Tousset (Anal. Chem. **44** [1972] 857/60).
[160] H. Chermette, C. Martelet, D. Sandino, J. Tousset (J. Inorg. Nucl. Chem. **34** [1972] 1627/38).

[161] M. Benmalek, H. Chermette, C. Martelet, D. Sandino, J. Tousset (J. Inorg. Nucl. Chem. **36** [1974] 1365/8).
[162] M. Benmalek, H. Chermette, C. Martelet, D. Sandino, J. Tousset (J. Inorg. Nucl. Chem. **36** [1974] 1359/63).
[163] C. Martelet (LYCEN-7307 [1973] 1/98; C.A. **81** [1970] No. 20454).
[164] H. Matsuda, A. Ninagawa, R. Nomura (Chem. Letters **1979** 1261/2).
[165] R. Nomura, A. Ninagawa, H. Matsuda (J. Org. Chem. **45** [1980] 3735/8).
[166] A. Ninagawa, H. Matsuda, R. Nomura (Kenkyu Hokoku Asahi Garasu Kogyo Gijutsu Shoreikai **40** [1982] 141/7).
[167] H. Matsuda (Japan. 80-122776 [1979/80]; C.A. **94** [1981] No. 139779).
[168] R. Harvey, H. M. Sachs, Halcon SD Group, Inc. (Ger. Offen. 3244456 [1983]; C. A. **99** [1983] No. 122444).
[169] H. Matsuda, T. Isaka, N. Iwamoto (Makromol. Chem. **179** [1978] 539/42; C.A. **88** [1978] No. 105862).
[170] J. Mleziva, V. Cermak (Plasty Kauc. **15** [1978] 129/35; C.A. **89** [1978] No. 90612).

[171] J. Havranek, J. Mleziva (Angew. Makromol. Chem. **84** [1980] 105/17).
[172] J. Havranek (Sb. Dokl. 1st Nats. Konf. Mladite Nauchni Rab. Spets. Neft Khim., Burgas, Bulg., 1976 [1977], pp. 152/9; C.A. **93** [1980] No. 187219).
[173] O. Horak, J. Havranek, J. Vladyka (Kunststoffe **72** [1982] 493/4).
[174] O. Horak, M. Pilny, V. Zvonar (Plasty Kauc. **18** [1981] 245/6).
[175] J. Sotiropoulos (Fr. 1157208 [1958]; C.A. **1960** 20312).
[176] J. Havranek, J. Muller, J. Mleziva (Sb. Ved. Pr. Vys. Sk. Chemickotechnol. Pardubice **42** [1980] 123/32; C.A. **94** [1981] No. 209645).
[177] B. O. Schoepfle, S. M. Burton, P. Robitschek (U.S. 2913428 [1959]; C.A. **1960** 5162).
[178] F. Challenger, L. R. Ridgway (J. Chem. Soc. **121** [1922] 104/20).
[179] Kali-Chemie A.-G. (Brit. 768765 [1957]; C.A. **1958** 421).
[180] C. Moureu, C. Dufraisse, M. Badoche (Compt. Rend. **187** [1928] 1092/6).

[181] H. Chermette, G. Reynaud, R. Chareyron (Analusis **4** [1976] 203/8; C.A. **85** [1976] No. 103262).
[182] J. R. Leebrick, M. & T. Chemicals Inc. (U.S. 3287210 [1962/66]; C.A. **66** [1967] No. 85070).
[183] R. E. Burrell, C. T. Corke, R. G. Goel (J. Agric. Food Chem. **31** [1983] 85/8).
[184] S. M. Manulkin, A. N. Tatarenko (Zh. Obshch. Khim. **21** [1951] 93/8; J. Gen. Chem. [USSR] **21** [1951] 103/8).

2.5.1.1.2.2.2 Other Triarylantimony Dichlorides, Trifuryl-, and Trithienylantimony Dichlorides

$(C_6F_5)_3SbCl_2$

Tris(perfluorophenyl)antimony dichloride is obtained by direct chlorination of $Sb(C_6F_5)_3$ in ethanol [1]. If the reaction is performed in CCl_4, the yield is 78% after 24 h standing and concentrating the solution [2]. The compound melts at 255 °C [2] or 255 to 257 °C after recrystallization from petroleum ether [1]. From $Sb(C_6F_5)_3$ and $TiCl_3$, 48 h in ether, or $Sb(C_6F_5)_3$ and $CuCl_2$, 1 h in acetone at 20 °C, the compound is also obtained after filtration of the mixture, evaporation of the solvent, and recrystallization from petroleum ether in yields of 70 and 80%, respectively [2]. $(C_6F_5)_3Sb(ON(CF_3)_2)_2$ and HCl react in a closed ampule for 6 d at 100 °C, forming the title compound with a melting point of 244 to 245 °C [3].

Complete IR spectra (between 4000 and 400 cm^{-1} in Nujol, between 400 to 40 cm^{-1} as wax discs) and Raman spectra (solid state between 4000 and 40 cm^{-1}) were measured, assigned, and discussed in view of the vibrations of the C_6F_5-substituent (see original). The ν_sSbCl in the Raman spectrum is assigned at 294 cm^{-1}, and the $\nu_{as}SbCl$ in the IR at 327 cm^{-1} [1]. The conductivity of a 5×10^{-4} M solution in $(CH_3)_2CO$ is $\Lambda = 40.08\ cm^2 \cdot \Omega^{-1} \cdot mol^{-1}$, and in CH_3NO_2 $\Lambda = 16.06\ cm^2 \cdot \Omega^{-1} \cdot mol^{-1}$ [2].

The title compound reacts with $AgNO_3$ or $AgClO_4$. Two hours reflux in C_6H_6 gives 40% of $(C_6F_5)_3Sb(NO_3)_2$ or 48% of $(C_6F_5)_3Sb(ClO_4)_2$. With $NaOCH_3$, 4 h at 20 °C in CH_3OH, it forms 48% $(C_6F_5)_3Sb(OCH_3)_2$. Hot aqueous C_2H_5OH and $(C_6F_5)_3SbCl_2$ give $[(C_6F_5)_3SbCl]_2O$ in 42% yield. Na-8-quinolinolate or Na-acetylacetonate, refluxed with the compound in $CHCl_3$ for 5 h, form 68% $(C_6F_5)_3SbCl(OC_9H_6N)$ and 54% $(C_6F_5)_3SbCl(acac)$, respectively [2].

$(2\text{-}ClC_6H_4)_3SbCl_2$

A suspension of iron powder (100 mesh) in dry $(CH_3)_2CO$ is treated first quickly then slowly with a solution of $[2\text{-}ClC_6H_4N_2]Cl$ (ratio 4:1) in the same solvent. After all the reactant is added, the mixture is stirred additionally for 45 min. All inorganic material is filtered, and the solvent is removed from the filtrate at reduced pressure. The residue is treated with 5 N HCl and with C_2H_5OH. The precipitate (73% yield) is recrystallized from heptane. It melts at 205 to 206 °C [4, 5]. The compound is also obtained in the same yield by reacting $(2\text{-}ClC_6H_4)_2SbCl$ with $[2\text{-}ClC_6H_4N_2]SbCl_4$ in cold $(CH_3)_2CO$ and workup as before [6, 7]. A reduction of $[2\text{-}ClC_6H_4N_2]SbCl_4$ with zinc dust (ratio 2:3) in ethyl acetate yields only 25% of the compound besides 17% $Sb(C_6H_4Cl\text{-}2)_3$ and 40% $2\text{-}ClC_6H_4SbO$. The given melting point of 185 °C [8] is an error [4].

$(4\text{-}ClC_6H_4)_3SbCl_2$

$Sb(C_6H_4Cl\text{-}4)_3$ reacts with chlorine gas in $CHCl_3$ at 0 °C in good (76%) [9] to quantitative [10] yields to form the compound. Chlorinating agents like $CuCl_2$ in ethanol [11] or SO_2Cl_2 in toluene at 0 °C [12] also react with the corresponding stibine to produce the compound in good yields. $[4\text{-}ClC_6H_4N_2]SbCl_4$ upon treatment with zinc dust (1:1.5) in ethyl acetate at 70 °C [8], or with iron powder (1:1) in $(CH_3)_2CO$ at 0 °C [13], gives only about 10% of the compound. The main products obtained with zinc dust are $(4\text{-}ClC_6H_4)_2SbO_2CCH_3$ and $4\text{-}ClC_6H_4SbO$, and with iron powder $(4\text{-}ClC_6H_4)_2SbO(OH)$. The same yield is obtained upon reacting $[4\text{-}ClC_6H_4N_2]Cl$ with antimony and $CaCO_3$ in $(CH_3)_2CO$ or ethyl acetate, refluxing for half an hour, and workup of the mixture [14]. $[4\text{-}ClC_6H_4N_2]SbCl_4$ and iron powder (ratio 1:4) may be reacted in $(CH_3)_2CO$ for 45 min. Subsequently, the filtrate is evaporated, and the residue treated with 5 N HCl and C_2H_5OH from which the compound precipitates

References on p. 58

in a yield of 30% [5]. Reaction of $(4\text{-}ClC_6H_4)_2ICl$ with $SbCl_3$ and Sb (ratio 2:1:3) for 3 h in boiling $(CH_3)_2CO$, filtering the inorganic residue, evaporating the solvent, taking up the remainder in C_6H_6, washing the latter with 5 N aqueous HCl, evaporating the C_6H_6, and taking up with C_2H_5OH give the compound upon cooling in a yield of 43% [15]. A similar yield is obtained by reacting the same iodonium chloride with $(4\text{-}ClC_6H_4)_2SbCl_3$ in concentrated HCl and reducing the resulting 1:1 complex in $(CH_3)_2CO$ at room temperature with zinc dust (ratio 1:4) after evaporation of the filtrate [16].

Melting points of the compound are given as 189.5 to 190.5 °C from $CH_3OH/CHCl_3$ [10], 193 to 193.5 °C from petroleum ether [11], 193 to 194 °C [13], 189.6 °C [9], 193 °C [5, 14], 193 °C from $C_2H_5OH/CHCl_3$ [12], 186 to 186.5 °C from the same mixture [16], and 102 °C [15]. In the ^1H NMR spectrum (in $CDCl_3$ or $(CD_3)_2CO$) resonances are found at $\delta=7.52$ (H-3,5) and 8.15 (H-2,6) ppm [12].

$(4\text{-}ClC_6H_4)_3SbCl_2$ reacts with KF in aqueous C_2H_5OH to form the corresponding difluoride [9, 12] in yields of about 80% [9]. The chlorination of the compound with $Cl_2/SbCl_3$ occurs with cleavage of the Sb-C bonds and formation of dichlorobenzenes. In the presence of $PO(N(CH_3)_2)_3$ the ratio of para/ortho product increases to 56% compared to 6% without $PO(N(CH_3)_2)_3$. It is proposed that a hexacoordinated intermediate is formed with the O-donor [12]. Substituted phenols XC_6H_4OH and substituted benzoic acids $XC_6H_4CO_2H$ with X=H, 4-NO_2, 4-Cl, 4-CH_3, and 4-CH_3O were reacted with the compound in C_6H_6 in the presence of $N(C_2H_5)_3$ as HCl acceptor to yield, after evaporation of the solvent from the filtrate, the corresponding diphenolates and dibenzoates in yields of 40 to 70% [17]. $(4\text{-}ClC_6H_4)_3SbCl_2$ and $4\text{-}CH_3C_6H_4MgBr$ react in dimethoxyethane for 8 h at 20 °C to form $(4\text{-}ClC_6H_4)_3Sb(C_6H_4CH_3\text{-}4)_2$ [12].

$(2,4\text{-}Cl_2C_6H_3)_3SbCl_2$

The compound is obtained in 60% yield by treating $[2,4\text{-}Cl_2C_6H_3N_2]SbCl_6$ with iron powder in a ratio of 1:4 in $(CH_3)_2CO$ for 45 min. Evaporation of the solvent from the filtrate, treating the residue twice with 5 N HCl, and then with ethanol give crystals which melt at 235 °C [5].

$(C_6Cl_5)_3SbCl_2$

Stoichiometric amounts of Cl_2 in CCl_4 are added to a suspension of $Sb(C_6Cl_5)_3$ in $CHCl_3$. After stirring for 30 min at room temperature, the solution is concentrated whereby the compound precipitates in a yield of 75% [18].

The compound is soluble in $CHCl_3$, CCl_4, and THF. No electrolytic conductivity is observed in CH_3NO_2 solution. An IR spectrum in Nujol shows νSbC at 300 cm^{-1} [18].

$(4\text{-}BrC_6H_4)_3SbCl_2$

The usual method for obtaining the compound is to treat the corresponding stibine, dissolved in $CHCl_3$, with Cl_2; 60 to 80% yields are obtained by concentrating the solution [9, 19]. $CuCl_2$ in a $C_2H_5OH/CHCl_3$ solution may be also used for chlorination of the stibine [15]. Preparations, based on diazonium and iodonium salts (for details see former compounds) lead to the following results. $[4\text{-}BrC_6H_4N_2]SbCl_4$ and zinc dust (ratio 1:1.5) in ethyl acetate at 80 °C give 21% yield [8]. The same diazonium adduct with iron powder (1:1) in $(CH_3)_2CO$ at 0 °C gives 9% yield [13]. $[4\text{-}BrC_6H_4N_2]Cl$, $CaCO_3$, and Sb in acetone or ethyl acetate refluxed for half an hour form about 10% of the compound [14]. $(4\text{-}BrC_6H_4)_2ICl$, $SbCl_3$, and Sb (2:1:3), 3 h in boiling $(CH_3)_2CO$, lead to 18% yield [15]. The reduction of the $(4\text{-}BrC_6H_4)_2SbCl_3 \cdot (4\text{-}BrC_6H_4)_2ICl$ complex with zinc (1:4) in $(CH_3)_2CO$ for 2 h at room temperature gives 36% of the title compound [16].

References on p. 58

Melting points for the compound are given as 192 °C [21], 184 to 185 °C [19], 185 °C [9], 194 °C [15], 197 °C from $C_2H_5OH/CHCl_3$ [16], 198 °C [8], 200 °C [14], and 200 to 201 °C from $C_2H_5OH/CHCl_3$ [11].

The compound reacts with KF in aqueous C_2H_5OH to form 72% $(4\text{-}BrC_6H_4)_3SbF_2$ [9]. With $N_2H_2 \cdot H_2O$ it can be reduced to the corresponding stibine [19]. The compound reacts in boiling alkaline C_2H_5OH with HgO to give $Hg(C_6H_4Br\text{-}4)_2$. This reaction is not observed in neutral C_2H_5OH [20].

$(4\text{-}IC_6H_4)_3SbCl_2$

$[4\text{-}IC_6H_4N_2]SbCl_4$ is treated with zinc dust (ratio 1:1.5) in ethyl acetate at 80 °C or in amyl alcohol at 120 °C. After workup of the mixture 23% of the compound is obtained in the first case, and 5.7% in the second case. The main products of the reaction are $4\text{-}IC_6H_4SbO$ and $(4\text{-}IC_6H_4)_2SbO_2CCH_3$ [8].

$(2\text{-}CH_3OC_6H_4)_3SbCl_2$

The compound is prepared by oxidation of the corresponding stibine with $CuCl_2$ in C_2H_5OH. After filtrating the CuCl deposit, the filtrate is concentrated. The compound melts at 237 to 238 °C, recrystallized from $CHCl_3$/petroleum ether [11]. Another method for preparation is the reaction of $[2\text{-}CH_3OC_6H_4N_2]SbCl_4$ with zinc dust (ratio 1:5) in $(CH_3)_2CO$. The yield is 20.1%, m.p. 245 °C [8]. The yield increases to 67%, if iron powder is used instead of zinc [13].

^{121}Sb Mössbauer parameters are given as $\delta = 3.1 \pm 0.1$ (vs. InSb at 80 K) and $e^2qQ = -18 \pm 0.4$ mm/s. The structure of the compound is discussed on the basis of these values [23]. The data given in a previous communication [22] are less accurate [23].

$(3\text{-}CH_3OC_6H_4)_3SbCl_2$

The compound is prepared like the previous one, m.p. 81.5 to 82.5 °C, recrystallized from $CHCl_3$/petroleum ether [11].

$(4\text{-}CH_3OC_6H_4)_3SbCl_2$

The compound is prepared by reacting the corresponding stibine with $CuCl_2$ in $C_2H_5OH/CHCl_3$. After filtrating the CuCl deposits the filtrate is extracted with C_6H_6, and the benzene is evaporated. The compound melts at 116 to 117 °C, recrystallized from $CHCl_3$/petroleum ether [21]. Reaction of $[4\text{-}CH_3OC_6H_4N_2]SbCl_4$ with zinc dust (ratio 1:5) in ethyl acetate gives the compound in 9.5% yield with a melting point of 180 °C [8].

The compound is soluble in C_6H_6, $CHCl_3$, and ether, slightly soluble in alcohol, and insoluble in petroleum ether. It crystallizes from C_6H_6 as large prismatic crystals which contain 1 mol C_6H_6. The melting point is 82 to 83 °C upon fast heating. Upon slow heating the C_6H_6 escapes [21].

$(2\text{-}C_2H_5OC_6H_4)_3SbCl_2$

The compound is obtained from the corresponding stibine and $CuCl_2$ in C_2H_5OH [11]. It may be also synthesized from $[2\text{-}C_2H_5OC_6H_4N_2]SbCl_4$ by adding iron powder (ratio 1:1.1) in $(CH_3)_2CO$. The compound is obtained in 49% yield besides 42% of $(2\text{-}C_2H_5OC_6H_4)_2SbCl_3$ [13]. It melts with decomposition at 231 to 232 °C, recrystallized from $C_2H_5OH/CHCl_3$ [11, 13].

References on p. 58

$(4\text{-}C_2H_5OC_6H_4)_3SbCl_2$

The compound is obtained by heating the corresponding stibine and $CuCl_2$ in C_2H_5OH. CuCl is filtered and the filtrate is concentrated yielding the compound with a melting point of 84 °C [21].

$(4\text{-}C_6H_5OC_6H_4)_3SbCl_2$

The compound is prepared by reacting the corresponding stibine with $CuCl_2$ in C_2H_5OH. It melts at 106 to 107 °C, recrystallized from $C_2H_5OH/CHCl_3$ [11].

$(4\text{-}(CH_3)_2NC_6H_4)_3SbCl_2$

$Sb(C_6H_4N(CH_3)_2)_3$ is reacted with $C_6H_5ICl_2$ in CH_2Cl_2 at −78 °C or in $CHCl_3$ at 0 °C. Warming to room temperature and addition of ether or petroleum ether to the solution yields 88% of colorless needles which melt with decomposition at 195 to 200 °C [24].

A ^1H NMR spectrum of the compound leads to the following resonances (in $CDCl_3$): δ=3.43 (CH_3), 7.23 (H-3,5), and 8.53 (H-2,6) ppm. The coupling constants J(H-2,3) and J(H-2,5) are 9 Hz. A figure of a UV spectrum in THF accompanied by a discussion is given. The maximum absorption is at λ=285 nm with log ε=4.8 [24].

Reaction with 2 mol $Li[C_6H_4N(CH_3)_2\text{-}4]$ gives $Sb[C_6H_4N(CH_3)_2\text{-}4]_5$ [58]. The compound reacts in 1,2-dichloroethane with $SbCl_5$ (0.5 h reflux) or $AlCl_3$ (24 h reflux). Upon addition of CCl_4 deeply colored precipitates of the composition $[(4\text{-}(CH_3)_2NC_6H_4)_3Sb][SbCl_6]_2$ (olive-green) and $[(4\text{-}(CH_3)_2NC_6H_4)_3Sb][AlCl_4]_2$ (light green-blue), respectively, are formed. The cations of the compounds are isoelectronic with that of crystal violet. The UV absorption of the $SbCl_6$ salt was measured by reflection and discussed (λ_{max}=625 nm). A ^1H NMR spectrum of this salt (in C_5D_5N) shows resonances at δ=2.47 (CH_3), 6.37 (H-3,5), and 7.03 (H-2,6) ppm. The coupling constants are J(H-2,3) and J(H-2,5)=9 Hz. The salts are sensitive towards air and moisture and are insoluble in the common solvents except C_5H_5N, but this solution is unstable [25].

$(4\text{-}CH_3CONHC_6H_4)_3SbCl_2$

$[4\text{-}CH_3CONHC_6H_4N_2]SbCl_6$ is treated in $(CH_3)_2CO$ for 45 min with iron powder (1:4). The solvent from the filtrate is evaporated, and the oily residue is treated twice with 5 N HCl and then with C_2H_5OH. The resulting precipitate is isolated in a yield of 76%. The compound sinters at 170 °C [5].

$(NO_2C_6H_4)_3SbCl_2$

A compound of this composition is mentioned in the older literature. It is obtained by reaction of $(C_6H_5)_3SbCl_2$, dissolved in concentrated H_2SO_4, with HNO_3 at 40 °C. Workup of the mixture with ice-water gives a small amount of crystals which melt at 157 °C, recrystallized from CH_3CO_2H. The compound is also formed, together with $(NO_2C_6H_4)_3Sb(OH)Cl$, by boiling $(NO_2C_6H_4)_3Sb(OH)_2$ with alcoholic HCl, but the product is not obtained in pure form [26].

$(2\text{-}CH_3C_6H_4)_3SbCl_2$

The compound is prepared by chlorination of $Sb(C_6H_4CH_3\text{-}2)_3$ with Cl_2 [27] in ether, THF, or $CHCl_3$ [28]. Colorless needles, which melt at 178 to 179 °C, crystallize from $C_2H_5OH/CHCl_3$ [27]. $Sb(C_6H_4CH_3\text{-}2)_3$ reacts with $HgCl_2$ in boiling C_6H_6 within 2 h to give the title compound. The precipitate Hg_2Cl_2 is filtered, the filtrate is concentrated, and the residue

References on p. 58

is washed with ether. The title compound is identified by IR [30]. The compound can be also obtained by extraction of an aqueous solution of $(2\text{-}CH_3C_6H_4)_3SbI_2$ and HCl with $CHCl_3$ [28, 30].

The solubility in $CHCl_3$ is 0.1(±50%) mol/L at 25 °C and the distribution coefficient between $CHCl_3$ and H_2O is 2.5×10^5. Hydrolysis constants obtained by titration with 0.1 N NaOH in $CHCl_3$ are given as log $K_1=8$ and log $K_2=6.5$ [28]. $(2\text{-}CH_3C_6H_4)_3SbCl_2$ reacts with KF in aqueous C_2H_5OH, refluxed for one hour, to form $(2\text{-}CH_3C_6H_4)_3SbF_2$ in a yield of 76% [9]. The difluoride is also obtained by extracting an aqueous solution of the title compound and HF with $CHCl_3$ [28]. The extraction of F^- with $CHCl_3$ from aqueous solutions containing the title compound is studied. From this a stability constant $[(2\text{-}CH_3C_6H_4)_3SbCl_2]/[2\text{-}CH_3C_6H_4)_3Sb^{2+}]\cdot[Cl^-]^2$ of 1.5 ± 1 L^2/mol^2 is derived [30]. $(2\text{-}CH_3C_6H_4)_3SbCl_2$ reacts with substituted phenols XC_6H_4OH (X=H, 4-NO_2, 4-Cl, 4-CH_3, and 4-CH_3O) and substituted benzoic acids $XC_6H_4CO_2H$ (X=H, 4-NO_2, 4-Cl, 4-CH_3, 4-NH_2) in a ratio of 1:2 in C_6H_6 in the presence of $N(C_2H_5)_3$ to form the corresponding diphenolates or dibenzoates, respectively [17].

$(3\text{-}CH_3C_6H_4)_3SbCl_2$

The compound is obtained by chlorination of the corresponding stibine with Cl_2 [27]. Treatment of $(3\text{-}CH_3C_6H_4)_3SbCl(OH)$ with ethanolic HCl leads to the compound [31]. As a by-product (25%), the compound is obtained by workup of the reaction mixture of $LiC_6H_4CH_3$-3 and $SbCl_3$ in ether with ice water and C_2H_5OH. The main product from this reaction is $Sb(C_6H_4CH_3\text{-}3)_3$ (71%) [9].

The compound crystallizes from ether/C_2H_5OH to form colorless, short thick needles [27]. It melts at 137 °C [9, 27, 31]. The crystals are monoclinic with a=11.2, b=12.6, c=15.6 Å, and β=112.5°, space group $P2_1/c\text{-}C^5_{2h}$ (No. 14); Z=4; $d_m=1.52$ and $d_c=1.54$ g/cm³ [9].

The compound is soluble in C_6H_6, ether, and $CHCl_3$, slightly soluble in C_2H_5OH and glacial acetic acid, and hardly soluble in petroleum ether. The better solubility in ether is in contrast to the corresponding ortho and para derivatives [27].

Reaction with H_2S in alcoholic NH_3 gives $(3\text{-}CH_3C_6H_4)_3SbS$ [27]. With KF in aqueous ethanolic solution, refluxed for one hour, the corresponding difluoride is formed in 85% yield [9]. It reacts with benzoic acid derivatives $XC_6H_4CO_2H$ (X=H, 4-NO_2, 4-Cl, 4-CH_3, 4-NH_2) in $C_6H_6/N(C_2H_5)_3$ forming the corresponding dibenzoates in yields of 40 to 70% [17]. The kinetics of isotope exchange in the system $(3\text{-}CH_3C_6H_4)_3SbCl_2/^{125}Sb\text{-}(C_6H_4CH_3\text{-}3)_3$ was studied in C_2H_5OH. The rate constants were found to be 0.65, 1.3, 2.6, and 4.8 $L\cdot mol^{-1}\cdot h^{-1}$ at 40, 50, and 70 °C, respectively; $E_a=14$ kcal/mol [32].

$(4\text{-}CH_3C_6H_4)_3SbCl_2$

Methods of preparation of this compound are summarized in Table 15.

The compound is obtained as large shiny crystals with a melting point of 156.5 °C, recrystallized from C_2H_5OH/C_6H_6 [27, 38]. Other melting points given range from 155 to 157 °C, recrystallized from C_6H_6/C_6H_{14} [39], $CHCl_3/C_2H_5OH$ [12, 16], or C_2H_5OH/HCl [43]. ^1H NMR spectra were measured: δ=2.42 (CH_3), 7.28 (H-3,5), 8.07 (H-2,6) ppm in $CDCl_3$ or $(CD_3)_2CO$ [12]; δ=2.43 (CH_3); 7.35, 8.11 (C_6H_4) ppm; J(H-2,3)=8 Hz in $CDCl_3$ [34]. The compound crystallizes in the cubic system with a=12.718 Å; space group $P4_332\text{-}O^6$ (No. 212) or $P4_132\text{-}O^7$ (No.213); Z=4; $d_m=1.44$ and $d_c=1.45$ g/cm³. The geometry of the molecule is discussed [9, 33].

References on p. 58

Table 15
Preparation of $(4\text{-}CH_3C_6H_4)_3SbCl_2$.

reactants	conditions (yield in %)	Ref.
$Sb(C_6H_4CH_3\text{-}4)_3 + Cl_2$	in petroleum ether, ether, THF, or $CHCl_3$	[27, 28, 38, 40]
	in CCl_4 with cooling	[39]
$Sb(C_6H_4CH_3\text{-}4)_3 + SO_2Cl_2$	in toluene at 0 °C	[12]
$Sb(C_6H_4CH_3\text{-}4)_3 + CuCl_2$ or $HgCl_2$	in refluxing $(CH_3)_2CO$ (70 to 85)	[41]
$Sb(C_6H_4CH_3\text{-}4)_3 + CuCl_2$	in C_2H_5OH/ether	[27]
$SbCl_5 + Hg(C_6H_4CH_3\text{-}4)_2$	ratio 1:3 in CCl_4, 1 h at 50 °C (90)	[42, 43]
$[4\text{-}CH_3C_6H_4N_2]SbCl_4 + Zn$ dust	ratio 1:1.5 in refluxing $(CH_3)_2CO$ (20)	[8]
$[4\text{-}CH_3C_6H_4N_2]SbCl_6 + Fe$ powder	ratio 1:4 in cold $(CH_3)_2CO$, 45 min (29)	[5]
$(4\text{-}CH_3C_6H_4)_2ICl + SbCl_3 + Sb$	ratio 2:1:3 in boiling $(CH_3)_2CO$, 3 h (8.5)	[15]
$(4\text{-}CH_3C_6H_4)_2ICl \cdot (4\text{-}CH_3C_6H_4)_2SbCl_3 + Zn$ dust	ratio 1:4 in $(CH_3)_2CO$, 2 h at 20 °C (34)	[16]
$(4\text{-}CH_3C_6H_4)_3Sb(ON{=}CC_5H_{10}\text{-}c)_2 + TeCl_4$	in refluxing C_6H_6, ca. 4 h (>80)	[44]
$(4\text{-}CH_3C_6H_4)_3Sb(ON{=}C(CH_3)C_6H_5)_2 + C_6H_5TeCl_3$	in refluxing C_6H_6, ca. 4 h (>80)	[44]
$(4\text{-}CH_3C_6H_4)_3Sb(N{=}P(C_6H_5)_3)_2 + TeCl_4$	in refluxing C_6H_6 (75)	[45]
$(4\text{-}CH_3C_6H_4)_3SbI_2 + HCl$	in H_2O, extracted with $CHCl_3$	[28, 30]

The dipole moment in C_6H_6 at 25 °C was measured to 0.83 D [35]. A molar susceptibility of $\chi_M = -249.2 \times 10^{-6}$ is given and compared with those of other organoantimony compounds [36, 37]. The molar solubility (in mol/L at 25 °C) is 2.5×10^{-3} in C_6H_6, 8×10^{-3} in CCl_4, and 9×10^{-3} in $CHCl_3$. Distribution coefficients between these organic phases and water are 5×10^2, 1.6×10^3, and 1.8×10^3, respectively. Hydrolysis constants were determined by titration with 0.1 N NaOH in $CHCl_3$ as log $K_1 = 9.5$ and log $K_2 = 7.5$ [28].

The compound decomposes upon heating under a vacuum of 5 to 7 Torr, forming $(4\text{-}CH_3C_6H_4)_2SbCl$ and $4\text{-}CH_3C_6H_4Cl$ [46]. With KF, refluxed for one hour in aqueous C_2H_5OH [9] or in water [12], it forms the corresponding difluoride. The same difluoride is obtained in 95% yield if an aqueous HF solution of the compound is extracted with $CHCl_3$ [28]. Concerning the extraction of fluoride from water to $CHCl_3$ with assistance of the compound see [30]. Chlorination of the compound with $SbCl_3/Cl_2$ mixtures in CH_2Cl_2 leads to a cleavage of the Sb-C bonds and formation of $CH_3C_6H_4Cl$ in 14.3% yield and a para/ortho ratio of 37.6. If the reaction is carried out in the presence of $PO(N(CH_3)_2)_3$, only $4\text{-}ClC_6H_4CH_3$ is formed in 42.8% yield. A hexacoordinated intermediate is proposed in the presence of the oxygen donor. A direct cleavage of the apical Sb-C bond is responsible for the selectivity of the para product [12]. $(4\text{-}CH_3C_6H_4)_3SbCl_2$ reacts with substituted phenols XC_6H_4OH (X = H, 4-NO_2, 4-Cl, 4-CH_3, 4-OCH_3) or carboxylic acids $XC_6H_4CO_2H$ (X = H, 4-NO_2, 4-Cl, 4-CH_3, 4-NH_2) in $C_6H_6/N(C_2H_5)_3$ to form the corresponding diphenolates and dibenzoates, respectively [17].

References on p. 58

Similarly, the reaction with oximes HON=CRR′ (R, R′ = CH_3, CH_3; CH_3, C_6H_5; CH_3, C_2H_5; CH_3, 4-$NO_2C_6H_4$; C_6H_5, C_6H_5; $(CH_2)_5$; H, 2-furyl) gives dioximates [44]. The reaction of the title compound with 4-$CH_3C_6H_4MgBr$ in dimethoxyethane, 8 h at 20 °C, leads to $Sb(C_6H_4CH_3\text{-}4)_5$ [12] while the same educts, refluxed in ether for 4 h and subsequent standing for 3 d give (4-$CH_3C_6H_4)_4SbOH$ after workup [39]. The reaction with $LiC_6H_4CH_3$-4 in ether gives also $Sb(C_6H_4CH_3\text{-}4)_5$ [40]. (4-$CH_3C_6H_4)_3SbCl_2$ reacts with $[(C_4H_9)_3Sn]_2S$ or $[(C_6H_5)_3Sn]_2S$ in $CHCl_3$ at −5 °C with formation of (4-$CH_3C_6H_4)_3SbS$ and the corresponding R_3SnCl (R = C_4H_9 or C_6H_5) [47]. The exchange reactions of the compound with $^{125}Sb(C_6H_4CH_3\text{-}2,\text{-}3,\text{ and }\text{-}4)_3$ was studied in C_2H_5OH at 40 to 70 °C. The rate of exchange is affected by the position of ^{125}Sb with respect to the CH_3 groups [48]. The chemical changes during the β-decay of ^{125}Sb in (4-$CH_3C_6H_5)_3{}^{125}SbCl_2$ were studied. (4-$CH_3C_6H_4)_2{}^{125m}TeCl_2$ is formed in 90±3% yield [49].

The title compound was tested as retarding agent for burning epoxy resins [50].

(5-Cl-2-$CH_3C_6H_3)_3SbCl_2$

The compound is obtained by oxidation of the corresponding stibine with Cl_2 in CCl_4. It crystallizes from petroleum ether as prisms, m.p. 238 °C. It is also formed in small amounts by reaction of [5-Cl-2-$CH_3C_6H_3N_2$]Cl with antimony powder in ethyl acetate [14].

(2,4-$(CH_3)_2C_6H_3)_3SbCl_2$

The corresponding stibine and Cl_2 in $CHCl_3$ give, after evaporation of the solvent, white crystals which melt at 189 °C [51]. Another way to the compound in a yield of 52% is the reaction of [2,4-$(CH_3)_2C_6H_3N_2$]$SbCl_6$ with iron powder (ratio 1:4) in $(CH_3)_2CO$. A melting point of 190 to 191.5 °C is given, recrystallized from $CHCl_3/C_2H_5OH$ [4, 5].

(2,5-$(CH_3)_2C_6H_3)_3SbCl_2$

This compound is obtained by reacting the corresponding stibine with $TlCl_3$ for 24 h in C_6H_6. After filtration, the solvent is concentrated. The resulting residue is recrystallized from petroleum ether with a yield of 50% and a melting point of 230 to 231 °C [52].

(3,5-$(CH_3)_2C_6H_3)_3SbCl_2$

(3,5-$(CH_3)_2C_6H_3)_2SbCl_3$ · (3,5-$(CH_3)_2C_6H_3)_2ICl$ reacts with zinc dust (ratio 1:4) in $(CH_3)_2CO$, 2 h at 20 °C, to form the compound in a yield of 33%. It melts at 189 to 190 °C [16].

(4-$C_2H_5O_2CC_6H_4)_3SbCl_2$

This compound results from the reaction of [4-$(C_2H_5O_2C)C_6H_4N_2$]$SbCl_6$ with iron powder (ratio 1:4) in $(CH_3)_2CO$. Evaporation of the filtrate and treatment of the residue with 5 N HCl and ethanol gives a precipitate in 20% yield which melts at 133 to 134 °C [5].

(4-CH_2=$CHC_6H_4)_3SbCl_2$

The preparation from the corresponding stibine and Cl_2 in ether, THF, or $CHCl_3$ is mentioned without details [28].

(4-$C_6H_5C_6H_4)_3SbCl_2$

$Sb(C_6H_4C_6H_5\text{-}4)_3$ and Cl_2 in $CHCl_3$ give small colorless prisms which melt at 273 to 274 °C under previous softening and decomposition. The crystals contain $CHCl_3$ in the ratio 1:1.

References on p. 58

The compound hydrolyzes quantitatively in 95% C_2H_5OH/NH_3, but no reaction occurs in cold or hot H_2O, or in warm 95% C_2H_5OH [53].

$(1\text{-}C_{10}H_7)_3SbCl_2$

Tris(1-naphthyl)stibine and chlorine are reacted in CCl_4 solution. Addition of petroleum ether precipitates crystals which are washed with hot $CHCl_3$ and dried [31, 54]; m.p. 260 °C [31], 256 °C [54]. The reaction may be also carried out in ether, THF, or $CHCl_3$ [28].

The solubility (in mol/L at 25 °C) of the compound is 1.5×10^{-3} in C_6H_6, 2.5×10^{-3} in CCl_4, and 2.5×10^{-2} in $CHCl_3$. Distribution coefficients between these solvents and water are 70, 7, and 6, respectively. Hydrolysis constants were determined by titration with 0.1 N NaOH in $CHCl_3$ as log $K_1 = 9$ and log $K_2 = 7.5$ [28]. The compound reacts with ethanolic KOH to form $(1\text{-}C_{10}H_7)_3SbO$ [54]. From a solution of the substance in aqueous HF, the corresponding difluoride can be extracted with $CHCl_3$ in yields of 95% [28]. The same difluoride is obtained from the compound and a mixture of KF, H_2O, and C_2H_5OH in a yield of 90% [9]. Concerning the extraction of fluoride from H_2O to the $CHCl_3$ phase in the presence of the compound at different pH values see [30].

$(2\text{-}C_{10}H_7)_3SbCl_2$

$[2\text{-}C_{10}H_7N_2]SbCl_6$ is reacted with iron powder (ratio 1:4) in $(CH_3)_2CO$ at 20 °C for 45 min. The filtrate of the mixture is evaporated and the remaining residue treated twice with 5 N HCl and then with C_2H_5OH. The precipitate (79% yield) is recrystallized from CH_3CO_2H or from $CH_3OH/(CH_3)_2CO$. The compound melts at 158 to 160 °C [4, 5].

$(C_4H_3O)_3SbCl_2$ (C_4H_3O = 2-furyl)

Tris(2-furyl)antimony dichloride is obtained from $Sb(OC_4H_3)_3$ (C_4H_3O = 2-furyl) and Cl_2 in petroleum ether. Recrystallization from C_2H_5OH gives 60% of the compound melting at 172 °C [55, 56].

$(C_4H_3S)_3SbCl_2$ (C_4H_3S = 2-thienyl)

Tris(2-thienyl)antimony dichloride is obtained from $Sb(C_4H_3S)_3$ (C_4H_3S = 2-thienyl) and Cl_2 in ether/CCl_4. It melts at 222.5 °C (corrected 229 °C), recrystallized from C_2H_5OH or a C_6H_6/petroleum ether mixture [57].

References:

[1] B. A. Nevett, A. Perry (Spectrochim. Acta A **31** [1975] 101/6).
[2] A. Otero, P. Royo (J. Organometal. Chem. **154** [1978] 13/19).
[3] H. G. Ang, W. S. Lien (J. Fluorine Chem. **9** [1977] 73/80).
[4] O. A. Reutov (Dokl. Akad. Nauk SSSR **87** [1952] 991/4).
[5] O. A. Reutov, V. V. Kondratyeva (Zh. Obshch. Khim. **24** [1954] 1259/65; J. Gen. Chem. [USSR] **24** [1954] 1245/9).
[6] O. A. Reutov, O. A. Ptitsyna (Dokl. Akad. Nauk SSSR **89** [1953] 877/80).
[7] A. N. Nesmeyanov, O. A. Reutov, O. A. Ptitsyna, P. A. Tsurkan (Izv. Akad. Nauk SSSR Otd. Khim. Nauk **1958** 1435/44; Bull. Acad. Sci. USSR Div. Chem. Sci. **1958** 1384/92).
[8] A. N. Nesmeyanov, K. A. Kocheshkov (Bull. Acad. Sci. USSR Div. Chem. Sci. **1944** 416/31).
[9] V. P. Glushkova, T. V. Talalaeva, Z. P. Razmanova, G. S. Zhdanov, K. A. Kocheshkov (Sb. Statei Obshch. Khim. **2** [1953] 992/6).
[10] L. W. Woods (Iowa State Coll. J. Sci. **19** [1944] 61/3).

[11] J. I. Harris, S. T. Bowden, W. J. Jones (J. Chem. Soc. **1947** 1568/71).
[12] B. Raynier, B. Waegell, R. Commandeur, H. Mathais (Nouv. J. Chim. **3** [1979] 393/401).
[13] A. N. Nesmeyanov, O. A. Reutov, O. A. Ptitsyna (Dokl. Akad. Nauk SSSR **91** [1953] 1341/4).
[14] F. B. Makin, W. A. Waters (J. Chem. Soc. **1938** 843/8).
[15] O. A. Ptitsyna, O. A. Reutov, G. Ertel (Izv. Akad. Nauk SSSR Otd. Khim. Nauk **1961** 265/70; Bull. Acad. Sci. USSR Div. Chem. Sci. **1961** 241/5).
[16] O. A. Reutov, A. N. Lovtsova (Vestn. Mosk. Univ. Ser. Mat. Mekhan. Astron. Fiz. Khim. **13** No. 3 [1958] 191/6).
[17] A. Ouchi, M. Nakatani, Y. Takahashi, S. Kitazima, T. Sugihara, M. Matsumoto, T. Uehiro, K. Kitano, K. Kawashima, H. Honda (Sci. Papers Coll. Gen. Educ. Univ. Tokyo **25** [1975] 73/99).
[18] A. Otero, P. Royo (J. Organometal. Chem. **171** [1979] 333/6).
[19] G. J. O'Donnell (Iowa State Coll. J. Sci. **20** [1945] 34/6).
[20] L. G. Makarova (J. Gen. Chem. [USSR] **7** [1937] 143/7).

[21] C. Löloff (Ber. Deut. Chem. Ges. **30** [1897] 2834/43).
[22] S. E. Gukasyan, V. S. Shpinel (Phys. Status Solidi **29** [1968] 49/52).
[23] S. E. Gukasyan, V. P. Gor'kov, P. N. Zaikin, V. S. Shpinel (Zh. Strukt. Khim. **14** [1973] 650/5; J. Struct. Chem. [USSR] **14** [1973] 603/7).
[24] J. M. Keck, G. Klar (Z. Naturforsch. **27b** [1972] 591/5).
[25] J. M. Keck, G. Klar (Z. Naturforsch. **27b** [1972] 596/9).
[26] P. May (Proc. Chem. Soc. **26** [1910] 218; J. Chem. Soc. **97** [1910] 1956/60).
[27] M. Michaelis, U. Genzken (Liebigs Ann. Chem. **242** [1887] 164/188).
[28] M. Benmalek, H. Chermette, C. Martelet, D. Sandino, J. Tousset (J. Organometal. Chem. **67** [1974] 53/9).
[29] G. Deganello, G. Dolcetti, M. Giustiniani, U. Belluco (J. Chem. Soc. A **14** [1969] 2138/40).
[30] M. Benmalek, H. Chermette, C. Martelet, D. Sandino, J. Tousset (J. Inorg. Nucl. Chem. **36** [1974] 1365/8).

[31] F. Challenger, F. Pritchard, J. R. A. Jinks (J. Chem. Soc. **125** [1924] 864/75).
[32] V. D. Nefedov, W.-C. Wang (Beijing Daxue Xuebao Ziran Kexueban **4** [1958] 315/9; C.A. **1961** 13007).
[33] G. S. Zhdanov, Z. P. Razmanova (Dokl. Akad. Nauk SSSR **72** [1950] 1055/7).
[34] G. L. Kuykendall, J. L. Mills (J. Organometal. Chem. **118** [1976] 123/8).
[35] L. M. Kataeva, Yu. V. Rydvanskii, N. I. Trofimova (Zh. Fiz. Khim. **50** [1976] 814/5; J. Phys. Chem. [USSR] **50** [1976] 486/7).
[36] N. K. Parab, D. M. Desai (Current Sci. [India] **26** [1957] 389).
[37] N. K. Parab, D. M. Desai (J. Indian Chem. Soc. **35** [1958] 569/72).
[38] A. Michaelis, U. Genzken (Ber. Deut. Chem. Ges. **17** [1884] 924/25).
[39] H. E. Affsprung, A. B. Gainer (Anal. Chim. Acta **27** [1962] 578/84).
[40] C. Brabant, J. Hubert, A. L. Beauchamp (Can. J. Chem. **51** [1973] 2951/7).

[41] S. N. Bhattacharya, M. Singh (Indian J. Chem. A **18** [1979] 515/6).
[42] A. N. Nesmeyanov, A. E. Borisov (Izv. Akad. Nauk SSSR Ser. Khim. **1969** 974/5; Bull. Acad. Sci. USSR Div. Chem. Sci. **1969** 895).
[43] A. N. Nesmeyanov, A. E. Borisov, N. V. Novikova (Izv. Akad. Nauk SSSR Ser. Khim. **1969** 1978/82; Bull. Acad. Sci. USSR Div. Chem. Sci. **1969** 1830/3).
[44] K. Bajpai, R. C. Srivastava (Syn. Reactiv. Inorg. Metal-Org. Chem. **11** [1981] 7/13).
[45] K. Bajpai, R. C. Srivastava (Syn. Reactiv. Inorg. Metal-Org. Chem. **12** [1982] 47/54).
[46] A. E. Goddard, V. E. Yarsley (J. Chem. Soc. **1928** 719/23).
[47] S. N. Bhattacharya, R. Prem, A. K. Saxena (Indian J. Chem. A **16** [1978] 1071/4).

[48] V. D. Nefedov, W.-C. Wang, H.-M. Hsiang, F.-S. Chang, K.-M. Ni (Beijing Daxue Xuebao Ziran Kexueban **1959** 377/81; C.A. **56** [1962] 10952).
[49] V. D. Nefedov, I. S. Kirin, V. M. Zaitsev (Radiokhimiya **4** [1962] 351/5; Soviet Radiochem. **4** [1962] 311/4; C.A. **59** [1963] 14829).
[50] J. Havranek, J. Mleziva (Angew. Makromol. Chem. **84** [1980] 105/17).
[51] A. E. Goddard (J. Chem. Soc. **123** [1923] 2315/23).
[52] A. E. Goddard (J. Chem. Soc. **123** [1923] 1161/72).
[53] D. E. Worrall (J. Am. Chem. Soc. **52** [1930] 2046/50).
[54] K. Matsumiya (Mem. Coll. Sci. Univ. Kyoto **8** [1925] 8/11; C.A. **1925** 1704).
[55] A. Etienne (Compt. Rend. **221** [1945] 562/4).
[56] A. Etienne (Bull. Soc. Chim. France **1947** 50/1).
[57] E. Krause, G. Renwanz (Ber. Deut. Chem. Ges. **65** [1932] 777/84).
[58] C. Cornselius, G. Klar (unpublished result from [24]).

2.5.1.1.3 Triorganoantimony Dibromides

2.5.1.1.3.1 R_3SbBr_2 Compounds with R = Alkyl and Alkenyl

2.5.1.1.3.1.1 Trimethylantimony Dibromide $(CH_3)_3SbBr_2$

Preparation. Trimethylantimony dibromide is prepared by reacting $Sb(CH_3)_3$ with Br_2 without a solvent or in ethereal solution [1 to 4]. The yield of the compound is 72 [1] to 89% [4]. In the most cases, $Sb(CH_3)_3$ was not isolated after preparation from $SbCl_3$ and Grignard compounds. The distillate from this reaction mixture, consisting of ether and $Sb(CH_3)_3$, was subsequently treated with Br_2 in CCl_4 until no further precipitate deposits and a light brown color of the solution appears [4 to 11]. The yields are about 50 to 60%, relative to the starting material $SbCl_3$ [8, 11]. Other methods for preparation of the compound are reaction of $(CH_3)_3SbO$ with aqueous HBr [2, 6], hydrolysis of $(CH_3)_3Sb(CN)Br$ [9], reaction of $(CH_3)_3Sb(NO_3)_2$ with KBr in water [12], or decomposition of $(CH_3)_3Sb \cdot BBr_3$ in $CHCl_3$ or CH_3CN [13]. The reaction of $(CH_3)_3SbS$ with $C_6H_5CH_2Br$ in $CHCl_3$ yields the compound in 89% yield besides $Sb(CH_3)_3$ and $(C_6H_5CH_2S)_2$ [14]. In a reaction of CD_3MgI with $SbCl_3$ and brominating the resulting mixture, $(CD_3)_3SbBr_2$ can be isolated in a yield of 20% after recrystallization from D_2O [9]. The crude compound may be recrystallized from ether/benzene [4], water [2, 6, 10] though addition of HBr is required to prevent hydrolysis [16], or 95% ethanol [2, 11].

Properties. Melting points with decomposition of the compound are given as 170 to 180 °C [2, 6], 192 to 193 °C [7], 198 °C [12], 195 °C [1], and 196 to 197 °C from ether/benzene [4].

IR and Raman spectra were published and discussed in several publications [9, 12, 17 to 21] and vibrations are compared with those of $(CD_3)_3SbBr_2$ [9, 19, 20]. The complete IR and Raman vibrations of the solid compound are given in Table 16 [21]. Force constants are calculated in [17 to 20, 22 to 24]. An approximate normal coordinate analysis assuming a molecular geometry of D_{3h} and force constant calculations using general valence, modified Urey-Bradley, and orbital valence force fields are given in [19, 20, 22]. Force constants, mean amplitudes of vibration, generalized mean square amplitudes, shrinkage constants, Coriolis coupling coefficients, and centrifugal distortion constants are evaluated in [24]. A correlation between fundamental vibrations and physical parameters is given in [25]. Further structural discussions based on some IR and Raman bands are given in [26, 27].

Table 16
IR and Raman Vibrations (in cm^{-1}) of Solid $(CH_3)_3SbBr_2$ [21].

IR	Raman	assignment	IR	Raman	assignment
3004 m	3024 mw, br	$\nu_{as}CH$		601 vw	
2925 w	2925 s	ν_sCH	569 vs	569 vs	ν_5, $\nu_{as}SbC$
2853 vw				526 vs	ν_1, ν_sSbC
2773 vw	2784 vw			481 mw	$(\nu_4+\nu_8)$ or $(\nu_6+\nu_8)$
2410 w	2417 w		394 vw	393 vvw	$\nu_5-\nu_2$
2356 mw			336 vw		
2332 mw			314 vw		$2\nu_6$
1780 vw			298 vw	305 vvw	
1732 w			275 w	265 vvw	
1627 mw			250 w		
	1465 w		228 vvw	233 vw	
1402 ms	1409 vw	$\delta_{as}CH$	215 mw		ν_3, $\nu_{as}SbBr$
	1237 m	δ_sCH		200 sh, br	ν_8, ϱ
1208 vw	1209 mw			168 vvs	ν_2, ν_sSbBr
1087 vw	1087 vw	$\nu_1+\nu_5$	172 vs		ν_4, δSbC out-of-plane
1019 vvw			160 vs, br		ν_6, δSbBr
875 vs		ϱCH_3	96 ms	86 mw	ν_7, δSbC in-plane
	635 vw			43 vs	lattice vibration

A UV spectrum of the compound is published, showing only a tail between 240 and 270 nm in CH_3CN solution (1.06×10^{-4} M) [14].

^{1}H NMR spectra were measured, and the following chemical shift values have been found: δ (in H_2O vs. H_2O) = 2.81 ppm [26], δ (D_2O) = 1.90 ppm [1], δ (CCl_4) = 2.61 ppm [28], δ ($CDCl_3$) = 2.72 ppm [29], δ ($CDCl_3$) = 2.66 (at −32 °C), 2.62 (at 30.5 °C), and 2.60 ppm (at 70 °C) [30], δ ($CHCl_3$) = 2.65 (at −65 °C), 2.59 ppm (at 30 °C) [31]. For ^{1}H NMR studies of mixtures of the compound with $(CH_3)_3SbX_2$ (X = F, Cl, I) [30] or with $(CH_3)_2SbCl_3$ [31] see originals. A correlation of the ^{1}H NMR chemical shifts with the inverse ionization potentials of the substituents in $(CH_3)_3SbX^1X^2$ (X^1, X^2 = all possible combinations of F, Cl, Br, I) is given in [32].

NQR spectra were measured and discussed on the basis of a D_{3h} symmetry of the molecule. At 300 K, the following resonances (in MHz) were found: $\nu(^{121}Sb)$ = 189.03(10) (5/2 ⇌ 3/2), 94.585(20) (3/2 ⇌ 1/2); $\nu(^{123}Sb)$ = 57.421(30) (3/2 ⇌ 1/2); e^2Qq/h = 630.57 (^{121}Sb), 803.89 (^{123}Sb); η = 0.00; $\nu(^{79}Br)$ = 115.07(60) and $\nu(^{81}Br)$ = 96.127(60) [33], see also [91]. The H/D isotope effect on quadrupole coupling was compared between $(CH_3)_3SbBr_2$ (e^2Qq/h = 803.9 ± 0.1 (^{123}Sb), 230.14 ± 0.5 (^{121}Sb)), and $(CD_3)_3SbBr_2$, where the corresponding values are 808.2 ± 0.1 and 229.86 ± 0.05, respectively [34]. Sb^V orbital population calculations on the basis of NQR and Mössbauer data are made in [35].

Mössbauer resonances from ^{121}Sb were measured at 80 K vs. $^{121}SnO_2$: δ = −6.0 ± 0.1, e^2qQ = −21 ± 1, Γ = 3.1 ± 0.2 mm/s [36]. The corresponding values at 4 K vs. $Ca^{121}SnO_3$ are: δ = −6.40 ± 0.02, e^2qQ = −22.1 ± 0.3, Γ = 5.58 ± 0.04 mm/s [37]. For a comparison of measured and calculated e^2qQ values see [38]. A discussion of the Goldanskii-Karyagin effect on Mössbauer quadrupole splittings of $(CH_3)_3SbBr_2$ is given in [39]. For a Hartree-Fock-Slater LCAO calculation of the Mössbauer parameters [40], the connection with the additive

References on p. 68

electric field gradient model [41], and the correlation of coupling values with those of Sn^{IV} compounds [42] see originals.

A high energy photoelectron spectrum with Al Kα as exciting radiation leads to binding energies of Sb 3d(3/2) = 540.8, Sb 3d(5/2) = 531.2 eV, Br 3p = 190.1 and 183.2 eV [43]. In a gas phase He(I) PE spectrum of the compound, the following ionization energies were found: 9.57, 10.02, 10.32, 11.7, and 14.5 eV. These values are discussed and compared with those of the other halides [44].

Trimethylantimony dibromide crystallizes in the hexagonal system [2]; space group $C\bar{6}2c$-D^4_{3h} (No. 190) [15, 16] with cell dimensions of a = 7.38 and c = 8.90 Å; Z = 2; d_c = 2.57 and d_m = 2.58 g/cm³ (by flotation). The geometry of the molecule is trigonal bipyramidal with bromine atoms in axial positions and an Sb-Br distance of 2.63 Å [16].

Mass spectra of the compound were measured and discussed in [45, 46]. At 70 eV, inlet 30 °C, source 150 °C, the following fragments were detected (relative intensity): $[(CH_3)_2SbBr_2]^+$ (3), $[CH_3SbBr_2]^+$ (1), $[SbBr_2]^+$ (12), $[(CH_3)_3SbBr]^+$ (100), $[(CH_3)_2SbBr]^+$ (50), $[CH_3SbBr]^+$ (20), $[SbBr]^+$ (18), $[(CH_3)_3Sb]^+$ (5), $[(CH_3)_2Sb]^+$ (20) [46].

Polarographic half-wave potentials (in V, at 25 °C vs. SCE) are: $E_{1/2} = -0.75$ (1 N HCl), −0.68 (pH 1), −0.82 (pH 7), −1.3 (pH 11), −1.24 (1 N NH_4OH), −0.73 (0.2 N KCl) [47]. About the potentiometric titration of the compound in aqueous solution with 0.1 N NaOH see [48]. A 0.05 M solution of the compound in H_2O shows a pH value of 1.55 [26], probably resulting from a hydrolysis of the compound to HBr and $(CH_3)_3Sb(OH)Br$ [49]. Redox processes in aqueous solution were studied by cyclic voltammetry; potentials $E_{ox} = 0.19$ V, $E_{red} = -0.08$ and −0.71 V vs. SCE have been reported. A figure is given in [26]. The compound dissolves in C_2H_5OH, $(CH_3)_2CO$, and CH_3CN, but is only slightly soluble in nonpolar organic solvents such as c-C_6H_{12} or CCl_4 [16].

Reactions. When combined, solutions of $(CH_3)_3SbBr_2$ and $(CH_3)_3SbX_2$ (X = F, Cl, I) in $CHCl_3$ produce a mixed halide, $(CH_3)_3SbBrX$ (X = F, Cl, I) [30]. The equilibrium constants were determined by ^1H or ^{19}F NMR. For X = F: $K_1 = 2.81 \pm 0.09$ at 0 °C, 3.24 ± 0.14 at 35 °C, 3.61 ± 0.19 at 60 °C; for X = Cl: $K_1 = 3.28 \pm 0.20$ at 0 °C, 3.50 ± 0.15 at 35 °C, 3.65 ± 0.13 at 60 °C; for X = I: $K_1 = 2.93 \pm 0.03$ at 0 °C, 3.02 ± 0.10 at 35 °C, 3.30 ± 0.16 at 60 °C [81, 82]. Equilibrium constants for the reactions $(CH_3)_3SbBr_2 + (C_6H_5CH_2)_3SbX_2 \rightleftharpoons (CH_3)_3SbX_2 + (C_6H_5CH_2)_3SbBr_2$ are $K_2 = 122 \pm 20$ for X = F, and 2.7 ± 0.4 for X = Cl in $CHCl_3$ at 32 °C [82].

Other reactions of trimethylantimony dibromide are summarized in Table 17.

Uses. The compound is claimed to be used as a catalyst for the reaction of epoxides with carbon dioxide to form cyclic carbonates [83 to 86] in yields of about 70 to 90% [83]. It is also a catalyst for the ring opening polymerization of alkene oxides [87]. The photo-initiated polymerization of vinyl monomers via a radical mechanism is promoted by the compound [88]. The use of the compound as a pharmaceutical and pest controlling agent is claimed [89].

References on p. 68

Table 17
Reactions of $(CH_3)_3SbBr_2$.

reactants	reaction conditions (ratio = $(CH_3)_3SbBr_2$: reactant)	products (yield in %)	Ref.
—	heating at 170 to 180 °C	dec. to oil, not identified	[2]
—	heating under CO_2, 200 °C/80 to 100 Torr	$(CH_3)_2SbBr$	[6, 50]
—	heating at 180 °C/90 Torr	$(CH_3)_2SbBr$ (70) + CH_3Br	[8]
Na	ratio 1:4 in liquid NH_3	red solution of $NaSb(CH_3)_2$	[51, 52]
AgF	in hot H_2O	$(CH_3)_3SbF_2$ (42)	[9]
AgF	in aqueous HF	$(CH_3)_3SbF_2$	[10]
$AgNO_3$	in hot H_2O or CH_3OH	$(CH_3)_3Sb(NO_3)_2$	[9, 10]
Ag_2SO_4	in H_2O	$(CH_3)_3SbSO_4$	[9, 10]
Ag_2CrO_4	in H_2O	$(CH_3)_3SbCrO_4$	[10]
Ag_2CO_3	in liquid SO_2	$(CH_3)_3SbCO_3$	[10]
$AgClO_4$	in H_2O	$[(CH_3)_3SbClO_4]_2O$	[9]
$AgClO_4$	in CH_3OH	*explosion!*	[10]
$AgBF_4$, $AgSiF_6$, $AgSbF_6$, or $AgB_{12}Cl_{12}$	in CH_3OH	impure products	[10]
Ag_2SeO_4	in H_2O	$(CH_3)_3SbSeO_4$	[53]
Ag_2O	some H_2O	$(CH_3)_3SbO$	[54, 55]
CH_3OH	$CH_3OH + NH_3$	$(CH_3)_3Sb(OCH_3)_2$	[56]
C_2H_5ONa	ratio 1:1 in $CHCl_3$ or C_6H_6	$(CH_3)_3SbBr(OC_2H_5)$	[56]
C_2H_5ONa	in C_2H_5OH, 30 min at 20 °C	$(CH_3)_3Sb(OC_2H_5)_2$	[57]
XC_6H_4OH, X = H, 4-Cl, 4-Br, 2-, 3-NO_2, 2-CH_3, 2-, 3-C_6H_5	in $C_6H_6 + N(C_2H_5)_3$	$(CH_3)_3Sb(OC_6H_4X)_2$ (40 to 70)	[58]

References on p. 68

Table 17 [continued]

reactants	reaction conditions (ratio = $(CH_3)_3SbBr_2$: reactant)	products (yield in %)	Ref.
C_6H_5OH	in $C_6H_6 + N(C_2H_5)_3$	$(CH_3)_3Sb(OC_6H_5)_2$	[59]
$C_6H_4(ONa)_2$-1,2	in C_6H_6 + some $(CH_3)_2CO$	$(CH_3)_3Sb(-OC_6H_4O-2-)$ (55)	[60]
t-C_4H_9OONa	in $C_6H_6/CHCl_3$	$(CH_3)_3Sb(OOC_4H_9\text{-t})_2$ (94)	[61]
t-C_4H_9OOH	via $(CH_3)_3Sb(OCH_3)_2$ in CH_3OH	$(CH_3)_3Sb(OOC_4H_9\text{-t})_2$ (83)	[61]
ROOH, R = $C_6H_5C(CH_3)_2$, C_9H_9O (= 3,4-dihydro-1H-benzopyranyl)	in $CHCl_3$/ether + NH_3 or amine	$(CH_3)_3Sb(OOR)_2$ (70 and 84, respectively)	[61]
$XC_6H_4CO_2H$, X = H, 2-, 3-, 4-NH_2, 2-, 3-, 4-CH_3, 2-, 3-, 4-NO_2	ratio 1:2 in $C_6H_6 + N(C_2H_5)_3$	$(CH_3)_3Sb(O_2CC_6H_4X)_2$	[58]
$XC_6H_4CO_2H$, X = H, 4-Cl	ratio 1:2 in $C_6H_6 + N(C_2H_5)_3$	$(CH_3)_3Sb(O_2CC_6H_4X)_2$	[59]
$CX_{3-n}H_nCO_2H + Ag_2O$, X = F, Cl, Br (n = 0 to 3)	ratio 1:2 in CH_3OH	$(CH_3)_3Sb(O_2CX_{3-n}H_n)_2$	[62]
$NCCH_2CO_2H + Ag_2O$	ratio 1:2 in CH_3OH	$(CH_3)_3Sb(O_2CCH_2CN)_2$	[62]
$CD_3CO_2H + Ag_2O$	ratio 1:2 in CH_3OH	$(CH_3)_3Sb(O_2CCD_3)_2$	[62]
RCO_2H, R = H, $(CH_3)_2CHCH_2$, $CH_3CH{=}CH$, $C_6H_5CH_2$, $C_6H_5CH{=}CH$	in refluxing $C_6H_6 + N(C_2H_5)_3$	$(CH_3)_3Sb(O_2CR)_2$	[63]
$Ag_2C_2O_4$	in H_2O	polymeric material (30) $(CH_3)_3SbC_2O_4$	[58] [10]
$AgO_2CCH_2CO_2Ag$	in H_2O	polymeric material (30)	[58]
RR'C=NOH, R = CH_3, R' = CH_3, C_2H_5, 4-$NO_2C_6H_4$	in refluxing $C_6H_6 + N(C_2H_5)_3$	$(CH_3)_3Sb(ON{=}CRR')_2$	[63]
RR'C=NOH, R, R' = $(CH_2)_5$; R = CH_3, R' = C_6H_5	3 h in refluxing $C_6H_6 + N(C_2H_5)_3$	$(CH_3)_3Sb(ON{=}CRR')_2$	[64]

References on p. 68

RR'C=NONa, R = CH_3, R' = CH_3, C_2H_5, C_3H_7, C_6H_5; R = R' = C_2H_5; R = C_6H_5, R' = NH_2	2 h in refluxing C_6H_6	$(CH_3)_3Sb(ON{=}CRR')_2$ (ca. 100)	[65]
8-hydroxyquinoline	via $(CH_3)_3Sb(OCH_3)Br$ in C_6H_6/CH_3OH	$(CH_3)_3Sb(Br)OC_9H_6N$	[66]
	same reaction, workup with aqueous $(CH_3)_2CO$	$(CH_3)_3Sb(OH)OC_9H_6N$	[67]
HX + 8-hydroxyquinoline, X = NO_3, 4-$NO_2C_6H_4O$, 2-,4-,6-$(NO_2)_3C_6H_2O$, CH_3CO_2, $ClCH_2CO_2$, Cl_2CHCO_2, Cl_3CCO_2, Br_3CCO_2, $NCCH_2CO_2$, HCO_2, $C_6H_5CO_2$, $(CH_3)_2CHCO_2$	via $(CH_3)_3Sb(OCH_3)_2$ in C_6H_6/CH_3OH, addition of HX, evaporation, addition of quinoline (ratio 1:1)	$(CH_3)_3Sb(X)OC_9H_6N$	[68]
2-methyl-8-hydroxyquinoline	$(CH_3)_3SbBr_2$ + $NaOCH_3$ + quinoline (ratio 1:1:2) in C_6H_6/CH_3OH, 30 min reflux	$(CH_3)_3SbBr(OC_9H_5NCH_3\text{-}2)$	[68]
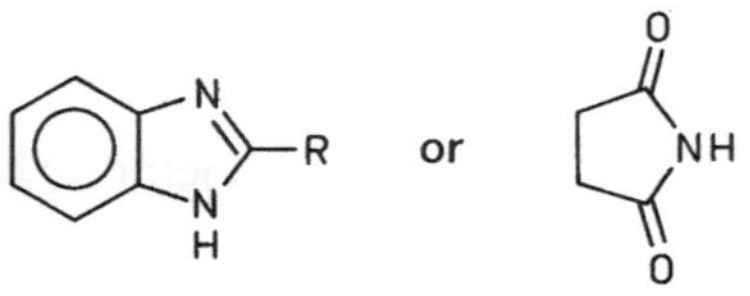 R = H, CH_3, C_2H_5	in refluxing C_6H_6 + $N(C_2H_5)_3$	$(CH_3)_3Sb(NC_7H_4NR)_2$ or $(CH_3)_3Sb(NC_4H_4O_2)_2$, respectively	[63]
R(H)P(O)OH, R = C_6H_5, $C_6H_5CH{=}CH$	in C_6H_6 + $N(C_2H_5)_3$	$(CH_3)_3Sb(OP(O)(H)R)_2$	[69]
PR_3, R = C_6H_5, C_4H_9	in C_6H_6 in a closed ampule, 16 h at 100 °C	$[R_3PCH_3]^+[(CH_3)_2SbBr_2]^-$ (75)	[70]
$(CH_3)_2NC(S)SNa$	in H_2O	$(CH_3)_3Sb(SC(S)N(CH_3)_2)_2$	[71]
$R_2NC(S)SNa$, R_2N = pyrrolidinyl, piperidinyl	in C_6H_6/CH_3CN, 1 h at 20 °C	$(CH_3)_3Sb(SC(S)NR_2)_2$ (50)	[72]
ROC(S)SNa, R = C_3H_7, i-C_3H_7, i-C_4H_9, $C_6H_5CH_2$	in aqueous i-C_3H_7OH	$(CH_3)_3Sb(SC(S)OR)_2$ (60)	[72]
$(CH_3)_3Sb(SC(O)R)_2$, R = CH_3, C_6H_5	in $CHCl_3$, 30 min at 20 °C	$(CH_3)_3Sb(SC(O)R)Br$ (ca. 100)	[73]
$(CH_3)_3SiONa$	organic solvent	$(CH_3)_3Sb(OSi(CH_3)_3)_2$	[74]
$LiBH_4$ or $LiAlH_4$	in ether or THF at −55 to −95 °C	H_2, B_2H_6, or AlH_3, LiBr, $Sb(CH_3)_3$	[75]

References on p. 68

Table 17 [continued]

reactants	reaction conditions (ratio = $(CH_3)_3SbBr_2$: reactant)	products (yield in %)	Ref.
$LiCH_3$	in ether, 3 h at 20 °C	$Sb(CH_3)_5$ (60)	[76, 90]
C_4H_9MgBr	in ether/THF, 3 h at 20 °C	$(CH_3)_3Sb(C_4H_9)_2$ (66), $(CH_3)_4SbC_4H_9$ (17), $(CH_3)_2Sb(C_4H_9)_3$ (17)	[76]
$LiCH_2Si(CH_3)_2$	in ether, 2 h at 20 °C	$(CH_3)_3Sb(CH_2Si(CH_3)_3)_2$ (83)	[77]
$Ga(CH_3)_3$	in CH_2Cl	$[Sb(CH_3)_4][(CH_3)_2GaBr_2]$	[78]
$(CH_3)_2GaBr$	in CH_2Cl	$[Sb(CH_3)_4][CH_3GaBr_3]$	[78]
$SbBr_3$	in boiling CH_2Cl_2	$(CH_3)_3SbBr_2 \cdot SbBr_3$ (74), see below	[79]
$Tl[Hg(Ge(C_6F_5)_3)_3]$	in $HCON(CH_3)_2/C_2H_5OH$	TlBr, $Hg(Ge(C_6F_5)_3)_2$, $(CH_3)_3Sb(OC_2H_5)_2$ (80)	[80]
$(CH_3)_3SbO$	in H_2O, evaporated	"oxybromide"	[2]

References on p. 68

$(CH_3)_3SbBr_2 \cdot SbBr_3$

The compound is prepared by reacting $(CH_3)_3SbBr_2$ and $SbBr_3$ in equimolar amounts in boiling CH_2Cl_2. Upon cooling, the compound crystallizes in the form of colorless small crystals in a yield of 74%. The compound melts at 122 to 126 °C [79].

The Raman spectra of the solid and of a CH_2Cl_2 solution below 600 cm^{-1} are given and assigned in the original. The Raman spectrum in CH_2Cl_2 is a composite of those of $(CH_3)_3SbBr_2$ and $SbBr_3$ showing that the adduct is dissociated in solution [79].

The compound crystallizes in the monoclinic system; space group $P2_1/n$-C^5_{2h} (No. 14) with a = 981.8(3), b = 1407.4(4), c = 1003.5(3) pm, and β = 93.95(2)°; d_m = 3.30, d_c = 3.304 g/cm³; Z = 4. The structure was solved and refined to an R value of 0.110. The coordination number of the Sb^V in the molecule is 5 and that of Sb^{III} is 6. The angles on $(CH_3)_3SbBr_2$ are practically the same as in the uncoordinated compound, but the lengths of the axial bromine bonds are different resulting from a bridging to the $SbBr_3$ moiety of the adduct. A projection of the elementary cell with important distances and angles is given in **Fig. 7** [79].

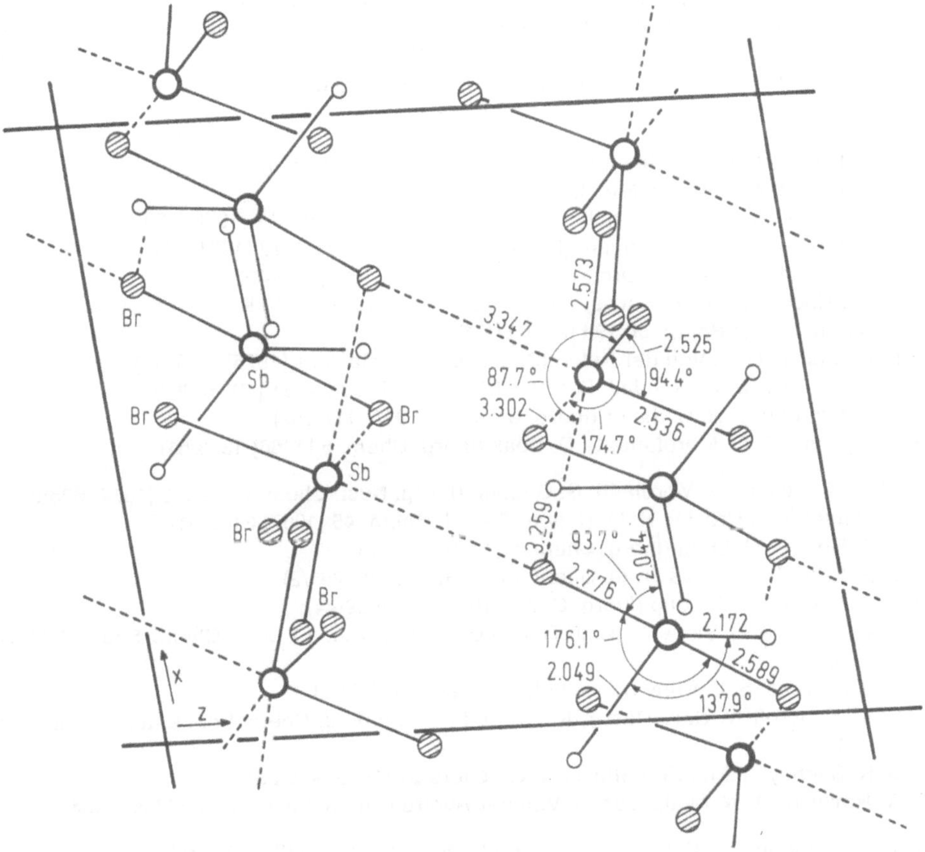

Fig. 7. Projection of the elementary cell of $(CH_3)_3SbBr_2 \cdot SbBr_3$ on the xz-plane (distances in Å) [79].

References on p. 68

References:

[1] A. G. Davies, S. C. W. Hook (J. Chem. Soc. C **1971** 1660/5).
[2] H. Landolt (J. Prakt. Chem. **84** [1861] 328/29).
[3] R. Müller, C. Dathe (Chem. Ber. **99** [1966] 1609/13).
[4] H. Schmidbaur, K. H. Mitschke (Chem. Ber. **104** [1971] 1837/41).
[5] H. Hibbert (Ber. Deut. Chem. Ges. **39** [1906] 160/2).
[6] G. T. Morgan, G. R. Davies (Proc. Roy. Soc. [London] A **110** [1926] 523/34).
[7] G. Wittig, H. Laib (Liebigs Ann. Chem. **580** [1953] 57/68).
[8] H. Hartmann, G. Kühl (Z. Anorg. Allgem. Chem. **312** [1961] 186/94).
[9] G. G. Long, G. O. Doak, L. D. Freedman (J. Am. Chem. Soc. **86** [1964] 209/13).
[10] H. C. Clar, R. G. Goel (Inorg. Chem. **5** [1966] 998/1003).

[11] G. O. Doak, G. G. Long, M. E. Key (Inorg.Syn. **9** [1967] 92/7).
[12] M. Shindo, R. Okawara (J. Organometal. Chem. **5** [1966] 537/44).
[13] M. L. Denniston, D. R. Martin (J. Inorg. Nucl. Chem. **36** [1974] 2175/6).
[14] J. Otera, R. Okawara (J. Organometal. Chem. **16** [1969] 335/8).
[15] G. S. Zhdanov, Z. P. Razmanova (Dokl. Akad. Nauk SSSR **72** [1950] 1055/7).
[16] A. F. Wells (Z. Krist. **99** [1938] 367/77).
[17] R. G. Goel, E. Maslowsky Jr., C. V. Senoff (Inorg. Chem. **10** [1971] 2572/7).
[18] R. G. Goel, E. Maslowsky Jr., C. V. Senoff (Inorg. Nucl. Chem. Letters **6** [1970] 833/5).
[19] C. Woods, G. G. Long (J. Mol. Spectrosc. **38** [1971] 387/95).
[20] C. Woods (Diss. North Carolina State Univ. at Raleigh 1971, pp. 1/100; Diss. Abstr. Intern. B **32** [1972] 3838).

[21] B. A. Nevett, A. Perry (Spectrochim. Acta A **33** [1977] 755/60).
[22] B. A. Nevett, A. Perry (J. Mol. Spectrosc. **66** [1977] 331).
[23] R. Namasivayam, S. Viswanathan (Bull. Soc. Chim. Belges **87** [1978] 733/6).
[24] A. Natarajan, K. Chockalingam (Indian J. Pure Appl. Phys. **19** [1981] 672/5).
[25] B. A. Nevett, A. Perry (J. Organometal. Chem. **71** [1974] 399/402).
[26] F. E. Brinckman, G. E. Parris, W. R. Blair, K. L. Jewett, W. P. Iverson, J. M. Bellama (Environ. Health Perspect. **19** [1977] 11/24).
[27] L. Verdonck, G. P. Van der Kelen (Spectrochim. Acta A **31** [1975] 1707/11).
[28] A. G. Massey, E. W. Randall, D. Shaw (Spectrochim. Acta **21** [1965] 263/73).
[29] R. L. McKenney, H. H. Sisler (Inorg. Chem. **6** [1967] 1178/82).
[30] G. G. Long, C. G. Moreland, G. O. Doak (Inorg. Chem. **5** [1966] 1358/61).

[31] W. A. Kustes, C. G. Moreland, G. G. Long (Inorg. Nucl. Chem. Letters **8** [1972] 695/9).
[32] T. Schaefer, F. Hruska, H. M. Hutton (Can. J. Chem. **45** [1967] 3143/51).
[33] T. B. Brill, G. G. Long (Inorg. Chem. **9** [1970] 1980/5).
[34] T. B. Brill, Z. Z. Hugus Jr. (J. Chem. Phys. **53** [1970] 1291/2).
[35] L. H. Bowen, G. G. Long (Inorg. Chem. **15** [1976] 1039/44).
[36] G. G. Long, J. G. Stevens, R. J. Tullbane, L. H. Bowen (J. Am. Chem. Soc. **92** [1970] 4230/5).
[37] J. G. Stevens, S. L. Ruby (Phys. Letters A **32** [1970] 91/2).
[38] G. M. Bancroft, V. G. K. Das, T. K. Sham, M. G. Clark (J. Chem. Soc. Dalton Trans. **1976** 643/54).
[39] G. K. Shenoy, J. M. Friedt (Phys. Rev. Letters **31** [1973] 419/22).
[40] W. Ravenek, J. W. M. Jacobs, A. Van der Avoird (Chem. Phys. **78** [1983] 391/404).

[41] J. N. R. Ruddick, J. R. Sams, J. C. Scott (Inorg. Chem. **13** [1974] 1503/7).
[42] G. M. Bancroft, V. G. K. Das, K. D. Butler (J. Chem. Soc. Dalton Trans. **1974** 2355/8).
[43] T. Birchall, J. A. Connor, I. H. Hillier (J. Chem. Soc. Dalton Trans. **1975** 2003/6).

[44] S. Elbel, H. Tom Dieck (Z. Anorg. Allgem. Chem. **483** [1981] 33/43).
[45] H. Preiss (Z. Anorg. Allgem. Chem. **389** [1972] 280/92).
[46] H. J. Breunig, W. Kanig (Phosphorus Sulfur **12** [1982] 149/59).
[47] M. K. Saikina (Uch. Zap. Kaz. Gos. Univ. **116** No. 2 [1956] 129/86; C.A. **1957** 7191).
[48] P. Nylen (Z. Anorg. Allgem. Chem. **246** [1941] 227/42).
[49] A. Hantzsch, H. Hibbert (Ber. Deut. Chem. Ges. **40** [1907] 1508/19).
[50] G. T. Morgan, G. R. Davies (Nature **116** [1925] 499).

[51] W. Levason, B. Sheikh (J. Organometal. Chem. **208** [1981] 1/9).
[52] W. Levason, B. Sheikh, W. E. Hill (J. Organometal. Chem. **219** [1981] 163/8).
[53] R. G. Goel, P. N. Joshi, D. R. Ridley, R. E. Beaumont (Can. J. Chem. **47** [1969] 1423/7).
[54] T. M. Lowry, J. H. Simons (Ber. Deut. Chem. Ges. **63** [1930] 1595/1602).
[55] W. Morris, R. A. Zingaro, J. Laane (J. Organometal. Chem. **91** [1975] 295/306).
[56] J. Dahlmann, A. Rieche (Chem. Ber. **100** [1967] 1544/9).
[57] Y. Matsumura, M. Shindo, R. Okawara (Inorg. Nucl. Chem. Letters **3** [1967] 219/22).
[58] A. Ouchi, M. Nakatani, Y. Takahashi, S. Kitazima, T. Sugihara, M. Matsumoto, T. Uehiro, K. Kitano, K. Kawashima, H. Honda (Sci. Papers Coll. Gen. Educ. Univ. Tokyo **25** [1975] 73/99).
[59] A. Ouchi, T. Uehiro, Y. Yoshino (J. Inorg. Nucl. Chem. **37** [1975] 2347/9).
[60] M. Shindo, R. Okawara (Inorg. Nucl. Chem. Letters **5** [1969] 77/80).

[61] A. Rieche, J. Dahlmann, D. List (Liebigs Ann. Chem. **678** [1964] 167/82).
[62] R. G. Goel, D. R. Ridley (J. Organometal. Chem. **38** [1972] 83/9).
[63] K. Bajpai, M. Srivastava, R. C. Srivastava (Indian J. Chem. A **20** [1981] 736/7).
[64] R. C. Srivastava (Syn. Reactiv. Inorg. Metal-Org. Chem. **11** [1981] 7/13).
[65] V. K. Jain, R. Bohra, R. C. Mehrotra (Inorg. Chim. Acta **51** [1981] 191/4).
[66] Y. Kawasaki (Inorg. Nucl. Chem. Letters **5** [1969] 805/10).
[67] Y. Kawasaki (Bull. Chem. Soc. Japan **49** [1976] 817/18).
[68] Y. Kawasaki, K. Hashimoto (J. Organometal. Chem. **99** [1975] 107/14).
[69] G. E. Graves, J. R. Van Wazer (J. Organometal. Chem. **131** [1977] 31).
[70] G. E. Graves, J. R. Van Wazer (J. Inorg.Nucl. Chem. **39** [1977] 1101).
[71] J. A. Cras, J. Willemse (Recl. Trav. Chim. **97** [1978] 28/9).
[72] A. Ouchi, M. Shimoi, F. Ebina, T. Uehiro, Y. Yoshino (Bull. Chem. Soc. Japan **51** [1978] 3511/3).
[73] J. Otera, R. Okawara (J. Organometal. Chem. **17** [1969] 353/7).
[74] H. Schmidbaur, M. Schmidt (Angew. Chem. **73** [1961] 655).
[75] E. Wiberg, K. Mödritzer (Z. Naturforsch. **11b** [1956] 750/1).
[76] A. N. Nesmeyanov, A. E. Borisov, N. V. Novikova, E. I. Fedin, P. V. Petrovskii (Izv. Akad. Nauk SSSR Ser. Khim. **1973** 1833/5; Bull. Acad. Sci. USSR Div. Chem. Sci. **1973** 1776/8).
[77] H. Schmidbaur, G. Hasslberger (Chem. Ber. **111** [1978] 2702/7).
[78] H. J. Widler, J. Weidlein (Z. Naturforsch. **34 b** [1979] 18/22).
[79] J. Werner, W. Schwarz, A. Schmidt (Z. Naturforsch. **36 b** [1981] 556/60).
[80] G. A. Razuvaev, M. N. Bochkarev, L. V. Pankratov (J. Organometal. Chem. **250** [1983] 135/43).

[81] C. G. Moreland, M. H. O'Brien, C. E. Douthit, G. G. Long (Inorg. Chem. **7** [1968] 834/6).
[82] C. G. Moreland, G. G. Long (Inorg. Nucl. Chem. Letters **8** [1972] 347/51).
[83] H. Matsuda, A. Ninagawa, R. Nomura (Chem. Letters **1979** 1261/2).
[84] R. Nomura, A. Ninagawa, H. Matsuda (J. Org. Chem. **45** [1980] 3735/8).
[85] R. Nomura, M. Kimura, S. Teshima, A. Ninagawa, H. Matsuda (Bull. Chem. Soc. Japan **55** [1982] 3200/3).
[86] H. Matsuda (Japan. 80-122776 [1979/80]; C.A. **94** [1981] No. 139779).

[87] R. Nomura, H. Hisada, A. Ninagawa, H. Matsuda (Makromol. Chem. Rapid Commun. **1** [1980] 135/8).
[88] H. Matsuda, T. Isaka, N. Iwamoto (Makromol. Chem. **179** [1978] 539/42).
[89] R. Müller, C. Dathe (Ger. 1249865 [1965/67]; C.A. **68** [1968] No. 49772).
[90] G. Wittig, K. Torssell (Acta Chem. Scand. **7** [1953] 1293/1301).
[91] D. J. Parker (Diss. Univ. Wisconsin 1959; Diss. Abstr. Intern. B **20** [1959/60] 2044).

2.5.1.1.3.1.2 Other Trialkylantimony Dibromides

$(C_2H_5)_3SbBr_2$

Triethylantimony dibromide is obtained by treating an ethereal solution of $Sb(C_2H_5)_3$ at 0 °C with Br_2. The mixture is poured into water [1] or an aqueous NH_4Cl solution [2]; the ether phase is dried and evaporated. The residue remains as a colorless oil with a density of $d^{17} = 1.953$ g/cm^3 which solidifies at −10 °C to white crystals [1]. Recrystallization at this temperature from C_2H_5OH leads to an impure product [3]. A given boiling point of 167 to 169 °C at 23 Torr [2] is probably incorrect, since the compound could not be distilled even at 0.2 Torr without some decomposition [3]. A boiling point of 111 °C at 0.1 Torr is given in [4]. The substance is directly obtained from C_2H_5MgBr and $SbCl_3$ with subsequent addition of Br_2 to the Grignard solution and workup of the mixture as before. The yield is about 96% in this case [5]. Another method of preparation is mentioned. The reactant, which has presumably the composition $(C_2H_5)_3SbSO_4$, is treated with $BaBr_2$ in an aqueous solution. $BaSO_4$ is filtered, and the water evaporated [47].

Far IR (in Nujol, c-C_6H_{12}, or C_6H_6) and Raman (solid or in C_6H_6) spectra show the following bands (IR (Raman) in cm^{-1}): 531 (534 dp, 525) ν_{as}SbCl, 480 (484 p) ν_sSbC, 280 (300 sh, 270 p, 237 p, 226 sh) δSbC, 190 ν_{as}SbBr, 165 sh (164 p) ν_sSbBr, 155 δSbC in-plane, 115 (118 dp) δSbC out-of-plane. From these data, a C_{3v} symmetry of the molecule is concluded [6]. A ^{1}H NMR spectrum in D_2O shows resonances at $\delta = 1.55$ (t, CH_3) and 2.78 (q, CH_2) ppm with a coupling constant $J = 8$ Hz [4].

The compound decomposes at heating to 220 to 240 °C/400 Torr for 1 h [2] or 190 to 210 °C/450 Torr for 1 h [5] or 170 to 180 °C/10 Torr [4] to form $(C_2H_5)_2SbBr$ and C_2H_5Br in yields of about 70 to 80% [4]. Triethylantimony dibromide is stable towards water [2], but is hydrolyzed on Amberlite 1R4B; 52% $(C_2H_5)_3SbO$ is obtained after treatment of the resulting hydroxide at 120 °C/0.1 Torr for 2 h [4]. HBr evolves with concentrated H_2SO_4 [1]. With AgF in CH_3OH, or CH_3CN, the corresponding difluoride is formed [6]. With Ag_2SO_4 in 70% C_2H_5OH, the compound forms $(C_2H_5)_3SbSO_4$ [3] and with $NaS \cdot xH_2O$ in CH_3OH one obtains $(C_2H_5)_3SbS$ [6]. Substituted benzoic acids $XC_6H_4CO_2H$ (X = 2-, 3-, 4-NH_2, 4-NO_2, 2-, 4-CH_3) and the compound in a ratio of 2:1 react in C_6H_6 in the presence of $N(C_2H_5)_3$ to give $(C_2H_5)_3Sb(O_2CC_6H_4X)_2$ in yields of 40 to 70%. With silver oxalate or silver terephthalate in water, 1:1 polymers in yields of about 30% are formed [8]. Sodium oximates NaON=CRR′ (R = R′ = CH_3; R = CH_3, R′ = C_2H_5; R = CH_3, R′ = C_6H_5) and the compound, refluxed in C_6H_6 for 2 h, form the corresponding triethylantimony dioximates in yields of 66 to 88% [9]. The compound is claimed to be a cocatalyst, together with $Al(C_2H_5)_3$ and $TiCl_3$, for the polymerization of 1-olefins [10].

$(C_3H_7)_3SbBr_2$

The compound is obtained by reaction of $Sb(C_3H_7)_3$ with Br_2 in CCl_4; after evaporation of the solvent, an oil remains. Treating this with ether gives crystals which melt at 45 °C [11]. From $(C_3H_7)_3Sb{=}NSO_2R$ (R = C_6H_5, 4-$CH_3C_6H_4$) and $SnBr_4$ dissolved in CH_2Cl_2, the compound is obtained by removal of the solvent, extraction with ether, evaporation and recrystallization from C_2H_5OH. The yield is 80 to 85%, m.p. 45 °C [12]. It decomposes upon heating to 200 °C for 4.5 h with formation of $(C_3H_7)_2SbBr$ and C_3H_7Br [13].

$(i\text{-}C_3H_7)_3SbBr_2$

$Sb(C_3H_7\text{-}i)_3$ and Br_2 react at −30 °C in pentane to form the compound. It is isolated in a yield of 99.5% by concentrating the solution and may be purified by recrystallization from CH_3OH or by sublimation at 120 °C/0.01 Torr [14]. The compound can also be synthesized directly from $i\text{-}C_3H_7MgBr$ and $SbCl_3$ by adding Br_2 to the Grignard solution without previous isolation of $Sb(C_3H_7\text{-}i)_3$. Workup in the usual manner gives the compound in a yield of 68.6% (relative to $SbCl_3$). It melts at 80 °C, recrystallized from CH_3OH [2]. In the latter preparation, a large amount of $(i\text{-}C_3H_7)_3SbCl_2$ was found to be present in the product [14].

Far IR (in Nujol, $c\text{-}C_6H_{12}$, or C_6H_6) and Raman (solid or in C_6H_6) spectra are given as (IR (Raman) in cm^{-1}): 499 (501 dp) $\nu_{as}SbC$, 478 (478) ν_sSbC, 408, 297, 261 (409 p, 260 p) δSbC, 180 (180 sh) $\nu_{as}SbBr$, 160 sh (163 p) ν_sSbBr, 135, 115 (138, 116) δSbC, 85 (85 sh) $\delta SbBr$. From these data a C_{3v} symmetry of the molecule is proposed [6]. IR (Nujol): 498 s, 478 m (νSbC). A ^1H NMR spectrum in C_6H_6 shows resonances at $\delta=1.32$(d) and 3.18(sept) ppm [14].

In the mass spectrum (20 eV, 40 °C inlet, 200 °C source the following fragments were found (relative intensity): $[(i\text{-}C_3H_7)_2SbBr_2]^+$ (70), $[(i\text{-}C_3H_7)_3SbBr]^+$ (100), $[(i\text{-}C_3H_7)_2SbBr]^+$ (28), $[(i\text{-}C_3H_7)_3Sb]^+$ (29), $[(i\text{-}C_3H_7)_2Sb]^+$ (9), $[i\text{-}C_3H_7SbH]^+$ (5), $[i\text{-}C_3H_7Sb]^+$ (4), $[i\text{-}C_3H_7]^+$ (100) [14]. The compound pyrolyzes at 160 to 170 °C/90 Torr for 2 h forming $(i\text{-}C_3H_7)_2SbBr$ and $i\text{-}C_3H_7Br$ in yields of about 80% [2, 14]. It is stable towards water [2].

$(C_4H_9)_3SbBr_2$

The compound is obtained by reacting $Sb(C_4H_9)_3$ with Br_2 in ether, and subsequent evaporation of the solvent. Antimony and C_4H_9Br, reacted for 30 h at 250 °C in a sealed tube, give the compound too. It remains as an oil [11]. It is formed in debromination reactions of phenacyl and arylmethyl bromides with $Sb(C_4H_9)_3$ [40].

IR absorptions were found at 519 and 460(νSbC) cm^{-1}. A figure of an IR spectrum is given in [15]. A ^{13}C NMR spectrum in $CHCl_3$ shows resonances at $\delta=17.2$(C-4), 26.8(C-3), 29.7(C-2), and 46.2(C-1) ppm [16]. The NQR spectrum at 77 K gives the following values (in MHz) [17]:

nucleus	$\nu(1/2 \rightleftharpoons 3/2)$	e^2Qq/h	η(%)
^{121}Sb	101.84	676.74	5.55
^{123}Sb	62.28	862.66	5.55
^{81}Br	94.48		

The compound decomposes within 4 h at 220 °C, forming 93% C_4H_9Br and 74% $(C_4H_9)_2SbBr$ [18]. It reacts with LiC_4H_9 to form $(C_4H_9)_4SbBr$ [19]. The same compound is obtained by reacting $(C_4H_9)_3SbBr_2$ with C_4H_9MgBr in THF/ether for 3.5 h at 20 °C. With an excess of the Grignard reagent (1:4) at the same conditions, 87% of $Sb(C_4H_9)_5$ is formed. With CH_3MgI under these conditions $(C_4H_9)_3Sb(CH_3)_2$ (75%), $(C_4H_9)_2Sb(CH_3)_3$ (20%), $C_4H_9Sb(CH_3)_4$ (4%), and $Sb(CH_3)_5$ (1%) were detected by ^1H NMR measurements [20]. $Ag_2C_2O_4$ and the compound, reacted in H_2O for one day, give a 30% yield of $(C_4H_9)_3SbC_2O_4$ [8]. The compound, dissolved in H_2O, gives $(C_4H_9)_3Sb(OH)_2$ on an anion exchange resin

References on p. 73

[21]. Together with $Al(C_2H_5)_3$ or $(C_2H_5)_2AlCl$ and $TiCl_3$ the compound is claimed to be a catalyst for ethylene polymerization [22].

$(i\text{-}C_4H_9)_3SbBr_2$

Triisobutylantimony dibromide is synthesized by reacting the corresponding stibine, dissolved in CCl_4, with Br_2. The compound remains after evaporation of the solvent [23 to 25]. The compound forms white rhombic plates, m.p. 88 °C (from $(CH_3)_2CO$) [23] or 95 °C (from ether) [24, 25].

An NQR spectrum at 77 K gives the following resonances (in MHz) [17]:

nucleus	ν(1/2 ⇌ 3/2)	ν(3/2 ⇌ 5/2)	e^2Qq/h	η(%)
^{121}Sb	100.28		666.64	5.2
^{123}Sb	60.84	120.36	843.79	5.2
^{79}Br	110.80			
^{81}Br	92.63			

The compound decomposes upon heating to 220 °C for 4 h to give 81% $(i\text{-}C_4H_9)_2SbBr$ and $i\text{-}C_4H_9Br$ [18].

$(C_5H_{11})_3SbBr_2$

The compound, which was characterized by Br analysis, was prepared by reacting $Sb(C_5H_{11})_3$ with Br_2 [11, 26, 27] or by treating $(C_5H_{11})_3SbO$ with aqueous HBr [26, 27, 38].

It is described as an oil [26, 27] or a white paste [11] with properties similar to the corresponding dichloride and diiodide [26]. It reacts with $AgNO_3$ or Ag_2SO_4 in C_2H_5OH to form the corresponding dinitrate or sulfate, respectively [26].

R_3SbBr_2 ($R = C_6H_{13}$, C_7H_{15}, C_8H_{17}, C_9H_{19}, and $C_{10}H_{21}$)

These compounds are obtained by reacting the corresponding stibines with Br_2 in $CHCl_3$ solution. Concentration of the solution and treating the residual oil with ether gives the compounds as white solid substances.

compound	yield	m.p.
$(C_6H_{13})_3SbBr_2$	63%	148 °C
$(C_7H_{15})_3SbBr_2$	65%	136 °C
$(C_8H_{17})_3SbBr_2$	58%	129 °C
$(C_9H_{19})_3SbBr_2$	44%	125 °C
$(C_{10}H_{21})_3SbBr_2$	50%	118 °C

They are stable to air, readily soluble in hot $(CH_3)_2CO$, C_6H_6, petroleum ether, $CH_3CO_2C_2H_5$, and $CH_3CO_2C_5H_{11}$, but sparingly soluble in C_2H_5OH [28].

$(c\text{-}C_5H_9)_3SbBr_2$

Tricyclopentylantimony dibromide is obtained from $Sb(C_5H_9\text{-}c)_3$ and Br_2 although no details are reported. It melts at 104 to 106 °C, recrystallized from C_6H_{14} [29, 30].

$(c\text{-}C_6H_{11})_3SbBr_2$

The compound is probably prepared by reaction of $Sb(C_6H_{11}\text{-}c)_3$ with Br_2. It reacts with $NaOCH_3/CH_3OH$, refluxed for one hour in C_6H_6, to give 88% $[(c\text{-}C_6H_{11})_3SbBr]_2O$ [31]. By addition of $HOCH_2CO_2H$ or $HSCH_2CO_2H$ to the filtrate of the former reaction mixture, $(c\text{-}C_6H_{11})_3Sb(\text{-}XCH_2CO_2\text{-})$ (X=O or S) is formed quantitatively [32]. With $t\text{-}C_4H_9OOH$ in C_6H_6 in the presence of NH_3 or $N(C_2H_5)_3$, 87% $(c\text{-}C_6H_{11})_3Sb(OOC_4H_9\text{-}t)_2$ is obtained [33]. $Na_2S \cdot xH_2O$ and the compound give in CH_3OH $(c\text{-}C_6H_{11})_3SbS$ [7]. A UV spectrum of the compound, dissolved in CH_3CN, shows a tail between 240 and 270 nm [34].

$(CF_3)_3SbBr_2$

An unstable product (m.p. −16 °C) of this chemical composition was obtained by cocondensation of $Sb(CF_3)_3$ and Br_2 at −30 °C. It was purified by recondensation in vacuum [35]. It is obtained in the form of the salt $[C_5H_5NH][(CF_3)_3SbBr_3]$ by reaction of $[C_5H_5NH][(CF_3)_3Sb(OH)_3]$ (C_5H_5NH=pyridinium) with excess of HBr in $(CH_3)_2CO$ [36].

$(CH_2Cl)_3SbBr_2$

The compound is obtained by reacting $Sb(CH_2Cl)_3$ with Br_2 in CCl_4 solution at 5 °C. The white shiny needles melt at 90 to 90.5 °C, recrystallized from petroleum ether [37, 39].

$[(CH_3)_3SiCH_2]_3SbBr_2$

A petroleum ether solution of $Sb(CH_2Si(CH_3)_3)_3$ is treated with Br_2 at 0 °C under N_2 atmosphere. The precipitating white solid melts at 158 to 160 °C [41]. The compound reacts with $LiCH_2Si(CH_3)_3$ in ether at 0 °C to give $Sb(CH_2Si(CH_3)_3)_5$ in 89% yield [42].

$(C_6H_5CH_2)_3SbBr_2$

Tribenzylantimony dibromide is presumably obtained by reacting $Sb(CH_2C_6H_5)_3$ with Br_2 [29, 30, 43], or by reacting $(C_6H_5CH_2)_3SbO$ with aqueous HBr (density 1.38 g/cm^3) in ether. The yield in the latter case is 60%. The compound melts at 107 to 109 °C, recrystallized from petroleum ether [44], or at 106 to 108 °C [29, 30].

An NQR spectrum of the substance at 300 K shows transitions (in MHz) for ^{121}Sb ($3/2 \rightleftharpoons 1/2$) at 88.195, for ^{79}Br at 120.37, and for ^{81}Br at 100.56 [45].

^1H NMR spectroscopic investigations of the exchange with $(C_6H_5CH_2)_3SbX_2$ (X=F, Cl) to give $(C_6H_5CH_2)_3SbBrX$ and with $(CH_3)_3SbI_2$ to give $(C_6H_5CH_2)_3SbI_2$ and $(CH_3)_3SbBr_2$ in $CHCl_3$ at 32 °C were performed and equilibrium constants were determined [46]. Tribenzylantimony dibromide reacts with $t\text{-}C_4H_9OOH$ in ether in the presence of NH_3 or $N(C_2H_5)_3$ to give $(C_6H_5CH_2)_3Sb(OOC_4H_9\text{-}t)_2$ in a yield of 91% [33].

References:

[1] C. Löwig, E. Schweizer (Liebigs Ann. Chem. **75** [1850] 315/55).
[2] H. Hartmann, G. Kühl (Z. Anorg. Allgem. Chem. **312** [1961] 186/94).
[3] G. G. Long, G. O. Doak, L. D. Freedman (J. Am. Chem. Soc. **86** [1964] 209/13).
[4] A. G. Davies, S. C. W. Hook (J. Chem. Soc. C **1971** 1660/5).
[5] K. Issleib, B. Hamann (Z. Anorg. Allgem. Chem. **339** [1965] 289/97).
[6] L. Verdonck, G. P. Van der Kelen (Spectrochim. Acta A **31** [1975] 1707/11).
[7] M. Shindo, Y. Matsumura, R. Okawara (J. Organometal. Chem. **11** [1968] 299/305).

[8] A. Ouchi, M. Nakatani, Y. Takahashi, S. Kitazima, T. Sugihara, M. Matsumoto, T. Uehiro, K. Kitano, K. Kawashima, H. Honda (Sci. Papers Coll. Gen. Educ. Univ. Tokyo **25** [1975] 73/99).
[9] V. K. Jain, R. Bohra, R. C. Mehrotra (Inorg. Chim. Acta **51** [1981] 191/4).
[10] Y. Takashi, I. Aijima, Y. Kobayashi, Y. Tsunoda (U.S. 3494910 [1965/70]; C.A. **72** [1970] No. 90967).

[11] W. J. C. Dyke, W. J. Jones (J. Chem. Soc. **1930** 1921/7).
[12] Z. I. Kuplennik, Z. N. Belaya, A. M. Pinchuk (Zh. Obshch. Khim. **51** [1981] 2711/15; J.Gen. Chem. [USSR] **51** [1981] 2339/43).
[13] E. A. Besolova, V. A. Foss, I. F. Lutsenko (Zh. Obshch. Khim. **38** [1968] 267/73; J. Gen. Chem. [USSR] **38** [1968] 270/5).
[14] H. J. Breunig, W. Kanig (Phosphorus Sulfur **12** [1982] 149/59).
[15] A. E. Borisov, N. V. Novikova, N. A. Chumaevskii, E. B. Shkirtil (Ukr. Fiz. Zh. **13** [1968] 75/82).
[16] A. Ouchi, T. Uehiro, Y. Yoshino (J. Inorg. Nucl. Chem. **37** [1975] 2347/9).
[17] V. I. Svergun, A. E. Borisov, N. V. Novikova, T. A. Babushkina, E. V. Bryukhova, G. K. Semin (Izv. Akad. Nauk SSSR Ser. Khim. **1970** 484/5; Bull. Acad. Sci. USSR Div. Chem. Sci. **1970** 443/4).
[18] S. Herbstman (J. Org. Chem. **29** [1964] 986/7).
[19] D. Hellwinkel, M. Bach (J. Organometal. Chem. **28** [1971] 349/58).
[20] A. N. Nesmeyanov, A. E. Borisov, N. V. Novikova, E. I. Fedin, P. V. Petrovskii (Izv. Akad. Nauk SSSR Ser. Khim. **1973** 1833/5; Bull. Acad. Sci. USSR Div. Chem. Sci. **1973** 1776/8).

[21] J. L. Zollinger (U.S. 3470176 [1967/9]; C.A. **71** [1969] No. 124513).
[22] H. Morita, Y. Takashi, I. Aijima; Asahi Chemical Industry Co., Ltd. (Japan. 70-13585 [1967/70]; C.A. **73** [1970] No. 46037).
[23] M. E. Brinnand, W. J. C. Dyke, W. H. Jones, W. J. Jones (J. Chem. Soc. **1932** 1815/19).
[24] L. I. Zakharkin, O. Yu. Okhlobystin (Dokl. Akad. Nauk SSSR **116** [1957] 236/8; Proc. Acad. Sci. USSR Chem. Sect. **112/117** [1957] 857/9).
[25] L. I. Zakharkin, O. Yu. Okhlobystin (Izv. Akad. Nauk SSSR Otd. Khim. Nauk **1959** 1942/7; Bull. Acad. Sci. USSR Div. Chem. Sci. **1959** 1853/8).
[26] F. Berlé (J. Prakt. Chem. **65** [1855] 385/418).
[27] C. Cramer (Verh. Naturforsch. Ges. Zürich **1851** May 12th; from Chem. Pharm. Centr. **26** [1855] 465/8; Jahresber. Fortschr. Chem. **1855** 590).
[28] A. N. Tatarenko, Z. M. Manulkin (Zh. Obshch. Khim. **34** [1964] 3462/5; J. Gen. Chem. [USSR] **34** [1964] 3503/5).
[29] N. S. Vyazankin, O. A. Kruglaya, G. A. Razuvaev, G. S. Semchikova (Dokl. Akad. Nauk SSSR **166** [1966] 99/102; Dokl. Chem. Proc. Acad. Sci. USSR **166/171** [1966] 8/11).
[30] N. S. Vyazankin, G. A. Razuvaev, O. A. Kruglaya, G. S. Semchikova (J. Organometal. Chem. **6** [1966] 474/83).

[31] Y. Kawasaki, Y. Yamamoto, M. Wada (Bull. Chem. Soc. Japan **56** [1983] 145/8).
[32] V. I. Bregadze, N. A. Chumaevskii, E. B. Shkirtil (Dokl. Akad. Nauk SSSR **181** [1968] 910/3).
[33] A. Rieche, J. Dahlmann, D. List (Liebigs Ann. Chem. **678** [1964] 167/82).
[34] J. Otera, R. Okawara (J. Organometal. Chem. **16** [1969] 335/8).
[35] J. W. Dale, H. J. Emeleus, R. N. Haszeldine, J. H. Moss (J. Chem. Soc. **1957** 3708/13).
[36] H. J. Emeleus, J. H. Moss (Z. Anorg. Allgem. Chem. **282** [1955] 24/8).
[37] A. Y. Yakubovich, V. A. Ginsburg, S. P. Makarov (Dokl. Akad. Nauk SSSR **71** [1950] 303/5).
[38] F. Berlé (Liebigs Ann. Chem. **97** [1856] 316/22).

[39] A. Y. Yakubovich, S. P. Makarov (Zh. Obshch. Khim. **22** [1952] 1528/34; J. Gen. Chem. [USSR] **22** [1952] 1569/74).
[40] K. Akiba, A. Shimidzu, H. Ohnari, K. Ohkata (Tetrahedron Letters **26** [1985] 3211/4).
[41] D. Seyferth (J. Am. Chem. Soc. **80** [1958] 1336/7).
[42] H. Schmidbaur, G. Hasslberger (Chem. Ber. **111** [1978] 2702/7).
[43] N. S. Vyazankin, G. S. Kalinina, O. A. Kruglaya, G. A. Razuvaev (Zh. Obshch. Khim. **39** [1969] 2005/11; J. Gen. Chem. [USSR] **39** [1969] 1964/8).
[44] J. P. Tsukervanik, D. Smirnov (Zh. Obshch. Khim. **7** [1937] 1527/31).
[45] T. B. Brill, G. G. Long (Inorg. Chem. **9** [1970] 1980/5).
[46] C. G. Moreland, G. G. Long (Inorg. Nucl. Chem. Letters **8** [1972] 347/51).
[47] W. Merck (J. Prakt. Chem. **66** [1855] 56/72).

2.5.1.1.3.1.3 Trialkenylantimony Dibromides

$(CH_2{=}CH)_3SbBr_2$

Trivinylantimony dibromide is obtained by reacting $Sb(CH{=}CH_2)_3$ with Br_2 in $CHCl_3$ at 0 to 5 °C. Distillation of the mixture at reduced pressure gives the compound in a yield of 74%. The boiling point is 117 to 119 °C at 1.5 Torr [1 to 3]. The compound can be prepared from $Sb(CH{=}CH_2)_5$ by two ways. Heating $Sb(CH{=}CH_2)_5$ for 40 min to 180 °C gives $Sb(CH{=}CH_2)_3$, which was not isolated, but reacted with Br_2 [1]. A solution of pentavinylantimony in $CHCl_3$ and Br_2 also gives the title compound in a yield of 77%. Similary, $(CH_2{=}CH)_4SbI$ and Br_2 in $CHCl_3$ at 0 °C form the compound in 70% yield. Boiling points are given as 118 to 119 °C at 1.5 Torr [4]. Only a yield of 15% is obtained by reacting SbF_3 with $(NH_4)_2[CH_2{=}CHSiF_5]$ in aqueous solution and treating the resulting oil, presumably $Sb(CH{=}CH_2)_3$, with Br_2 in $CHCl_3$. A boiling point is given as 128 °C at 7 Torr with partial decomposition [5]. The relative density of the liquid is $d_4^{20}=2.11529$, and its refractive index $n_D^{20}=1.6480$ [1 to 3].

The compound, dissolved in C_2H_5OH, reacts with KF in water to give 77% yield of $(CH_2{=}CH)_3SbF_2$ [4]. Reaction with $CH_2{=}CHMgBr$ in THF at −2 to −5 °C, 3.5 h in a N_2 atmosphere, gives upon cooling with dry ice 79% yield of $Sb(CH{=}CH_2)_5$ [1, 2]. C_2H_5MgBr in ether reacts with the compound, dissolved in THF, to give $(CH_2{=}CH)_3Sb(C_2H_5)_2$, which was isolated in a yield of 70% by evaporating the solvents in vacuum [6, 7]. Reaction of the compound with $LiCH_3$ (ratio 1:2) in ether at −20 °C gives, after distillation at 0.1 Torr with a bath temperature of 100 °C, a mixture of $(CH_2{=}CH)_nSb(CH_3)_{5-n}$ (n = 0 to 3) compounds, as shown by ^1H NMR spectroscopy [8].

$(cis\text{-}CH_3CH{=}CH)_3SbBr_2$ and $(trans\text{-}CH_3CH{=}CH)_3SbBr_2$

Both compounds are obtained from the corresponding stibines by reaction with Br_2 in $CHCl_3$ at 0 °C. After evaporation of the solvent, the *cis*-compound remains as a solid which is recrystallized from ether (yield 70%, melting point 85 to 86 °C), and the *trans*-compound remains as a liquid which is distilled at 166 to 167 °C/4 Torr (yield 75%, $d_4^{20}=1.8596$, $n_D^{20}=1.6270$ [3]) [9]. Both compounds may be prepared by two ways from $Sb(CH{=}CHCH_3\text{-}cis)_5$ or $Sb(CH{=}CHCH_3\text{-}trans)_5$. Heating the compounds to 100 to 160 °C gives, with evolution of gas, the corresponding tripropenylstibines which are further reacted with Br_2 to yield the *cis*- and *trans*-compounds, respectively. The direct reaction of the corresponding $Sb(CH{=}CHCH_3)_5$ and Br_2 in $CHCl_3$ at −5 °C and workup of the mixtures as before give 77% *cis*- and 64% *trans*-compound with physical properties as described. Reaction of $(cis\text{-}CH_3CH{=}CH)_4SbI$ and $(trans\text{-}CH_3CH{=}CH)_4SbI$ with Br_2 in $CHCl_3$ at 0 °C and the usual workup form the title compounds in 55 and 87% yield, respectively [4].

IR vibrations of the compounds are given in [9, 10]. The following absorptions were found and discussed in [12]: 1604, 1443, 1382, 1305, 1199, 1045, 939, 925, 818, 663, and 452 cm^{-1} for the *cis*-isomer in KBr; 1605, 1440, 1377, 1306, 1190, 1105, 1075, 1041, 951, 722, 655, and 620 cm^{-1} for the liquid *trans*-isomer.

The *cis*-compound is readily soluble in CH_3OH, C_2H_5OH, $(CH_3)_2CO$, $CHCl_3$, and C_6H_6, and moderately soluble in ether, whereas the *trans*-compound dissolves readily in CH_3OH, C_2H_5OH, $(CH_3)_2CO$, $CHCl_3$, and ether [9]. (*trans*-CH_3CH=$CH)_3SbBr_2$ becomes turbid in air [9]. Both compounds react with KF in aqueous C_2H_5OH to form the corresponding difluorides in yields of about 75% [4]. With the appropriate LiCH=$CHCH_3$, both compounds react in ether at −10 to −5 °C to give Sb(CH=$CHCH_3$-*cis*$)_5$ or Sb(CH=$CHCH_3$-*trans*$)_5$ in yields of about 80% [9 to 11]. LiC_6H_5 reacts with the title compounds under the same conditions to give (*cis*-CH_3CH=$CH)_3Sb(C_6H_5)_2$ (86%) or (*trans*-CH_3CH=$CH)_3Sb(C_6H_5)_2$ (74% yield) [6]. In THF solution, the compounds react with C_2H_5MgBr in ether to form (*cis*-CH_3CH=$CH)_3Sb(C_2H_5)_2$ (71%) and (*trans*-CH_3CH=$CH)_3Sb(C_2H_5)_2$ (80% yield) [6, 7].

[CH_2=$C(CH_3)]_3SbBr_2$

Triisopropenylantimony dibromide is prepared by reacting the corresponding stibine with Br_2 in $CHCl_3$ at 0 °C. Evaporation of the solvent and recrystallization of the residue from CCl_4 give 73% of the compound, melting at 138 °C. Heating of Sb(C(CH_3)=$CH_2)_5$ for 40 min at 180 °C and adding of Br_2 to the residue give 91% of the substance [1]. From Sb(C(CH_3)=$CH_2)_5$ or from [CH_2=$C(CH_3)]_4SbI$ and Br_2 in $CHCl_3$ at 0 °C the title compound is obtained after the usual workup (as above) in yields of 80 and 61%, respectively [4].

The compound reacts with KF in aqueous C_2H_5OH to give the corresponding difluoride in 76% yield [4]. Reaction with LiC(CH_3)=CH_2 in ether for 4 h at −5 to 0 °C, gives 69% yield of Sb(C(CH_3)=$CH_2)_5$ [1]; under similar conditions, LiC_6H_5 forms 93% [CH_2=$C(CH_3)]_3Sb(C_6H_5)_2$ [6]. The title compound, dissolved in THF, reacts with an ethereal solution of C_2H_5MgBr to give [CH_2=$C(CH_3)]_3Sb(C_2H_5)_2$ in a yield of 67% [6, 7].

(CH_2=$CHCH_2)_3SbBr_2$

Triallylantimony dibromide is prepared by reacting Sb(CH_2CH=$CH_2)_3$ with Br_2 in $CHCl_3$ at −60 °C. The solvent is evaporated at low temperatures, the residue dissolved in ether, cooled, filtered, and the ether driven off. A syrup-like mass remains in a yield of 51%. The compound hydrolyzes immediately in air [13].

(*cis*-ClCH=$CH)_3SbBr_2$

The compound is prepared from Sb(CH=CHCl-*cis*$)_3$ and Br_2 in $CHCl_3$ solution at 0 °C. Evaporation of the solvent and recrystallization of the residue from $CHCl_3$ gives the compound in 68% yield with a melting point of 186 to 187 °C [14].

IR spectra of the compound were measured [14, 15], and the following characteristic absorptions (in cm^{-1}) were found: 1605 s, 1265 s, 1130 s, 931 w, 903 s, and 690 to 670 s. A ^1H NMR spectrum shows resonances at δ=7.12 (H-1) and 7.56 (H-2) ppm with a coupling of J(H-1,2)=7.1 Hz [14]. An NQR spectrum recorded at 77 K gives the following resonances (in MHz) [16]:

nucleus	$\nu(^1/_2 \rightleftharpoons {^3/_2})$	$\nu(^3/_2 \rightleftharpoons {^5/_2})$	e^2Qq/h	η (%)	nucleus	$\nu(^1/_2 \rightleftharpoons {^3/_2})$
^{121}Sb	73.437	146.616	494.00	9.8	^{79}Br	140.640
	74.880	147.936	488.80	3.7		143.928
^{123}Sb	44.548	88.920	629.73	9.8	^{81}Br	117.500
	46.548	89.490	623.00	3.7		120.240
						105.400
						106.960
					^{35}Cl	34.710
						34.520
						34.446
						34.326

$[(CO)_3MnC_5H_4]_3SbBr_2$

$Sb(C_5H_4Mn(CO)_3)_3$ is titrated with Br_2 in CCl_4 solution. Removal of the solvent and recrystallization of the residue from C_6H_6/C_2H_5OH (3:1) give yellow crystals which decompose at 220 °C. A thermogravimetric analysis of the compound shows that the corresponding stibine and CO are formed at 220 to 285 °C and that at 285 to 430 °C quantitative decomposition to Sb, inorganic Mn^{2+}, and Sb^{3+} compounds occurs [17].

References:

[1] A. N. Nesmeyanov, A. E. Borisov, N. V. Novikova (Izv. Akad. Nauk SSSR Otd. Khim. Nauk **1961** 1578/82; Bull. Acad. Sci. USSR Div. Chem. Sci. **1961** 1473/6).

[2] A. N. Nesmeyanov, A. E. Borisov, N. V. Novikova (Izv. Akad. Nauk SSSR Otd. Khim. Nauk **1960** 952; Bull. Acad. Sci. USSR Div. Chem. Sci. **1960** 893).

[3] A. N. Nesmeyanov, A. E. Borisov, N. V. Novikova (Dokl. Akad. Nauk SSSR **134** [1960] 100/1; Proc. Acad. Sci. USSR Chem. Sect. **130/135** [1960] 995/6).

[4] A. N. Nesmeyanov, A. E. Borisov, N. V. Novikova (Izv. Akad. Nauk SSSR Ser. Khim. **1964** 1202/9; Bull. Acad. Sci. USSR Div. Chem. Sci. **1964** 1116/21).

[5] R. Müller, S. Reichel, C. Dathe (Inorg. Nucl. Chem. Letters **3** [1967] 125/33).

[6] A. N. Nesmeyanov, A. E. Borisov, N. V. Novikova (Izv. Akad. Nauk SSSR Ser. Khim. **1964** 1197/202; Bull. Acad. Sci. USSR Div. Chem. Sci. **1964** 1112/5).

[7] A. N. Nesmeyanov, A. E. Borisov, N. V. Novikova (Izv. Akad.Nauk SSSR Otd. Khim. Nauk **1961** 730; Bull. Acad. Sci. USSR Div. Chem. Sci. **1961** 678).

[8] H. A. Meinema, J. G. Noltes (J. Organometal. Chem. **22** [1970] 653/7).

[9] A. N. Nesmeyanov, A. E. Borisov, N. V. Novikova (Izv. Akad. Nauk. SSSR Otd. Khim. Nauk **1961** 612/17; Bull. Acad. Sci. USSR Div. Chem. Sci. **1961** 564/8).

[10] A. N. Nesmeyanov, A. E. Borisov, N. V. Novikova (Tetrahedron Letters **8** [1960] 23/4).

[11] A. N. Nesmeyanov, A. E. Borisov, N. V. Novikova (Izv. Akad. Nauk SSSR Otd. Khim. Nauk **1960** 147; Bull. Acad. Sci. USSR Div. Chem. Sci. **1960** 140).

[12] A. E. Borisov, N. V. Novikova, N. A. Chumaevskii (Dokl. Akad. Nauk SSSR **136** [1961] 129/32).

[13] A. E. Borisov, N. V. Novikova, A. N. Nesmeyanov (Izv. Akad. Nauk SSSR Ser. Khim. **1963** 1506/7; Bull. Acad. Sci. SSSR Div. Chem. Sci. **1963** 1368/9).

[14] A. N. Nesmeyanov, A. E. Borisov, N. V. Novikova (Izv. Akad. Nauk SSSR Ser. Khim. **1969** 1978/82; Bull. Acad. Sci. USSR Div. Chem. Sci. **1969** 1830/3).

[15] A. E. Borisov, V. V. Klinkova, N. A. Chumaevskii (Dokl. Akad. Nauk SSSR **200** [1971] 64/7).

[16] V. I. Svergun, A. E. Borisov, N. V. Novikova, T. A. Babushkina, E. V. Bryukhova, G. K. Semin (Izv. Akad. Nauk SSSR Ser. Khim. **1970** 484/5; Bull. Acad. Sci. USSR Div. Chem. Sci. **1970** 443/4).

[17] G. A. Razuvaev, G. A. Domrachev, V. V. Sharutin, O. N. Suvorova (Dokl. Akad. Nauk SSSR **237** [1977] 852/4; Dokl. Chem. Proc. Acad. Sci. USSR **232/237** [1977] 711/3).

2.5.1.1.3.2 R_3SbBr_2 Compounds with R = Aryl and Thienyl

2.5.1.1.3.2.1 Triphenylantimony Dibromide $(C_6H_5)_3SbBr_2$

Preparation. Methods for preparation and formation of the compound are summarized in Table 18.

Table 18
Preparation and Formation of $(C_6H_5)_3SbBr_2$.

reactants	reaction conditions (yield in %), other products	Ref.
$Sb(C_6H_5)_3 + Br_2$	in petroleum ether (90)	[1]
	in ether	[2]
	in glacial acetic acid	[2, 4]
	in CH_3CN, conductometric titration	[5]
$Sb(C_6H_5)_3 + SnBr_4$	in C_5H_{12} at evaporation of the solvent, obtained as mixture with $SnBr_2$	[6]
$Sb(C_6H_5)_3 + HgBr_2$	in refluxing $(CH_3)_2CO$ (70 to 85), Hg_2Br_2	[7]
$Sb(C_6H_5)_3 + CuBr_2$	in refluxing $(CH_3)_2CO$ (70 to 85), CuBr	[7, 8]
$(C_6H_5)_3SbO_{1.4} + Br_2$	in $CHCl_3$, O_2	[9]
$(C_6H_5)_3SbI_2 + Br_2$	conductometric titrations in CH_3CN, I_2	[5]
$(C_6H_5)_3SbIBr + {}^1/_2\ Br_2$		
$(C_6H_5)_3SbI_2 + 2\ IBr$		
$(C_6H_5)_3SbIBr + IBr$		
$(C_6H_5)_3Sb(-O(CHC_6H_5)_2O-) + Br_2$	in $CHCl_3$ (80), benzil, benzoin	[9]
$(C_6H_5)_3Sb(-O(CHC_6H_5)_2O-) +$ 1-Br-2,5-pyrrolidinedione	in $CHCl_3$, benzil, 2,5-pyrrolidinedione	[9]
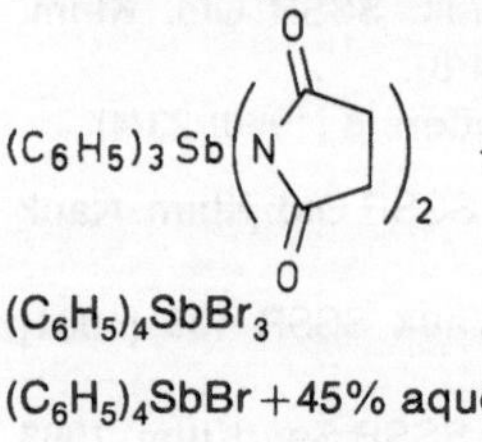 $+ Br_2$	in $CHCl_3$ as mixture with 1-Br-2,5-pyrrolidinedione	[10]
$(C_6H_5)_4SbBr_3$	without solvent at ≥ 135 °C (82), C_6H_5Br	[11, 12]
$(C_6H_5)_4SbBr$ + 45% aqueous HBr	at boiling (92), C_6H_6	[11]
$(C_6H_5)_5Sb \cdot c\text{-}C_6H_{12}$ + 45% aqueous HBr	at boiling (90), $C_6H_6 + c\text{-}C_6H_{12}$	[11]
$SbCl_3 + C_6H_5Br/Na$	in C_6H_6 (small amount), $Sb(C_6H_5)_3$	[2]
$Sb(C_6H_5)_3$ + 4-$BrC_6H_4COCH_2Br$ or other α-bromoketones	in CD_3CN at 35 °C, 10 d (70), 4-$BrC_6H_4COCH_3$	[58]

Properties. Melting points for the compound are given as 221 to 222 °C (from C_6H_6 or CH_3CO_2H [15], 217 to 218 °C (from $C_5H_{11}OH$ and C_6H_6/C_2H_5OH) [11, 14], 216 °C (from C_6H_6) [2, 4], 215 to 216 °C (from C_6H_6/C_2H_5OH) [11, 12], 214 to 215 °C (from petroleum ether), 214 °C [7 to 9], and 212 to 214 °C [13]. The solid is obtained from C_6H_6 as beautiful glass-like shiny crystals [2] or needles [4] which become easily turbid [2]. From glacial acetic acid broad needles crystallize [2].

The dipole moment in C_6H_6 is 1.15 D at 25 °C [35], and a molar susceptibility of $\chi_M = -232.4 \times 10^{-6}$ is given [36, 37]. IR spectra were measured, assigned, and discussed in several publications [13, 16 to 22]. Raman spectra are published in [20 to 22]. Complete IR and Raman vibrations of the solid compound from 4000 to 40 cm^{-1} are listed and assigned with Whiffen's nomenclature in [22]. IR (Nujol): 456 vs, 294 vs, 248 s; 188 vs (ν_{as}SbBr), 162 vw cm^{-1}; Raman (solid): 457 w, 400 vw, 293 m, 219 s; 161 vs (ν_sSbBr) cm^{-1} [20].

^{1}H NMR spectra show the following resonances: δ($CDCl_3$ at 26 °C) = 5.57 (m) and 8.17 (m) ppm [23], or δ ($CHCl_3$) = 7.58 ppm [24]. ^{13}C NMR spectra give δ ($CHCl_3$ vs. C_6H_6) = 2.2(C-3), 4.3(C-4), 6.2(C-2), 13.6 (C-1) ppm [24], or δ($CDCl_3$) = 129.44 (C-3; J(C,H) = 164.6 Hz), 131.48 (C-4; J(C,H) = 161.9 Hz), 133.48 (C-2; J(C,H) = 166.0 Hz), and 140.79 (C-1) ppm [25].

^{121}Sb Mössbauer spectra gave the following data (in mm/s):

	δ	e^2qQ	Γ	Ref.
at 80 K vs. $^{121}SnO_2$	-6.1 ± 0.1	-18 ± 1	2.1 ± 0.1	[28]
at 4 K vs. $Ca^{121}SnO_3$	-6.32 ± 0.02	-19.8 ± 0.3	2.75 ± 0.04	[29]

A quadrupole coupling constant of $e^2qQ = -19.9$ mm/s was calculated, and is consistent with the measured value [30]. Correlation of the coupling constant with that of $[(C_6H_5)_3SnBr_2]^-$ is given in [31]. An additive model for the electric field gradient on Sb is applied. The data are consistent with a trigonal bipyramidal geometry of the molecule [32]. The electron populations in the orbitals directed towards the ligands are calculated by a method which uses both ^{121}Sb Mössbauer isomer shifts and quadrupole coupling constants [33].

NQR spectra were measured at 300 K [26] and 77 K [27]. The transition frequency for ^{79}Br is given as 127.05 MHz and for ^{81}Br as 106.14 MHz at 300 K [26]. The values for ^{121}Sb and ^{123}Sb are (in MHz):

T (K)	ν ^{121}Sb		ν ^{123}Sb		e^2Qq/h		η (%)
	$^5/_2 \rightleftharpoons {}^3/_2$	$^3/_2 \rightleftharpoons {}^1/_2$	$^5/_2 \rightleftharpoons {}^3/_2$	$^3/_2 \rightleftharpoons {}^1/_2$	^{121}Sb	^{123}Sb	
300	169.97	85.193	–	52.219	565.78	720.93	4.3
77	178.16	89.22	108.16	54.28	594.00	757.59	3.4

A high-energy photoelectron spectrum with Al Kα exciting radiation gives core-binding energies of 539.8 (Sb $3d^3/_2$), 530.4 (Sb $3d^5/_2$), 189.6 and 183.0 (Br 3p) eV [34].

The compound is insoluble or nearly insoluble in petroleum ether, alcohol, and ether, slightly soluble in glacial acetic acid, easily soluble in hot glacial acetic acid, C_6H_6, and CS_2 [2]. It forms monomers in C_6H_6 [1]. The molar conductivity in CH_3CN solution is smaller than 1 $cm^2 \cdot \Omega^{-1} \cdot mol^{-1}$ [5]. Polarographic half-wave potentials were obtained in glyme with $[N(C_4H_9)_4]ClO_4$ as supporting electrolyte against 10^{-3} M $Ag/AgClO_4$. The value of −1.4 V

References on p. 85

shows that a one-step two electron reduction occurs to give $Sb(C_6H_5)_3$ and $2\,Br^-$, and a second step at -3.3 V gives $Sb(C_6H_5)_2^-$ and $C_6H_5^-$ [38]. The 70 eV mass spectrum gave the following fragments (source 250 °C, direct inlet 25 to 150 °C): $[(C_6H_5)_3SbBr_n]^+$ (n = 0, 1), $[(C_6H_5)_2SbBr_n]^+$ (n = 0, 1), $[C_6H_5SbBr_n]^+$ (n = 0, 1), $[SbBr]^+$, $[Sb(C_6H_4)_n]^+$ (n = 0 to 2), $[C_{12}H_n]^+$ (n = 8 to 10), $[C_6H_n]^+$ (n = 5, 6), $[C_4H_3]^+$, $[BrH]^+$, and $[Br]^+$. The spectrum is discussed and compared with those of other triphenylantimony dihalides [3].

Reactions. The exchange reaction of $(C_6H_5)_3SbBr_2$ with $(C_6H_5)_3SbF_2$ to give $(C_6H_5)_3SbFBr$ was studied by ^1H and ^{19}F NMR in $CHCl_3$ or $CDCl_3$. The equilibrium constant K was determined as 4.0 ± 0.7 at 32 °C (35 °C in [70]) [70, 71]. Reaction of the title compound with $(CH_3)_3SbF_2$ gives $(C_6H_5)_3SbF_2$ and $(CH_3)_3SbBr_2$; $K = 600 \pm 200$ at 32 °C, measured by ^{19}F NMR [71]. The kinetics of the isotope exchange reaction in the system $^{125}Sb(C_6H_5)_3/(C_6H_5)_3SbBr_2$ were studied in $n\text{-}C_3H_7OH$. The mechanism is analogous to the corresponding dichloride derivative [83]. Other reactions of the title compound are summarized in Table 19.

Uses. The compound is useful as a catalyst for the reaction of epoxides with CO_2 to cyclic carbonates [73 to 76]. The kinetics and the mechanism of this reaction are discussed in [76]. The compound catalyzes the ring-opening polymerization of ethylene oxide [77]. A binary catalyst, containing triphenylantimony dibromide and $P(C_6H_5)_3$ [78, 79], $N(C_2H_5)_3$, $N(C_4H_9)_3$, or pyridine is used for the polymerization of ethylene oxide. A mechanism for this reaction is postulated in [79]. The title compound may be also used as a catalyst for the polycondensation of glycol esters to give a transparent polyester [80]. It was tested as a retarding agent in the burning of epoxy resins and compared in this connection with other triphenylantimony dihalides [81, 82].

Formulas for Tables 19 and 20:

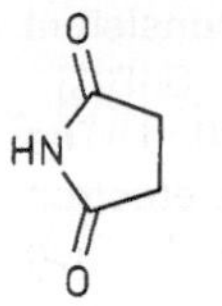

I

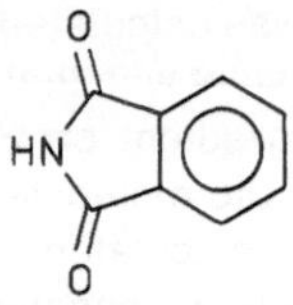

II

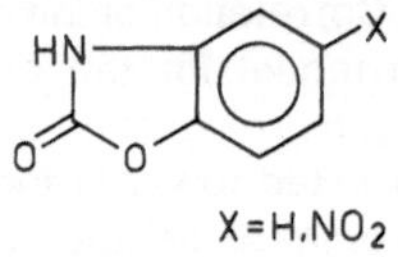

III

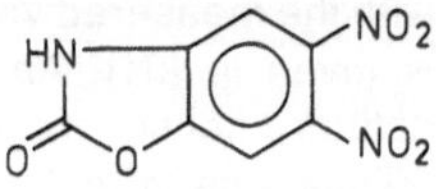

IV

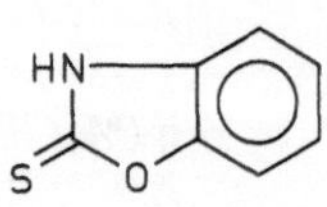

V

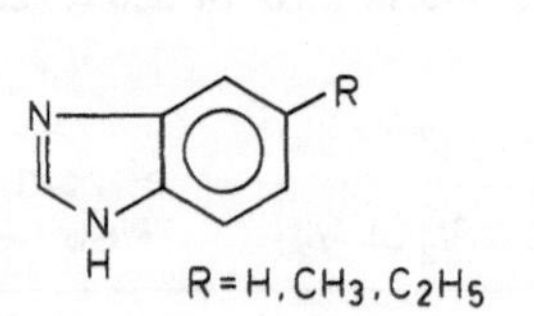

VI

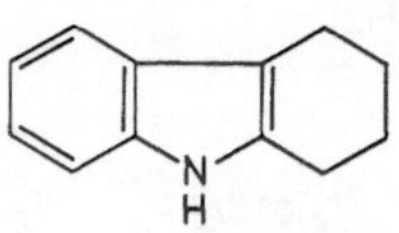

VII

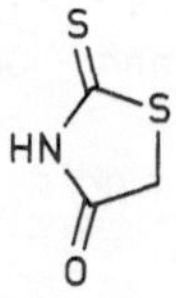

VIII

References on p. 85

Table 19
Reactions of $(C_6H_5)_3SbBr_2$.

reactants	reaction conditions (ratio = $(C_6H_5)_3SbBr_2$: reactant)	products (yield in %)	Ref.
—	0.5 h heating to 200 to 250 °C in a vacuum	C_6H_5Br, $(C_6H_5)_2SbBr$ (92)	[39]
—	heating to 220 °C	C_6H_5Br, $(C_6H_5)_2SbBr$, $C_6H_5SbBr_2$, $Sb(C_6H_5)_3$	[40]
H_2O	2 h reflux in C_2H_5OH/H_2O	$(C_6H_5)_3Sb(OH)Br$ (89)	[11]
H_2O	in $H_2O + H_2NC_2H_4NH_2$	$(C_6H_5)_3Sb(OH)_2$	[1]
$H_2NC_2H_4NH_2$	in C_2H_5OH/ether or C_6H_6	$[H_3NC_2H_4NH_3]Br_2$	[1]
KOH	in boiling C_2H_5OH	$(C_6H_5)_3Sb(OH)_2$	[2, 14]
KOH	in $(CH_3)_2CO/C_2H_5OH$ 1 h at 30 °C	$(-(C_6H_5)_3SbO-)_n$	[41]
H_2O_2	in C_6H_6/ether + NH_3 or amine	$(-(C_6H_5)_3SbO_2-)_n$	[42]
$NaOCH_3$	ratio 1:2 in CH_3OH	$(C_6H_5)_3Sb(OCH_3)_2$	[43]
$NaOCH_3$	ratio 1:2 in C_6H_6/CH_3OH, 1 h at 20 °C	$(C_6H_5)_3Sb(OCH_3)_2$ (95)	[44]
$NaOCH_3$	ratio 1:1 in $CHCl_3$ or C_6H_6	$(C_6H_5)_3Sb(OCH_3)Br$	[45]
$NaOC_2H_5$ in excess	in $C_2H_5OH + NH_3$	$(C_6H_5)_3Sb(OC_2H_5)_2$	[45]
8-hydroxyquinoline	in refluxing C_6H_6/CH_3OH, via $(C_6H_5)_3Sb(OCH_3)Br$, ratio 1:1	$(C_6H_5)_3Sb(Br)OC_9H_6N$	[46]
8-hydroxyquinoline, 2-CH_3-8-hydroxyquinoline, or 5-Cl-8-hydroxyquinoline	in refluxing C_6H_6/CH_3OH, via $(C_6H_5)_3Sb(OCH_3)_2$, workup with $(CH_3)_2CO/H_2O$	$(C_6H_5)_3Sb(OH)OC_9H_6N$, $(C_6H_5)_3Sb(OH)OC_{10}H_8N$, or $(C_6H_5)_3Sb(OH)OC_9H_5ClN$	[48]
Na-8-quinolinolate	ratio 1:3 in refluxing C_6H_6	$(C_6H_5)_3Sb(OC_9H_6N)_2$ (90)	[4]
$HOCC_6H_4ONa$-2	—	$(C_6H_5)_3Sb(C_7H_5O_2)_2$	[4]
$C_6H_5(C_6H_5CO)NONa$	—	$(C_6H_5)_3Sb(C_{13}H_{10}NO_2)_2$	[4]

References on p. 85

Table 19 [continued]

reactants	reaction conditions (ratio = $(C_6H_5)_3SbBr_2$: reactant)	products (yield in %)	Ref.
Na salt of dehydroacetic acid	—	$(C_6H_5)_3Sb(C_8H_7O_4)_2$	[4]
$[CH_3COCH_2CONC_6H_5]Na$	—	$(C_6H_5)_3Sb(C_{10}H_{10}NO_2)_2$	[4]
$HXCH_2CO_2H$ (X = O, S)	in C_6H_6/CH_3OH, via $(C_6H_5)_3Sb(OCH_3)_2$	$(C_6H_5)_3Sb(-XCH_2CO_2-)$	[47]
$RCOCH_2COR'$, R = CF_3, R' = CH_3, C_6H_5	in $C_6H_6 + N(C_2H_5)_3$	$[(C_6H_5)_3Sb(OCR{=}CHCOR')]_2O$	[49]
Na(RCOCHCOR'), R = R' = CH_3 or C_6H_5; R = CH_3, R' = C_6H_5, 4-$CH_3C_6H_4$, 4-$CH_3OC_6H_4$, 4-ClC_6H_4, 4-BrC_6H_4; R = CF_3, R' = 2-thienyl	ratio 1:1 in C_6H_6, 2 h reflux	$(C_6H_5)_3SbBr(OCR{=}CHCOR')$ (50 to 70)	[50]
XC_6H_4OH, X = H, 2-, 3-, 4-Cl, 4-Br, 2-, 3-, 4-CH_3, 4-OCH_3, 4-C_4H_9-t	ratio 1:2 in $C_6H_6 + N(C_2H_5)_3$	$(C_6H_5)_3Sb(OC_6H_4X)_2$ (40 to 70)	[51]
XC_6H_4OH, X = H, 4-Cl, 4-CH_3	as above	$(C_6H_5)_3Sb(OC_6H_4X)_2$	[24]
XC_6H_4OH, X = 2-, 3-, 4-NO_2	ratio 1:2 in $C_6H_6 + N(C_2H_5)_3$	$[(C_6H_5)_3Sb(OC_6H_4X)]_2O$	[51]
RCO_2H, R = H, CH_2Cl, CH_2Br, CCl_3, C_2H_5, C_3H_7, $C_6H_5CH_2$, $C_6H_5CH{=}CH$, $CH_3CH{=}CH$, C_8H_{17}, $C_{17}H_{35}$, $CH_3COCH_2CH_2$, 2-, 3-, 4-$CH_3C_6H_4OCH_2$	ratio 1:2 in refluxing $C_6H_6 + N(C_2H_5)_3$	$(C_6H_5)_3Sb(O_2CR)_2$ (70 to 90)	[52]
$AgO_2CCF_{3-n}H_n$ (n = 0 to 2)	in C_6H_6	$(C_6H_5)_3Sb(O_2CCF_{3-n}H_n)_2$	[53]
$Ag_2O + H_nX_{3-n}CCO_2H$ in H_2O (n = 0 to 2), X = Cl, Br	in C_6H_6	$(C_6H_5)_3Sb(O_2CCX_{3-n}H_n)_2$	[53]
$C_6H_5OCH_2CO_2H$	ratio 1:2.5 in $C_6H_6 + N(C_2H_5)_3$, 1 h at 20 °C	$(C_6H_5)_3Sb(O_2CCH_2OC_6H_5)_2$	[54]
$RSCH_2CO_2H$, R = CH_3, C_2H_5, C_3H_7, i-C_3H_7, C_4H_9, $C_6H_5CH_2$, C_6H_5	ratio 1:2.5 in $C_6H_6 + N(C_2H_5)_3$, 1 h at 20 °C	$(C_6H_5)_3Sb(O_2CCH_2SR)_2$ (60)	[54]

References on p. 85

$XC_6H_4CO_2H$, X = H, 2-, 3-, 4-Cl, 2-, 3-, 4-NO_2, 2-, 4-OH, 2-, 3-, 4-NH_2, 2-, 3-, 4-CH_3, 4-CH_3O	ratio 1:2 in $C_6H_6 + N(C_2H_5)_3$	$(C_6H_5)_3Sb(O_2CC_6H_4X)_2$ (40 to 70)	[51]
$C_6H_5CO_2H$	as above	$(C_6H_5)_3Sb(O_2CC_6H_5)_2$	[24]
$RR'C(OH)_2$, R = CF_3, R' = CF_3; R = H, R' = CF_3 or CCl_3	in $C_6H_6 + N(C_2H_5)_3$, 1 h at 20 °C	$(C_6H_5)_3Sb(-OCRR'O-)$ (75)	[23]
$NaOC_6H_4ONa$-2	in $C_6H_6/(CH_3)_2CO$, 20 min at 20 °C	$(C_6H_5)_3Sb(-OC_6H_4O-2-)$	[55]
$HO_2C(CH_2)_nCO_2H$ (n = 0 to 4)	in $H_2O + Ag_2O$	$(C_6H_5)_3Sb(-O_2C(CH_2)_nCO_2-)$ (30)	[51]
$Ag_2C_2O_4$	in CH_3OH	$(C_6H_5)_3Sb(-O_2CCO_2-)$	[65]
$XC_6H_4CO_2H$, X = 2-, 3-, 4-CO_2H	in H_2O with Ag_2O	corresponding $(C_6H_5)_3Sb(-O_2CC_6H_4CO_2-)$ (30)	[51]
t-C_4H_9OOH	in $C_6H_6 + NaNH_2$ in C_6H_6/ether + CH_3ONa in $C_6H_6 + NH_3$ or amine	$(C_6H_5)_3Sb(OOC_4H_9\text{-t})_2$ (77) (85) (75 to 79)	[42]
$C_6H_5(CH_3)_2COOH$	in $C_6H_6 + NH_3$ or amine	$(C_6H_5)_3Sb(OOC(CH_3)_2C_6H_5)_2$ (85)	[42]
ROOH, R = 1, 2, 3, 4-tetrahydronaphthalinyl or 3,4-dihydro-1H-benzoisopyranyl	in $C_6H_6/NaNH_2$ or NH_3 or amine	$(C_6H_5)_3Sb(OOR)_2$ (ca. 100)	[42]
$NaOSi(CH_3)_3$	organic solvent	$(C_6H_5)_3Sb(OSi(CH_3)_3)_2$	[56]
RR'C=NOH, R = CH_3, R' = C_6H_5, 4-$NO_2C_6H_4$; R = R' = C_6H_5; R = H, R' = 4-$CH_3OC_6H_4$, 2-furyl; R, R' = $(CH_2)_5$	in refluxing $C_6H_6 + N(C_2H_5)_3$	$(C_6H_5)_3Sb(ON{=}CRR')_2$	[57]
NHR_2, Formulas I to III, V, or VI, p. 80	ratio 1:2 in $C_6H_6 + N(C_2H_5)_3$, 2 h at 20 °C	$(C_6H_5)_3Sb(NR_2)_2$ (ca. 100)	[10]
Ag salt of Formula I, p. 80	in $CHCl_3$, 0.5 h at 20 °C	$(C_6H_5)_3Sb(NC_4H_4O_2)_2$ (70)	[59]
NHR_2, Formulas VII or VIII, p. 80	in $C_6H_6 + N(C_2H_5)_3$, 2 h reflux	$(C_6H_5)_3Sb(NR_2)_2$	[60]
2,6-diamino-8-purinol or 4-$H_2NC_6H_4C_6H_4NH_2$-4'	in $CCl_4/H_2O/NaOH$ at 20 °C	polymers (62 and 51, respectively)	[61]

References on p. 85

Table 19 [continued]

reactants	reaction conditions (ratio = $(C_6H_5)_3SbBr_2$: reactant)	products (yield in %)	Ref.
$RPO(H)OH$, R = C_6H_5, $CH{=}CHC_6H_5$	ratio 1:2 in $C_6H_6 + N(C_2H_5)_3$ for 1 h	$(C_6H_5)_3Sb(OP(O)(H)R)_2$ (61, 69)	[62]
H_2S	in C_2H_5OH saturated with NH_3	$(C_6H_5)_3SbS$ (100)	[1, 63, 64]
Ag_2SeO_4	in CH_3OH	$(C_6H_5)_3SbSeO_4$	[65]
Ag_2CrO_4	in $C_6H_5NO_2$	$(C_6H_5)_3SbCrO_4$	[65]
$[(C_4H_9)_3Sn]_2S$	in $CHCl_3$ at −5 °C	$(C_6H_5)_3SbS + (C_4H_9)_3SnBr$ (80 to 90)	[66]
C_6H_5MgBr	in THF, first at 0 °C, then 3 h reflux	$Sb(C_6H_5)_5$ (90)	[15]
C_5H_5MgBr	in ether	$Sb(C_6H_5)_3$ and tar	[67]
2-$LiC_6H_4C_6H_4Li$-2′	in ether at −70 °C for 12 h	$(C_6H_5)_3Sb(-2\text{-}C_6H_4C_6H_4\text{-}2'-)$ (55)	[68]
2-$LiC_6H_4OC_6H_4Li$-2′	in ether	$(C_6H_5)_3Sb(-2\text{-}C_6H_4OC_6H_4\text{-}2'-)$ (34)	[68]
$Pb_2(C_6H_5)_6$	ratio 1:1 in $CHCl_3$, 6 h reflux	$Sb(C_6H_5)_3$ (50), $Pb(C_6H_5)_4$ (78), $(C_6H_5)_2PbBr_2$ (70)	[69]
$SbBr_3$	in $CHCl_3$, concentration of the solution	$(C_6H_5)_3SbBr_2 \cdot SbBr_3$ (m.p. 102 °C)	[72]

References:

[1] W. J. Lile, R. J. Menzies (J. Chem. Soc. **1950** 617/21).
[2] A. Michaelis, A. Reese (Liebigs Ann. Chem. **233** [1886] 39/60).
[3] H. Preiss (Z. Anorg. Allgem. Chem. **389** [1972] 280/92).
[4] S. Gopinathan, C. Gopinathan (Indian J. Chem. A **15** [1977] 660/2).
[5] A. D. Beveridge, G. S. Harris, F. Inglis (J. Chem. Soc. A **1966** 520/8).
[6] D. Cunningham, M. J. Frazer, J. D. Donaldson (J. Chem. Soc. A **1971** 2049).
[7] S. N. Bhattacharya, M. Singh (Indian J. Chem. A **18** [1979] 515/6).
[8] D. Makanova, G. Ondrejovic (Proc. 9th Conf. Coord. Chem., Bratislava, Czech., 1983, pp. 277/81).
[9] F. Nerdel, J. Buddrus, K. Höher (Chem. Ber. **97** [1964] 124/31).
[10] K. Bajpai, R. C. Srivastava (Syn. Reactiv. Inorg. Metal-Org. Chem. **9** [1979] 557/64).

[11] G. Wittig, K. Clauss (Liebigs Ann. Chem. **577** [1952] 26/39).
[12] G. Wittig, D. Hellwinkel (Chem. Ber. **97** [1964] 789/93).
[13] G. O. Doak, G. G. Long, L. D. Freedman (J. Organometal. Chem. **4** [1965] 82/91).
[14] K. A. Jensen (Z. Anorg. Allgem. Chem. **250** [1943] 257/67).
[15] T. C. Thepe, R. J. Garascia, M. A. Selvoski, A. N. Patel (Ohio J. Sci. **77** [1977] 134).
[16] K. A. Jensen, P. H. Nielsen (Acta Chem. Scand. **17** [1963] 1875/85).
[17] A. E. Borisov, N. V. Novikova, N. A. Chumaevskii, E. B. Shkirtil (Dokl. Akad. Nauk SSSR **173** [1967] 855/8).
[18] A. E. Borisov, N. V. Novikova, N. A. Chumaevskii, E. B. Shkirtil (Ukr. Fiz. Zh. **13** [1968] 75/82).
[19] K. M. Mackay, D. B. Sowerby, W. C. Young (Spectrochim. Acta A **24** [1968] 611/31).
[20] R. G. Goel, E. Maslowsky Jr., C. V. Senoff (Inorg. Chem. **10** [1971] 2572/7).

[21] R. G. Goel, E. Maslowsky Jr., C. V. Senoff (Inorg. Nucl. Chem. Letters **6** [1970] 833/5).
[22] B. A. Nevett, A. Perry (Spectrochim. Acta A **33** [1977] 755/60).
[23] A. Ouchi, F. Ebina, T. Uehiro, Y. Yoshino (Bull. Chem. Soc. Japan **51** [1978] 2427/8).
[24] A. Ouchi, T. Uehiro, Y. Yoshino (J. Inorg. Nucl. Chem. **37** [1975] 2347/9).
[25] J. Havranek, A. Lycka (Sb. Ved. Pr. Vys. Sk. Chemickotechnol. Pardubice **43** [1980] 123/7).
[26] T. B. Brill, G. G. Long (Inorg. Chem. **9** [1970] 1980/5).
[27] V. I. Svergun, A. E. Borisov, N. V. Novikova, T. A. Babushkina, E. V. Bryukhova, G. K. Semin (Izv. Akad. Nauk SSSR. Ser. Khim. **1970** 484/5; Bull. Acad. Sci. USSR Div. Chem. Sci. **1970** 443/4).
[28] G. G. Long, J. G. Stevens, R. J. Tullbane, L. H. Bowen (J. Am. Chem. Soc. **92** [1970] 4230/5).
[29] J. G. Stevens, S. L. Ruby (Phys. Letters A **32** [1970] 91/2).
[30] G. M. Bancroft, V. G. K. Das, T. K. Sham, M. G. Clark (J. Chem. Soc. Dalton Trans. **1976** 643/54).

[31] G. M. Bancroft, V. G. K. Das, K. D. Butler (J. Chem. Soc. Dalton Trans. **1974** 2355/8).
[32] J. N. R. Ruddick, J. R. Sams, J. C. Scott (Inorg. Chem. **13** [1974] 1503/7).
[33] L. H. Bowen, G. G. Long (Inorg. Chem. **15** [1976] 1039/44).
[34] T. Birchall, J. A. Connor, I. H. Hillier (J. Chem. Soc. Dalton Trans. **1975** 2003/6).
[35] L. M. Kataeva, Yu. V. Rydvanskii, N. I. Trofimova (Zh. Fiz. Khim. **50** [1976] 814/5; Russ. J. Phys. Chem. **50** [1976] 486/7).
[36] N. K. Parab, D. M. Desai (Current Sci. [India] **26** [1957] 389).
[37] N. K. Parab, D. M. Desai (J. Indian Chem. Soc. **35** [1958] 573/5).
[38] R. E. Dessy, T. Chivers, W. Kitching (J. Am. Chem. Soc. **88** [1966] 467/70).
[39] G. B. Reinert (Prax. Naturwiss. III **22** [1973] 169/82).

[40] T. Severengiz, H. J. Breunig (Chemiker-Ztg. **104** [1980] 202/3).

[41] D. L. Venezky, C. W. Sink, B. A. Nevett, W. F. Fortescue (J. Organometal. Chem. **35** [1972] 131/42).
[42] A. Rieche, J. Dahlmann, D. List (Liebigs Ann. Chem. **678** [1964] 167/82).
[43] G. H. Briles, W. E. McEwen (Tetrahedron Letters **1966** 5191/6).
[44] W. E. McEwen, G. H. Briles, B. E. Giddings (J. Am. Chem. Soc. **91** [1969] 7079/84).
[45] J. Dahlmann, A. Rieche (Chem. Ber. **100** [1967] 1544/9).
[46] Y. Kawasaki (Inorg. Nucl. Chem. Letters **5** [1969] 805/10).
[47] Y. Matsumura, M. Shindo, R. Okawara (J. Organometal. Chem. **27** [1971] 357/63).
[48] Y. Kawasaki (Bull. Chem. Soc. Japan **49** [1976] 817/8).
[49] F. Ebina, T. Uehiro, T. Iwamoto, A. Ouchi, Y. Yoshino (J. Chem. Soc. Chem. Commun. **1976** 245/6).
[50] V. K. Jain, R. Bohra, R. C. Mehrotra (J. Organometal. Chem. **184** [1980] 57/62).

[51] A. Ouchi, M. Nakatani, Y. Takahashi, S. Kitazima, T. Sugihara, M. Matsumoto, T. Uehiro, K. Kitano, K. Kawashima, H. Honda (Sci. Papers Coll. Gen. Educ. Univ. Tokyo **25** [1975] 73/99).
[52] K. Bajpai, R. Singhal, R. C. Srivastava (Indian J. Chem. A **18** [1979] 73/5).
[53] R. G. Goel, D. R. Ridley (J. Organometal. Chem. **38** [1972] 83/9).
[54] A. Ouchi, H. Honda, S. Kitazima (J. Inorg. Nucl. Chem. **37** [1975] 2559/61).
[55] M. Shindo, R. Okawara (Inorg. Nucl. Chem. Letters **5** [1969] 77/80).
[56] H. Schmidbaur, M. Schmidt (Angew. Chem. **73** [1961] 655).
[57] K. Bajpai, R. C. Srivastava (Syn. Reactiv. Inorg. Metal-Org. Chem. **11** [1981] 7/13).
[58] K. Akiba, A. Shimizu, H. Ohnari, K. Okata (Tetrahedron Letters **26** [1985] 3211/4).
[59] J. Dahlmann, K. Winsel (J. Prakt. Chem. **321** [1979] 370/8).
[60] P. Raj, A. Ranjan, A. K. Saxena (Indian J. Chem. A **22** [1983] 120/3).

[61] C. E. Carraher Jr., M. D. Naas, D. J. Giron, D. R. Cerutis (J. Macromol. Sci. Chem. A **19** [1983] 1101/20).
[62] G. E. Graves, J. R. Van Wazer (J. Organometal. Chem. **131** [1977] 31).
[63] L. Kaufmann (Ber. Deut. Chem. Ges. **41** [1908] 2762/6).
[64] L. Kaufmann (C. **1908** II 1260/1).
[65] R. G. Goel, P. N. Joshi, D. R. Ridley, R. E. Beaumont (Can. J. Chem. **47** [1969] 1423/7).
[66] S. N. Bhattacharya, P. Raj, A. K. Saxena (Indian J. Chem. A **16** [1978] 1071/4).
[67] N. A. Nesmeyanov, V. V. Pravdina, O. A. Reutov (Dokl. Akad. Nauk SSSR **155** [1964] 1364/7; Proc. Acad. Sci. USSR **154/159** [1964] 424/7).
[68] D. Hellwinkel, M. Bach (J. Organometal. Chem. **17** [1969] 389/403).
[69] S. N. Bhattacharya, A. K. Saxena (Indian J. Chem. A **17** [1979] 307/9).
[70] C. G. Moreland, M. H. O'Brien, C. E. Douthit, G. G. Long (Inorg. Chem. **7** [1968] 834/6).

[71] C. G. Moreland, G. G. Long (Inorg. Nucl. Chem. Letters **8** [1972] 347/51).
[72] M. Hall, D. B. Sowerby (J. Chem. Soc. Dalton Trans. **1983** 1095/9).
[73] H. Matsuda, A. Ninagawa, R. Nomura (Chem. Letters **1979** 1261/2).
[74] R. Nomura, A. Ninagawa, H. Matsuda (J. Org. Chem. **45** [1980] 3735/8).
[75] H. Matsuda (Japan. 80-122776 [1979/80]; C.A. **94** [1981] No. 139779).
[76] A. Ninagawa, H. Matsuda, R. Nomura (Kenkyu Hokoku Asahi Garasu Kogyo Gijutsu Shoreikai **39** [1981] 117/23).
[77] R. Nomura, H. Hisada, A. Ninagawa, H. Matsuda (Makromol. Chem. Rapid Commun. **1** [1980] 135/8).
[78] R. Nomura, H. Hisada, A. Ninagawa, H. Matsuda (Makromol. Chem. **183** [1982] 1073/80).
[79] R. Nomura, H. Hisada, A. Ninagawa, H. Matsuda (Makromol. Chem. Rapid Commun. **1** [1980] 705/7).

[80] T. Ozeki, I. Kanzaki, R. Kamiya, Mitsubishi Rayon Co., Ltd. (Japan. 71-06633 [1967/71]; C.A. **75** [1971] No. 64624).

[81] J. Havranek, J. Mleziva (Angew. Makromol. Chem. **84** [1980] 105/17).

[82] J. Havranek (Sb. Dokl. 1st Nats. Konf. Mladite Nauchni Rab. Spets. Neft Khim., Burgas, Bulg., 1976 [1977], pp. 152/9).

[83] N. I. Trofimova, V. E. Zhuravlev, E. N. Sinotova, N. E. Shchepina, M. V. Moshkovskaya (Tr. Estestvennonauchn. Inst. Permsk. Gos. Univ. **13** [1975] 187/93; from Ref. Zh. Khim. B **22** [1976] 658).

2.5.1.1.3.2.2 Other Triarylantimony Dibromides and Tris(2-thienyl)antimony Dibromide

$(C_6F_5)_3SbBr_2$

The compound is prepared by reaction of $Sb(C_6F_5)_3$ with Br_2 in CCl_4. The resulting precipitate, recrystallized from petroleum ether, melts at 207 to 209 °C. The observed IR and Raman vibrations of the solid between 4000 and 40 cm^{-1} are given together with their assignments; ν_sSbBr is observed at 188 cm^{-1} in the Raman, and $\nu_{as}SbBr$ at 240 cm^{-1} in the IR [1].

$(4\text{-}ClC_6H_4)_3SbBr_2$

The compound is obtained by reacting the corresponding stibine with $CuBr_2$ in C_2H_5OH. Evaporating the filtrate and recrystallizing from CCl_4/petroleum ether give the compound which melts at 189.5 to 190 °C [2].

A ^1H NMR spectrum in $CDCl_3$ at 26 °C shows two quartets at $\delta=7.58$ and 8.12 ppm [3].

The compound reacts with $CF_3COCH_2COCH_3$ in C_6H_6 in the presence of $N(C_2H_5)_3$ to give $[(C_6H_5)_3SbOC(CF_3)\text{=}CHCOCH_3]_2O$ [4]. Under similar conditions, reactions with $(CF_3)_2C(OH)_2$ or $CCl_3CH(OH)_2$ give the heterocycle $(C_6H_5)_3Sb(\text{-}OCRR'O\text{-})$ (R, $R'=CF_3$ or R=H, $R'=CCl_3$) [3].

$(C_6Cl_5)_3SbBr_2$

$Sb(C_6Cl_5)_3$, dissolved in $CHCl_3$, is reacted with Br_2 in CCl_4 for 30 min at room temperature. With concentration of the solution, the compound precipitates in a yield of 75%. The crystals were washed with hexane or ether [5].

The vibrations of the C_6Cl_5 groups are observed in the IR spectrum. The compound is soluble in CCl_4, $CHCl_3$, and THF. A solution of the compound in CH_3NO_2 shows no electrolytic conductivity. The solid is indefinitely stable in air at room temperature. It is thermally stable and melts without decomposition [5].

$(4\text{-}BrC_6H_4)_3SbBr_2$

The compound is obtained by the reaction of the corresponding stibine with $CuBr_2$ in C_2H_5OH. After evaporation of the solvent the residue is recrystallized; m.p. 182 °C with dec. (from $CHCl_3/C_2H_5OH$) [2].

$(2\text{-}CH_3OC_6H_4)_3SbBr_2$

The compound is prepared like the previous one; m.p. 225 to 226 °C with dec. (from $CHCl_3$/petroleum ether) [2].

$(3\text{-}CH_3OC_6H_4)_3SbBr_2$

The compound is prepared like the 4-bromo derivative described above; m.p. 74.5 to 75.5 °C with dec. (from $CHCl_3/C_2H_5OH$) [2].

$(4\text{-}CH_3OC_6H_4)_3SbBr_2$

$Sb(C_6H_4OCH_3\text{-}4)_3$ and Br_2 react in $CHCl_3$ to give the compound as a precipitate upon treating the solution with petroleum ether; m.p. 123 °C (from $CHCl_3/C_2H_5OH$) [6].

The compound is soluble in $CHCl_3$, C_6H_6, and ether. It crystallizes from C_6H_6 as a 1:1 adduct. These rhombohedral prisms melt at 81 to 82 °C. With $AgNO_3$ in C_2H_5OH it forms the corresponding dinitrate, with NaOH in the same solvent the oxide, and with H_2S in ethanolic NH_3 it is reduced to the corresponding stibine [6].

$(2\text{-}C_2H_5OC_6H_4)_3SbBr_2$

The corresponding stibine and $CuBr_2$ are reacted in C_2H_5OH to give, after evaporation of the solvent from the filtrate, a residue which melts at 237 to 238 °C with dec. after recrystallization from $CHCl_3/C_2H_5OH$ [2].

$(4\text{-}C_2H_5OC_6H_4)_3SbBr_2$

From the corresponding stibine and Br_2 in petroleum ether an oily phase is obtained which solidifies after separation. Recrystallization from $CHCl_3/C_2H_5OH$ gives the compound which melts at 110 to 111 °C. It is soluble in ether, C_6H_6, and $CHCl_3$. Reaction with $AgNO_3$ in C_2H_5OH gives the corresponding dinitrate [6].

$(4\text{-}C_6H_5OC_6H_4)_3SbBr_2$

The compound is prepared like the $2\text{-}C_2H_5O$ derivative described above; m.p. 151 to 152 °C (from $CHCl_3/C_2H_5OH$) [2].

$(4\text{-}(CH_3)_2NC_6H_4)_3SbBr_2$

$Sb(C_6H_4N(CH_3)_2\text{-}4)_3$, dissolved in CH_2Cl_2 or $CHCl_3$, reacts with Br_2 in the same solvents at 0 to −78 °C to give the compound as a precipitate in a yield of 48% after treating the reaction solution with ether or petroleum ether; dec. at 160 to 170 °C [7].

A UV spectrum is shown in [7]. The ^1H NMR spectrum (in $CDCl_3$) shows the following resonances: $\delta=3.43$ (CH_3), 7.21 (H-3,5), 8.50 (H-2,6) ppm; J (H-2,3) = 9 Hz. The structure is discussed on the basis of these values. The compound reacts with KF in aqueous C_2H_5OH (1:5) under 2 h reflux to form 72% of the corresponding difluoride [7].

$(2\text{-}CH_3C_6H_4)_3SbBr_2$

The compound is obtained from the corresponding stibine with Br_2 in petroleum ether. The small shiny crystals melt at 209 to 210 °C [8].

The compound is easily soluble in $CHCl_3$ and C_6H_6, and nearly insoluble in ether, alcohol, and petroleum ether. It reacts with alcoholic KOH to form $(2\text{-}CH_3C_6H_4)_3SbO$, and is reduced by H_2S in alcoholic NH_3 to the stibine [8]. With substituted phenols XC_6H_4OH (X = H, 4-NO_2, 4-Cl, 4-CH_3, and 4-CH_3O) in C_6H_6 in the presence of $N(C_2H_5)_3$, the diphenolates $(2\text{-}CH_3C_6H_5)_3Sb(OC_6H_4X)_2$ are obtained in yields of about 40 to 60%. Under the same condi-

References on p. 92

tions, reaction with substituted benzoic acids $XC_6H_4CO_2H$ (X=H, 4-NO_2, 4-Cl, 4-CH_3, and 4-NH_2) forms the corresponding dibenzoates [9].

$(3\text{-}CH_3C_6H_4)_3SbBr_2$

The compound is obtained from the corresponding stibine with Br_2 in petroleum ether. It crystallizes from ether as shiny, only slightly colored crystals which melt at 113 °C [8].

It is easily soluble in $CHCl_3$, ether, C_6H_6 and glacial acetic acid, and rather insoluble in petroleum ether. Reaction with alcoholic KOH gives $(3\text{-}CH_3C_6H_4)_3SbO$ [8]. It reacts with substituted benzoic acids $XC_6H_4CO_2H$ (X=H, 4-NO_2, 4-Cl, 4-CH_3, and 4-NH_2) in C_6H_6 in the presence of $N(C_2H_5)_3$ to form the corresponding dibenzoates [9].

$(4\text{-}CH_3C_6H_4)_3SbBr_2$

Tris(4-methylphenyl)antimony dibromide is obtained from the corresponding stibine and Br_2 in petroleum ether [8, 10], ether [8, 10, 11], or $CHCl_3$. In the last case, the yield is 80% after concentrating the reaction mixture [12]. The same yield is obtained by reacting the stibine with $CuBr_2$ or $HgBr_2$ in refluxing $(CH_3)_2CO$ and concentrating the filtrate from the mixture [13].

Recrystallization from C_6H_6/C_2H_5OH gives small shiny crystals which melt at 233 to 234 °C [8, 10]. Other melting points given are: 230 °C (from C_6H_6/C_2H_5OH) [11], 232 °C [13], 233 °C (from C_6H_6 or $CHCl_3$) [12], and 234 °C [14]. A molar susceptibility of $\chi_M = -267.4 \times 10^{-6}$ is given [18, 19]. ^1H NMR spectra show the following resonances in CCl_4: δ=2.43 (s, CH_3), 7.35 (d, H-2,6), 8.07 (d, H-3,5) ppm, J(H-2,3)=8 Hz [12]. In $CDCl_3$ the corresponding values are: δ=2.40, 7.32, 8.04 ppm, and J=8 Hz [16, 17]. The compound crystallizes in the cubic system with a=12.817 Å; Z=4; d=1.70 g/cm³; space group $P4_332\text{-}O^6$ (No. 212) or $P4_132\text{-}O^7$ (No. 213) [14, 15].

The compound is easily soluble in C_6H_6 and $CHCl_3$, soluble in glacial acetic acid, and rather insoluble in ether, C_2H_5OH, and petroleum ether [8]. Reactions of the compound are summarized in Table 20. The title compound was tested as a retarding agent in the burning of epoxy resins [20].

$(4\text{-}CH_2BrCHBrCH_2C_6H_4)_3SbBr_2$

$Sb(C_6H_4CH_2CH{=}CH_2\text{-}4)_3$, dissolved in $CHCl_3$, is titrated with Br_2 in the same solvent until a light brown color persists. Two hours standing and concentrating of the mixture give the compound with a melting point of 122 to 125 °C [29].

$(2,4\text{-}(CH_3)_2C_6H_3)_3SbBr_2$

The corresponding stibine reacts with stoichiometric amounts of Br_2 in $CHCl_3$ solution. The compound precipitates upon concentration of the solution and addition of C_2H_5OH; m.p. 195 °C, recrystallized from $CHCl_3/C_2H_5OH$ [18, 19, 30]. The molar susceptibility is $\chi_M = -300.7 \times 10^{-6}$ [18, 19]. The compound is moderately soluble in petroleum ether [30].

$(2,4\text{-}(CH_3)_2\text{-}5\text{-}NO_2\text{-}6\text{-}BrC_6H)_3SbBr_2$

The compound is prepared by treating a $CHCl_3$ solution of $Sb(C_6H_2NO_2\text{-}5\text{-}(CH_3)_2\text{-}2,4)_3$ with Br_2. It is precipitated by light petroleum ether, and it is recrystallized from a mixture of $CHCl_3$ and petroleum ether, to form a white crystalline powder [30].

References on p. 92

Table 20
Reactions of $(4\text{-}CH_3C_6H_4)_3SbBr_2$.

reactants	reaction conditions (ratio = $(4\text{-}CH_3C_6H_4)_3SbBr_2$: reactant)	products (yield in %)	Ref.
—	heating at 5 to 7 Torr	$(4\text{-}CH_3C_6H_4)_2SbBr + CH_3C_6H_4Br$	[21]
ethanolic NaOH	heated, then treated with warm H_2O	$(4\text{-}CH_3C_6H_4)_3SbO$	[8, 10]
H_2S	in $C_2H_5OH + NH_3$	$(4\text{-}CH_3C_6H_4)_3SbS$ (82)	[14]
8-hydroxyquinoline	ratio 1:1 in C_6H_6/CH_3OH, via $(4\text{-}CH_3C_6H_4)_3SbBr(OCH_3)$	$(4\text{-}CH_3C_6H_4)_3Sb(Br)OC_9H_6N$	[22]
8-hydroxyquinoline	ratio 1:1 in C_6H_6/CH_3OH, via $(4\text{-}CH_3C_6H_4)_3Sb(OCH_3)_2$, and workup with aqueous $(CH_3)_2CO$	$(4\text{-}CH_3C_6H_4)_3Sb(OH)OC_9H_6N$	[23]
XC_6H_4OH, X = H, 4-NO_2, 4-Cl, 4-CH_3, 4-CH_3O	ratio 1:2 in $C_6H_6 + N(C_2H_5)_3$	$(4\text{-}CH_3C_6H_4)_3Sb(OC_6H_4X)_2$	[9]
$RR'C(OH)_2$, R = CF_3, R' = CF_3; R = H, R' = CF_3, CCl_3	in $C_6H_6 + N(C_2H_5)_3$, 1 h at 20 °C	$(4\text{-}CH_3C_6H_4)_3Sb(\text{-}OCRR'O\text{-})$	[17]
RCO_2H, R = H, CH_2Cl, CH_2Br, CCl_3, C_2H_5, C_3H_7, $C_6H_5CH_2$, $C_6H_5CH{=}CH$, $CH_3CH{=}CH$, C_8H_{17}, $C_{17}H_{35}$, $CH_3COCH_2CH_2$, 2-,3-$CH_3C_6H_4OCH_2$, 4-$ClC_6H_4OCH_2$	ratio 1:2 in $C_6H_6 + N(C_2H_5)_3$, 2 h at 20 °C, 2 h reflux, concentration of the filtrate in vacuum	$(4\text{-}CH_3C_6H_4)_3Sb(O_2CR)_2$ (70 to 90)	[24]
$XC_6H_4CO_2H$, X = H, 4-NO_2, 4-Cl, 4-CH_3, 4-NH_2	ratio 1:2 in $C_6H_6 + N(C_2H_5)_3$	$(4\text{-}CH_3C_6H_4)_3Sb(O_2CC_6H_4X)_2$	[9]
$RR'C{=}NOH$, R = CH_3, R' = C_6H_5, $C_6H_4NO_2$-4, C_2H_5, CH_3; R = C_6H_5, R' = C_6H_5; R = H, R' = 2-furyl; R, R' = $(CH_2)_5$	ratio 1:2 in $C_6H_6 + N(C_2H_5)_3$, 3 h reflux	$(4\text{-}CH_3C_6H_4)_3Sb(ON{=}CRR')_2$ (ca. 100)	[25]
NHR_2, Formulas I to VI, p. 80	ratio 1:2 in $C_6H_6 + N(C_2H_5)_3$, 2 h at 20 °C	$(4\text{-}CH_3C_6H_4)_3Sb(NR_2)_2$ (ca. 100)	[26]
4-$CH_3C_6H_4MgBr$	in C_6H_6/ether, 3 h at 20 °C, + ice-cold HBr	$(4\text{-}CH_3C_6H_4)_4SbBr$ (45)	[12]
$Pb_2(C_6H_5)_6$	ratio 1:1, 6 h reflux in $CHCl_3$	$Sb(C_6H_4CH_3\text{-}4)_3$ (75), $Pb(C_6H_5)_4$ (80), $(C_6H_5)_2PbBr_2$ (60)	[27]
$(R_3Sn)_2S$, R = C_4H_9, C_6H_5	in $CHCl_3$ at −5 °C	$(4\text{-}CH_3C_6H_4)_3SbS + R_3SnBr$ (80 to 90)	[28]

References on p. 92

$(2,4-(CH_3)_2-5-(2',4',6'-(NO_2)_3C_6H_2NH)-6-Br-C_6H)_3SbBr_2$

Addition of Br_2 in light petroleum ether to a solution of $Sb(C_6H_2(NHC_6H_2(NO_2)_3-2',4',6')-5-(CH_3)_2-2,4)_3$ in $CHCl_3$ gives the compound, which precipitates upon addition of light petroleum ether as a lemon yellow, crystalline powder. It blackened at 183 °C and melted at 188 °C. The reaction with alcoholic potash is described [30].

$(4-CH_3CH{=}CHC_6H_4)_3SbBr_2$

The compound is prepared from the corresponding stibine and Br_2 in $CHCl_3$. After removal of the solvent, the residual oil was dissolved in a little dioxane. Addition of C_2H_5OH gave the light yellow compound in a yield of 90.3%; m.p. 174 °C [31].

In the IR spectrum a characteristic frequency for νC=C appears at 960 cm^{-1}. The air stable compound is soluble in ether, $CHCl_3$, dioxane, $(CH_3)_2CO$, and C_6H_6, sparingly soluble in C_2H_5OH [31].

$(4-CH_2{=}C(CH_3)C_6H_4)_3SbBr_2$

The compound is prepared by reaction of the corresponding stibine with Br_2 in $CHCl_3$ to yield 88% of a cream colored solid, m.p. 168 °C. See the preceding compound for the solubility. It is air stable [31].

$(4-C_2H_5CH{=}CHC_6H_4)_3SbBr_2$

The compound is prepared like the previous one in a yield of 70%. The cream colored solid, which is stable in air, melts at 185 °C [31].

$(2-C_6H_5C_6H_4)_3SbBr_2$

The compound is prepared by reaction of the corresponding stibine with Br_2 in $CHCl_3$. Addition of petroleum ether and concentration until crystallization starts gives needle-like clusters, which are a 1:1 adduct with $CHCl_3$; m.p. 152 to 154 °C. It reacts with ethanolic NH_3 to give the corresponding dihydroxide [32].

$(4-C_6H_5C_6H_4)_3SbBr_2$

The compound is prepared by reaction of the corresponding stibine with Br_2 in $CHCl_3$. It precipitates as a 1:1 solvate with $CHCl_3$. The white narrow plates melt at 259 to 260 °C under preliminary softening and subsequent decomposition. It does not react with hot or cold water, and only slowly with warm 95% C_2H_5OH. Boiling for 1 h with ethanolic NH_3 gives the corresponding dihydroxide quantitatively [33].

$(1-C_{10}H_7)_3SbBr_2$

This compound is prepared from tris(1-naphthyl)stibine and Br_2 in CCl_4. Addition of petroleum ether to the reaction mixture precipitates yellow crystals which melt at 229 °C [34] or 232 °C. The compound reacts with ethanolic KOH to give $(1-C_{10}H_7)_3SbO \cdot C_6H_6$ when recrystallized from C_6H_6 [35].

$(C_4H_3S)_3SbBr_2$ (C_4H_3S = 2-thienyl)

Tris(2-thienyl)antimony dibromide is obtained by reaction of the corresponding stibine, dissolved in ether, with Br_2 dissolved in CCl_4. It melts at 178.5 °C, recrystallized from C_2H_5OH or C_6H_6/petroleum ether. The compound reacts with hot ethanolic KOH to give the corre-

References on p. 92

sponding oxide. With aqueous Ag_2O at 80 °C Sb-C cleavage occurs, yielding antimony oxide [36].

References:

[1] B. A. Nevett, A. Perry (Spectrochim. Acta A **31** [1975] 101/6).
[2] J. I. Harris, S. T. Bowden, W. J. Jones (J. Chem. Soc. **1947** 1568/71).
[3] A. Ouchi, F. Ebina, T. Uehiro, Y. Yoshino (Bull. Chem. Soc. Japan **51** [1978] 2427/8).
[4] F. Ebina, T. Uehiro, T. Iwamoto, A. Ouchi, Y. Yoshino (J. Chem. Soc. Chem. Commun. **1976** 245/6).
[5] A. Otero, P. Loyo (J. Organometal. Chem. **171** [1979] 333/6).
[6] C. Löloff (Ber. Deut. Chem. Ges. **30** [1897] 2834/43).
[7] J. M. Keck, G. Klar (Z. Naturforsch. **27b** [1972] 591/5).
[8] A. Michaelis, U. Genzken (Liebigs Ann. Chem. **242** [1887] 164/188).
[9] A. Ouchi, M. Nakatani, Y. Takahashi, S. Kitazima, T. Sugihara, M. Matsumoto, T. Uehiro, K. Kitano, K. Kawashima, H. Honda (Sci. Papers Coll. Gen. Educ. Univ. Tokyo **25** [1975] 73/99).
[10] A. Michaelis, U. Genzken (Ber. Deut. Chem. Ges. **17** [1884] 924/5).

[11] P. Pfeiffer, I. Heller (Ber. Deut. Chem. Ges. **37** [1904] 4620/3).
[12] K. W. Shen, W. E. McEwen, S. J. La Placa, W. C. Hamilton, A. P. Wolf (J. Am. Chem. Soc. **90** [1968] 1718/23).
[13] S. N. Bhattacharya, M. Singh (Indian J. Chem. A **18** [1979] 515/6).
[14] V. P. Glushkova, T. V. Talalaeva, Z. P. Razmanova, G. S. Zhdanov, K. A. Kocheshkov (Sb. Statei Obshch. Khim. Akad. Nauk SSSR **2** [1953] 992/6).
[15] G. S. Zhdanov, Z. P. Razmanova (Dokl. Akad. Nauk SSSR **72** [1950] 1055/7).
[16] G. L. Kuykendall, J. L. Mills (J. Organometal. Chem. **118** [1976] 123/8).
[17] A. Ouchi, F. Ebina, T. Uehiro, Y. Yoshino (Bull. Chem. Soc. Japan **51** [1978] 2427/8).
[18] N. K. Parab, D. M. Desai (Current Sci. [India] **26** [1957] 389).
[19] N. K. Parab, D. M. Desai (J. Indian Chem. Soc. **35** [1958] 69/72).
[20] J. Havranek, J. Mleziva (Angew. Makromol. Chem. **84** [1980] 105/17).

[21] A. E. Goddard, V. E. Yarsley (J. Chem. Soc. **1928** 719/23).
[22] Y. Kawasaki (Inorg. Nucl. Chem. Letters **5** [1969] 805/10).
[23] Y. Kawasaki (Bull. Chem. Soc. Japan **49** [1976] 817/8).
[24] K. Bajpai, R. Singhal, R. C. Srivastava (Indian J. Chem. A **18** [1979] 73/5).
[25] K. Bajpai, R. C. Srivastava (Syn. Reactiv. Inorg. Metal-Org. Chem. **11** [1981] 7/13).
[26] K. Bajpai, R. C. Srivastava (Syn. Reactiv. Inorg. Metal-Org. Chem. **9** [1979] 557/64).
[27] S. N. Bhattacharya, A. K. Saxena (Indian J. Chem. A **17** [1979] 307/9).
[28] S. N. Bhattacharya, P. Raj, A. K. Saxena (Indian J. Chem. A **16** [1978] 1071/4).
[29] F. Yu. Yusupov, Z. M. Manulkin (Tr. Tashkent. Farm. Inst. **4** [1966] 531/7; C.A. **68** [1968] No. 59676).
[30] A. E. Goddard (J. Chem. Soc. **123** [1923] 2315/23).

[31] A. N. Tatarenko, Z. M. Manulkin (Zh. Obshch. Khim. **38** [1968] 273/5; J. Gen. Chem. [USSR] **38** [1968] 276/7).
[32] D. E. Worrall (J. Am. Chem. Soc. **62** [1940] 2514/5).
[33] D. E. Worrall (J. Am. Chem. Soc. **52** [1930] 2046/50).
[34] F. Challenger, F. Pritchard, J. R. A. Jinks (J. Chem. Soc. **125** [1924] 864/75).
[35] K. Matsumiya (Mem. Coll. Sci. Univ. Kyoto Imp. **8** [1925] 11/8; C.A. **1925** 1704).
[36] E. Krause, G. Renwanz (Ber. Deut. Chem. Ges. **65** [1932] 777/84).

2.5.1.1.4 Triorganoantimony Diiodides

2.5.1.1.4.1 R_3SbI_2 Compounds with R = Alkyl and Alkenyl

$(CH_3)_3SbI_2$

Trimethylantimony diiodide is obtained by reaction of $Sb(CH_3)_3$ with I_2 in C_2H_5OH [1], in ether [2, 3, 22], or in a mixture of $Sb(CH_3)_3$ dissolved in ether and I_2 dissolved in CCl_4 [4] at 0 °C [3]. The yield of the compound is between 60 and 85% [4]. In some cases, the compound is directly obtained in yields of 45 to 62% [3] by treating the azeotropic distillate of the Grignard mixture of $SbCl_3$ and CH_3MgI with I_2 [2, 3]. The compound was obtained directly by reacting antimony with CH_3I at 140 °C [5, 6], but the resulting crystals were very impure [2]. $(CH_3)_3SbO$ and aqueous HI give the compound as a precipitate which melts at 107 °C with decomposition [1]. $Sb(CH_3)_3$ and AsI_3 react in toluene in a ratio of 3:2 to form the substance and metallic arsenic [7]. Upon decomposition of a $(CH_3)_3Sb \cdot BI_3$ adduct in $CHCl_3$ or CH_3CN solution, the compound is also obtained [8].

The compound can be recrystallized from CH_3OH [2], from C_2H_5OH [1, 3] as long thin needles [1], or from H_2O [1, 4] as prisms [1] upon addition of HI [17]. The crystals slowly become yellow and opaque [1]. It melts at 128 °C with decomposition [17]. The melting point is not reproducible because the compound loses CH_3I upon heating [3].

Table 21
IR and Raman Vibrations (in cm^{-1}) of Solid $(CH_3)_3SbI_2$ [9].

IR	Raman	assignment	IR	Raman	assignment
3004 m	3006 vw	$\nu_{as}CH$	317 vw		$2\nu_6$
2924 w	2914 ms	ν_sCH	311 vw	306 vvw	
2851 vw			290 vw	292 vvw	
2405 w			283 vw		
2325 vw			277 vw	274 vvw	
1777 vw			265 vw	263 vvw	
1730 w				247 w	$2\nu_2$
1630 mw			229 vw	224 vvw	$\nu_2+\nu_7$
	1402 ms	$\delta_{as}CH$	201 vw		
	1232 w	δ_sCH		196 vw	ν_8, ϱ
	1207 vw		173 sh		ν_4, δSbC out-of-plane
1020 vw	1030 vw			163 mw	ν_6, δSbI
867 vs		ϱCH_3	144 vs, br		ν_3, ν_{as}SbI
563 s	555 mw	ν_5, ν_{as}SbC		122 vvs	ν_2, ν_sSbI
	508 s	ν_1, ν_sSbC	80 m	81 vw	ν_7, δSbC in-plane
396 vw		$3\nu_3$ first overtone		70 vw	
328 vw				43 s	lattice vibration

Complete IR and Raman vibrations are shown in Table 21 [9]. Calculations of force constants and frequencies are made in [10 to 12]. A detailed analysis of force constants, mean amplitudes of vibration, generalized mean-square amplitudes, shrinkage constants, Coriolis coupling coefficients, and centrifugal distortion constants have been evaluated from Raman and infrared vibrations and the structural parameters, using the general valence force field approach. The variation of these constants with the changing halogen atom (F, Cl, Br, I) is discussed [13].

References on p. 97

The ^1H NMR spectrum in $CDCl_3$ shows the CH_3 singlet at $\delta=3.01$ (−32 °C), 2.98 (30.5 °C), and 2.96 (70 °C) ppm [14]. A ^{121}Sb Mössbauer spectrum vs. InSb at 4.2 K gives values of $\delta=2.16$, $e^2qQ=-19.28$, and $\Gamma=2.5$ mm/s. These values are compared with those of other trialkylantimony dihalides [15]. For a Hartee-Fock-Slater LCAO calculation of the parameters see [16]. An NQR spectrum is given in [48].

A UV spectrum in the region of 240 to 320 nm shows no characteristic maxima [17]. The He(I) photoelectron spectrum gives ionization energies of 8.89, 9.56, 9.95, 11.18, 12.0, and 14.8 eV. These values are discussed and compared with those of the other halides in [18]. A correlation of the ^1H NMR chemical shifts of the compounds $(CH)_3SbX_2$ and $(CH_3)_3SbXX'$ (X and X′=F, Cl, Br, I) with the inverse ionization potentials of the ligands X and X′ is established and discussed in [25].

The compound crystallizes in the hexagonal space group $P\bar{6}m2-D^3_{3h}$ (No. 189) or $P\bar{6}c2-D^4_{3h}$ (No. 190) [20, 21] with two molecules in the elementary cell. The cell dimensions are a=7.53 and c=9.50 Å; $d_c=2.95$ and $d_m=2.97$ g/cm³. The Sb-I distances in the molecule with trigonal bipyramidal geometry (I in axial positions) were determined to 2.88 Å [20].

The compound is readily soluble in hot H_2O and C_2H_5OH [1, 17], in $(CH_3)_2CO$, CH_3CN, and C_6H_5CN [17], but only slightly soluble in ether [1], c-C_6H_{12}, and CCl_4 [17]. The conductivity in CH_3CN at 25 °C is between 2.35 and 7.46 $cm^2\cdot\Omega^{-1}\cdot mol^{-1}$. However, no exact value can be given since the solution turns yellow, and the resistance increases [17]. A mass spectrum (70 eV, 30 °C inlet, 100 °C source) shows the fragments $[(CH_3)_3SbI_n]^+$ (n=0, 1), $[(CH_3)_2SbI_n]^+$ (n=0 to 2), $[CH_3SbI_n]^+$ (n=0 to 2), and SbI_n (n=1, 2) [19]. The compound pyrolyzes [23] at 140 °C/70 Torr to give 95% $(CH_3)_2SbI$ and CH_3I [19], at 100 to 140 °C/50 Torr [2] or 60 to 80 Torr [22] to give the same products.

It reacts with iodine in hot water; upon cooling, green-black needle-like crystals of the composition $(CH_3)_3SbI_6$ precipitate, which melt at 68 to 70 °C after workup. The crystals are not very stable and decompose even in a sealed tube. Nonpolar solvents abstract iodine. With water, a dark red solution is obtained which turns clear upon boiling due to loss of iodine. Concentrated HCl gives $(CH_3)_3SbCl_2$ [17].

$(CH_3)_3SbI_2$ reacts with $AgNO_3$ or Ag_2SO_4 in aqueous solution to form the corresponding dinitrate and sulfate, respectively. With $(CH_3)_3SbO$ in H_2O in a ratio of 1:1 the compound gives $(CH_3)_3SbI(OH)$ [1]. In an inert atmosphere at 100 °C, the title compound and $Zn(CH_3)_2$ form $Sb(CH_3)_3$ and $Sb(CH_3)_5$, which were isolated by distillation of the mixture [5, 6]. $(CH_3)_3SbI_2$ gives exchange reactions with $(C_6H_5CH_2)_3SbX_2$ (X=F, Cl, Br) to give $(CH_3)_3SbX_2$ and $(C_6H_5CH_2)_3SbI_2$. The equilibrium constants for these reactions were determined in $CDCl_3$ by NMR at 32 °C as 350±60, 7.8±1.0, and 2.9±0.8, respectively [24]. Exchange reactions with $(CH_3)_3SbX_2$ (X=F, Cl, Br) lead to $(CH_3)_3SbIX$ [14]. The following equilibrium constants were obtained by ^1H NMR or ^{19}F NMR in $CDCl_3$ for X=F: 1.09±0.10 at 0 °C, 1.50±0.06 at 35 °C (32 °C in [24]), 1.68±0.13 at 60 °C; for X=Cl: 2.05±0.07 at 0 °C, 2.25±0.10 at 35 °C (2.3±0.1 at 32 °C in [24]), 2.85±0.01 at 60 °C; for X=Br: 2.93±0.03 at 0 °C, 3.02±0.10 at 35 °C, and 3.30±0.16 at 60 °C [31]. No antimony polyesters are formed in a reaction with the Na salts of 1,1′-ferrocenedicarboxylic acid or terephthalic acid in CCl_4/H_2O at 25 °C [32].

$(C_2H_5)_3SbI_2$

Triethylantimony diiodide can be prepared from $Sb(C_2H_5)_3$ and I_2 in water or ether; it is best prepared in cooled C_2H_5OH. The compound crystallizes in long colorless needles which melt at 70.5 °C. It can be recrystallized from C_2H_5OH and from ether [26]. It is mentioned that a reaction of $SbCl_5$ with C_2H_5MgI gives $(C_2H_5)_3SbCl_2$, which was isolated as

References on p. 97

the diiodide [27]. Coarsely powdered antimony reacts with C_2H_5I in a sealed ampule at 140 °C. An oil is obtained which probably contains the title compound and $(C_2H_5)_2SbI_3$ [5, 6].

The far IR (in Nujol, c-C_6H_{12}, or C_6H_6) and Raman spectra show the following absorptions (IR (Raman) in cm^{-1}): 525, 510 sh (529 dp, 515) $\nu_{as}SbC$, 468 (475 p) ν_sSbC, 280 to 250 (ca. 280, 250 p) δSbC, 170 ϱ, 140 (140) $\nu_{as}SbI$, 115 (118 p) ν_sSbI. The experimental data are in good agreement with a C_{3v} geometry [28].

The compound is soluble in ethanol, ether, and water. Most of it is recovered from a hot saturated aqueous solution. If the solid is heated for a longer time at 100 °C, part of it sublimes. A slight increase of temperature gives decomposition, and a thick white steam evolves. Melted triethylantimony diiodide is immediately reduced by potassium. With HCl, the corresponding dichloride is formed. I_2 is evolved upon reaction with Cl_2, Br_2, and HNO_3. Concentrated H_2SO_4 reacts with formation of HI, I_2, and SO_2 [26]; see also [29]. The compound reacts with Ag_2SO_4, $AgNO_3$, Ag_2CO_3 [29, 30], or $HgCl_2$ [26, 29, 30] in aqueous medium to form the corresponding $(C_2H_5)_3SbX_2$ ($X_2=SO_4$, $(NO_3)_2$, CO_3, Cl_2) compounds. With $Zn(C_2H_5)_2$ in an inert atmosphere, $Sb(C_2H_5)_3$, $Sb(C_2H_5)_5$, and other compounds are obtained upon distillation [5, 6].

The compound is useful as a catalyst [33] or a cocatalyst together with $TiCl_3$ and $Al(C_3H_7)_3$ [34] for the polymerization of olefins.

$(C_3H_7)_3SbI_2$

The compound is mentioned as a product from the reaction of the corresponding stibine with I_2 in ether. It is obtained as a yellow mass [35].

$(i\text{-}C_3H_7)_3SbI_2$

$Sb(C_3H_7\text{-}i)_3$ and I_2, reacted at 0 °C in ether, give the compound as a pale yellow solid in a yield of 87% after removal of the ether. It decomposes above 125 °C [19].

The far IR (in Nujol, c-C_6H_{12}, or C_6H_6) and Raman (solid) spectra show the following absorptions (IR (Raman) in cm^{-1}): 494 (499dp) $\nu_{as}SbC$, 473 (476p) ν_sSbC; 408, 298, 255 (410p, 252p) δSbC; 175, 165 (175, 164p) ϱ, 136 (131dp) $\nu_{as}SbI$, 116 (117p) ν_sSbI, 70 δSbI. The experimental data are in good agreement with a C_{3v} geometry of the molecule [28]. In [19] the νSbC bands are given as 489s and 469 m in Nujol. A ^1H NMR spectrum in C_6H_6 shows resonances at δ = 1.27 (d) and 3.28 (hept) ppm with J = 6.9 Hz.

The following fragments were found in the mass spectrum (35 eV, inlet 35 °C, source 150 °C): $[(C_3H_7)_3SbI_n]^+$ (n = 0, 1), $[(C_3H_7)_2SbIH]^+$, $[(C_3H_7)_2SbI_n]^+$ (n = 0, 1), $[C_3H_7SbI_n]^+$ (n = 0, 1), $[CH_3SbI_n]^+$ (n = 0, 1), SbI, and C_3H_7. The compound pyrolyzes at 130 °C/60 Torr to give 5% $Sb(C_3H_7\text{-}i)_3$ and 85% $(i\text{-}C_3H_7)_2SbI$ [19].

It is useful together with $TiCl_3$ and $Al(C_2H_5)_3$ or $(C_2H_5)_2AlCl$ as a catalyst for ethylene polymerization [36].

$(C_4H_9)_3SbI_2$

The compound is obtained as an oily solid by reaction of the corresponding stibine with I_2 in $CHCl_3$ solution [35]. The reaction of $Sb(C_4H_9)_3$ with CH_3AsI_2 gives a 65% yield of the compound and 85% $(CH_3As)_n$. With $C_6H_5AsI_2$, the yield of $(C_6H_5As)_6$ is 54%. With $(C_6H_5)_2AsI$ in ether solution, the stibine reacts to form the title compound and $(C_6H_5)_2As\text{-}As(C_6H_5)_2$ (82% yield) [7].

References on p. 97

$(i\text{-}C_4H_9)_3SbI_2$

The yellow crystals resulting from the reaction of $Sb(C_4H_9\text{-}i)_3$ and I_2 melt at 70 °C [37].

$(C_5H_{11})_3SbI_2$

The compound is prepared by reacting $Sb(C_5H_{11})_3$ with I_2 in ether and precipitation with alcohol [38 to 40], or by treating $(C_5H_{11})_3SbO$ with aqueous HI and precipitation with water [38, 40].

The oily compound reacts with $AgNO_3$ or Ag_2SO_4 in C_2H_5OH to give the corresponding dinitrate and sulfate [38].

$(C_7H_{15})_3SbI_2$

The substance is formed as a half-solid residue by the reaction of the corresponding stibine with I_2 in ether with subsequent evaporation of the solvent [41].

$(C_6H_5CH_2)_3SbI_2$

Exchange reactions of the compound with the corresponding difluorides and dichlorides in $CHCl_3$ solution at 32 °C were followed by ^1H NMR spectroscopy. Equilibrium constants of $K=7.1\pm0.8$ (X = F) and 3.3 ± 0.5 (X = Cl) were found for the reaction $(C_6H_5CH_2)_3SbI_2+(C_6H_5CH_2)_3SbX_2 \rightleftharpoons 2(C_6H_5CH_2)_3SbFX$ [24].

$(CH_2{=}CH)_3SbI_2$

$Sb(CH{=}CH_2)_3$ is reacted with a stoichiometric amount of I_2 in CCl_4. Evaporation of the solvent gives a yellow oily residue [42]. The same reaction performed in ether at 0 °C yields 63% light yellow crystals after concentration and cooling of the mixture. These crystals melt at 35 to 37 °C [43]. In a complex reaction, the title compound with the same melting point is obtained in a yield of 15% from $(CH_2{=}CH)Si(OC_2H_5)_3$, SbF_3, and NH_4HF_2 in water and subsequent addition of I_2 in ether solution [44].

The compound dissolves well in $CHCl_3$, poorly in ether, and is insoluble in petroleum ether [43]. It decomposes upon heating under reduced pressure with formation of SbI_3 and some $(CH_2{=}CH)_2SbI$ [42].

$[CH_2{=}C(CH_3)]_3SbI_2$

Light yellow crystals, which melt at 162 to 164 °C, are obtained in a yield of 63% from the reaction of the corresponding stibine with stoichiometric amounts I_2 in ether at −5 °C. The solubility is the same as for the preceding compound [43].

$(cis\text{-}CH_3CH{=}CH)_3SbI_2$

The compound is prepared in the usual way by reacting the corresponding stibine with I_2, dissolved in ether at 0 °C. Evaporation of the solvent in an inert atmosphere gives the compound as a residue [45, 46] in a yield of 80.5%; m.p. 122 to 123 °C [46]. IR(KBr): 1600, 1425, 1378, 1297, 1196, 1100, 1040, 937, 925, 660, 610, and 452 cm^{-1} [47]. The spectrum was also recorded in [45].

(*trans*-CH_3CH=$CH)_3SbI_2$

The compound is prepared like the previous one [45, 46]. It is obtained as a liquid in a yield of 84% after washing with aqueous $NaHSO_3$ [46]. IR (neat liquid): 1598, 1437, 1375, 1302, 1185, 1105, 1065, 1039, 945, 718, 660, and 615 cm^{-1} [47]. IR data are also given in [45].

References:

[1] H. Landolt (J. Prakt. Chem. **84** [1861] 328/339).
[2] K. Brodersen, R. Palmer, D. Breitinger (Chem. Ber. **104** [1971] 360/4).
[3] G. O. Doak, G. G. Long, M. E. Key (Inorg. Syn. **9** [1967] 92/7).
[4] G. G. Long, G. O. Doak, L. D. Freedman (J. Am. Chem. Soc. **86** [1964] 209/13).
[5] G. B. Buckton (Z. Chem. Pharm. **1860** 611/6; Jahresber. Fortschr. Chem. **1860** 371/374).
[6] G. B. Buckton (Quart. J. Chem. Soc. **13** [1860] 115/21).
[7] J. C. Summers, H. H. Sisler (Inorg. Chem. **9** [1970] 862/9).
[8] M. L. Denniston, D. R. Martin (J. Inorg. Nucl. Chem. **36** [1974] 2175/6).
[9] B. A. Nevett, A. Perry (Spectrochim. Acta A **33** [1977] 755/60).
[10] B. A. Nevett, A. Perry (J. Mol. Spectrosc. **66** [1977] 331).

[11] R. Namasivayam, S. Viswanathan (Bull. Soc. Chim. Belges **87** [1978] 733/6).
[12] B. A. Nevett, A. Perry (J. Organometal. Chem. **71** [1974] 399/402).
[13] A. Natarajan, K. Chockalingam (Indian J. Pure Appl. Phys. **19** [1981] 672/5).
[14] G. G. Long, C. G. Moreland, G. O. Doak, M. Miller (Inorg. Chem. **5** [1966] 1358/61).
[15] K. Dehnicke, K. Fleck, K. Schmidt, J. Pebler (Z. Anorg. Allgem. Chem. **451** [1979] 109/14).
[16] W. Ravenek, J. W. M. Jacobs, A. Van der Avoird (Chem. Phys. **78** [1983] 391/404).
[17] T. M. Lowry, J. H. Simons (Ber. Deut. Chem. Ges. **63** [1930] 1595/1602).
[18] S. Elbel, H. Tom Dieck (Z. Anorg. Allgem. Chem. **483** [1981] 33/43).
[19] H. J. Breunig, W. Kanig (Phosphorus Sulfur **12** [1982] 149/59).
[20] A. F. Wells (Z. Krist. **99** [1938] 367/77).

[21] G. S. Zhdanov, Z. P. Razmanova (Dokl. Akad. Nauk SSSR **72** [1950] 1055/7).
[22] G. T. Morgan, G. R. Davies (Proc. Roy. Soc. [London] B **110** [1926] 523/34).
[23] G. T. Morgan, G. R. Davies (Nature **116** [1925] 499).
[24] C. G. Moreland, G. G. Long (Inorg. Nucl. Chem. Letters **8** [1972] 347/51).
[25] T. Schaefer, F. Hruska, H. M. Hutton (Can. J. Chem. **45** [1967] 3143/51).
[26] C. Löwig, E. Schweizer (Liebigs Ann. Chem. **75** [1850] 315/55).
[27] P. Pfeiffer, K. Schnurmann (Ber. Deut. Chem. Ges. **37** [1904] 319/22).
[28] L. Verdonck, G. P. Van der Kelen (Spectrochim. Acta A **31** [1975] 1707/11).
[29] W. Merck (J. Prakt. Chem. **66** [1855] 56/72).
[30] W. Merck (Liebigs Ann. Chem. **97** [1856] 329/33).

[31] C. G. Moreland, M. H. O'Brien, C. E. Douthit, G. G. Long (Inorg. Chem. **7** [1968] 834/7).
[32] C. Carraher Jr., H. Blaxall (Angew. Makromol. Chem. **83** [1979] 37/45).
[33] Y. Takashi, I. Aiiima, Y. Kobayashi, Y. Tsunoda, Asahi Chemical Industry Co., Ltd. (Japan. 9439 [1961/63]; C.A. **59** [1963] 10256).
[34] Y. Takashi, I. Aiiima, Y. Kobayashi, Y. Tsunoda, Asahi Chemical Industry Co., Ltd. (U.S. 3494910 [1965/67/70]; C.A. **72** [1970] No. 90967).
[35] W. J. C. Dyke, W. J. Jones (J. Chem. Soc. **1930** 1921/7).
[36] H. Morita, Y. Takashi, I. Aiiima, Asahi Chemical Industry Co., Ltd. (Japan. 7013585 [1967/70]; C.A. **73** [1970] No. 46037).
[37] M. E. Brinnand, W. J. C. Dyke, W. H. Jones, W. J. Jones (J. Chem. Soc. **1932** 1815/19).
[38] F. Berlé (J. Prakt. Chem. **65** [1855] 385/418).

[39] C. Cramer (Verh. Naturf. Ges. Zürich **1851** May 12th from Chem. Pharm. Centr. **26** [1855] 465/8; Jahresber. Fortschr. Chem. **1855** 590).
[40] F. Berlé (Liebigs Ann. Chem. **97** [1856] 316/22).
[41] Chao-Lun Tseng, Wen-Yu Shih (J. Chinese Chem. Soc. **4** [1936] 183/6).
[42] L. Maier, D. Seyferth, F. G. A. Stone, E. G. Rochow (J. Am. Chem. Soc. **79** [1957] 5884/9).
[43] A. N. Nesmeyanov, A. E. Borisov, N. V. Novikova (Izv. Akad. Nauk SSSR Otd. Khim. Nauk **1961** 1578/82; Bull. Acad. Sci. USSR Div. Chem. Sci. **1961** 1473/6).
[44] R. Müller, S. Reichel, C. Dathe (Inorg. Nucl. Chem. Letters **3** [1967] 125/33).
[45] A. N. Nesmeyanov, A. E. Borisov, N. V. Novikova (Tetrahedron Letters **1960** 23/4).
[46] A. N. Nesmeyanov, A. E. Borisov, N. V. Novikova (Izv. Akad. Nauk SSSR Otd. Khim. Nauk **1961** 612/7; Bull. Acad. Sci. USSR Div. Chem. Sci. **1961** 564/8).
[47] A. E. Borisov, N. V. Novikova, N. A. Chumaevskii (Dokl. Akad. Nauk SSSR **136** [1961] 129/32).
[48] D. J. Parker (Diss. Univ. Wisconsin 1959; Diss. Abstr. Intern. B **20** [1959/60] 2044).

2.5.1.1.4.2 R_3SbI_2 Compounds with R = Aryl

$(C_6H_5)_3SbI_2$

Triphenylantimony diiodide is prepared by reacting triphenylstibine and iodine in organic solvents like petroleum ether [1, 2], CH_3CN [7], or cyclohexane [3]; see also [41]. The stibine and HgI_2 in refluxing $(CH_3)_2CO$ give the compound in a yield of 70 to 85% after evaporation of the filtrate [4]. Irradiation of $Sb(C_6H_5)_3$ and C_6H_5I for 60 h in CH_3OH, C_6H_6, or $CHCl_3$ gives 35, 70, or 55%, respectively, of the compound; a radical mechanism is proposed for the reaction. Similarly, the stibine reacts with CH_3I in CH_3OH for 20 min to form 60% of the title compound besides 76% CH_4 [5, 6]. Preparation from the sulfide with I_2 is reported in [29]. For the conductometric titration of $Sb(C_6H_5)_3$ with I_2 in CH_3CN see [7]. A spot test for $Sb(C_6H_5)_3$ or I_2 in $CHCl_3$, C_6H_6, or CS_2 is based on the formation of $(C_6H_5)_3SbI_2$ and a further reaction with excess of I_2 to brown triphenylantimony polyiodide [8].

Melting points of the compound are given as 153 °C [1 to 3] for shiny white plates, recrystallized from petroleum ether/C_6H_6 [1], 150 °C [4], or 163 to 164 °C precipitated from C_6H_6 solution with petroleum ether [9].

IR and Raman spectra of the compound are given and discussed in several publications between 4000 and 40 cm^{-1} [9 to 11]. The Raman band at 117 cm^{-1} is assigned to $\nu_s SbI$, and the IR band at 143 cm^{-1} to $\nu_{as} SbI$. For a complete listing of the bands with assignment according to Whiffen's nomenclature see [11]. In the UV spectrum of the compound in $CHCl_3$ a weak absorption is observed at 265 nm [3]. A ^{13}C NMR spectrum in $CDCl_3$ shows resonances at $\delta = 129.48$ (C-3; J(C, H) = 164.6), 131.43 (C-4; J(C, H) = 162.1), 132.99 (C-2; J(C, H) = 165.0), 141.22 (C-1) [12]. ^{121}Sb Mössbauer spectra gave the following values (in mm/s):

	δ	e^2qQ	Γ	Ref.
80 K vs. $^{121}SnO_2$	−6.3 ± 0.2	−16 ± 3	2.4 ± 0.5	[13]
4 K vs. $Ca^{121}SnO_3$	−6.72 ± 0.05	−18.1 ± 0.4	2.58 ± 0.14	[14]

From these values and an additive model for the electric field gradient, a trigonal bipyramidal geometry of the molecule was established [15]. The quadrupole coupling constant was calculated as −17.7 mm/s and compared with the experimental value [16]. An Sb^V orbital population analysis is made using the Mössbauer data in [17].

The compound is insoluble in ether, alcohol, and petroleum ether, but easily soluble in benzene [1]. The dipole moment of the compound in C_6H_6 at 25 °C is 2.07 D [18], and the molar susceptibility $\chi_M = -261.1 \times 10^{-6}$ [19, 20]. For electrical conductance and refractometric measurements for detection of separated ion pairs in CCl_4, $CHCl_3$, and dioxane/THF mixtures, see [21]. $(C_6H_5)_3SbI_2$ reacts with alkali in alcohol to form $(C_6H_5)_3Sb(OH)_2$ [1]. The same compound is obtained with KOH in boiling C_2H_5OH and treating the reaction products with ether, CH_3CO_2H, and H_2O [6]. $NaOCH_3$ in CH_3OH and $(C_6H_5)_3SbI_2$ in C_6H_6 are refluxed for 20 min, and 8-hydroxyquinoline is subsequently added to give $(C_6H_5)_3SbI(OC_9H_6N)$ [22]. Conductometric titration of the compound with Br_2 or with IBr in CH_3CN gives $(C_6H_5)_3SbBr_2$ and I_2 [7]. The exchange reaction with $(C_6H_5)_3SbF_2$ in $CDCl_3$ solution which yields $(C_6H_5)_3SbFI$ was followed by ^{1}H NMR spectroscopy. An equilibrium constant of 1.0 ± 0.3 was determined at 35 °C [23]. The kinetics of the isotope exchange reaction in the system $^{125}Sb(C_6H_5)_3$/$(C_6H_5)_3SbI_2$ was studied in C_3H_7OH. The mechanism is analogous to that of the corresponding dichloride derivative [42].

The compound was tested as a retarding agent of burning epoxy resins [24, 25]. It may be used for extraction of chloride from aqueous to CCl_4 or C_6H_6 phases [26, 27]. It is a catalyst for the decomposition of $[(C_6H_5)_3Sb(X)O]_2$ (X = Cl, Br) in C_6H_5Cl with evolution of singlet oxygen [28]. The use of the compound in iodine therapy and for treatment of pleuritic exudations is described in a patent [29], see also [41].

$(4\text{-}CH_3C_6H_4)_3SbI_2$

Tris(4-methylphenyl)antimony diiodide is obtained by reaction of the corresponding stibine with iodine in ether/ethanol [33] or in petroleum ether [36]. In refluxing $(CH_3)_2CO$, the stibine reacts with HgI_2. After filtration from the precipitated Hg_2I_2, the compound is isolated in 70 to 85% yield [4]. $(4\text{-}CH_3C_6H_4)_3SbO$, treated with an excess of 48% aqueous HI in acetone, gives the compound which precipitates upon addition of cold water [37].

Recrystallization from $CHCl_3$ gives small shiny crystals of m.p. 182 to 183 °C. Large crystals are obtained from C_6H_6 [33]. Other melting points given are 182.5 °C (from C_6H_6/C_2H_5OH) [36] and 181 °C [4, 38, 39]. A molar susceptibility of $\chi_M = -295.3 \times 10^{-6}$ was found [19, 20], and ^{1}H NMR signals in $CDCl_3$ at $\delta = 2.42$ (CH_3); 7.32, 8.02 (H-2,3; J(H-2,3) = 8 Hz) ppm [37].

The compound is easily soluble in $CHCl_3$ and C_6H_6, and only slightly soluble in ether, C_2H_5OH, and petroleum ether [33]. It decomposes upon heating under 5 to 7 Torr to form $4\text{-}CH_3C_6H_4I$ and $(4\text{-}CH_3C_6H_4)_2SbI$ quantitatively [40]. It is useful for the extraction of chloride ions from water into $CHCl_3$ [34].

Other R_3SbI_2 Compounds (R = substituted phenyl and 1-naphthyl)

The compounds are collected in Table 22. They are prepared by the following methods:

Method I: The corresponding stibine reacts with CuI_2 in C_2H_5OH. After filtration of the mixture, the solution is concentrated, and the resulting precipitate is recrystallized.

Method II: The corresponding stibine reacts with I_2 in an organic solvent. Details are given in the table.

References on p. 101

Table 22
R_3SbI_2 Compounds with R = Substituted Phenyl and 1-Naphthyl.
For explanations, abbreviations, and units, see p. X.

No.	compound method of preparation, conditions (yield in %)	properties and remarks	Ref.
1	$(4\text{-}ClC_6H_4)_3SbI_2$ I	yellow plates m.p. 137 to 138° (dec., from C_6H_6/C_2H_5OH)	[30]
2	$(4\text{-}BrC_6H_4)_3SbI_2$ I	pale yellow needles m.p. 155 to 156° (dec., from petroleum ether)	[30]
3	$(2\text{-}CH_3OC_6H_4)_3SbI_2$ I	yellow solid m.p. 141 to 143° (dec., from CCl_4/petroleum ether)	[30]
4	$(3\text{-}CH_3OC_6H_4)_3SbI_2$ I	pale yellow plates m.p. 99.5 to 100° (dec., from $CHCl_3/C_2H_5OH$)	[30]
5	$(2\text{-}C_2H_5OC_6H_4)_3SbI_2$ I	pale yellow plates m.p. 143° (from $CHCl_3/C_2H_5OH$)	[30]
6	$(4\text{-}C_6H_5OC_6H_4)_3SbI_2$ I	pale yellow plates m.p. 140° (dec., from $CHCl_3/C_2H_5OH$)	[30]
7	$(4\text{-}CH_3OC_6H_4)_3SbI_2$ II in $CHCl_3$	yellow monoclinic plates m.p. 116° (from C_2H_5OH) soluble in ether, C_6H_6, $CHCl_3$ stable towards H_2O dec. with C_2H_5OH/H_2O	[31]
8	$(4\text{-}C_2H_5OC_6H_4)_3SbI_2$ II in $CHCl_3$	prismatic crystals m.p. 121 to 122° (from C_2H_5OH)	[31]
9	$(4\text{-}(CH_3)_2NC_6H_4)_3SbI_2$ II in CH_2Cl_2 at −78°, in $CHCl_3$ at 0°, precipitated with petroleum ether or ether (91)	fine yellow crystals dec. at ca. 135° UV(THF): $\lambda = 285$ ($\log\varepsilon = 4.8$), 375 (3.8) ^{1}H NMR($CDCl_3$): $\delta = 3.45$ (CH_3), 7.18 (H-3,5), 8.43 (H-2,6); J (H-2,3) + J (H-2,5) = 9	[32]
10	$(2\text{-}CH_3C_6H_4)_3SbI_2$ II in ether/C_2H_5OH	crystalline powder of light yellow color m.p. 175 to 176° (dec.) useful for extraction of Cl^- from H_2O into $CHCl_3$	[33] [34]
11	$(3\text{-}CH_3C_6H_4)_3SbI_2$ II in petroleum ether	white powder m.p. 138 to 139° (from $CHCl_3$/petroleum ether; washed with $CHCl_3/C_2H_5OH$) soluble in $CHCl_3$, C_6H_6, ether, C_2H_5OH	[33]
12	$(4\text{-}C_6H_5C_6H_4)_3SbI_2$ II in $CHCl_3$	small plates, contains $CHCl_3$ of crystallization m.p. 176 to 178° reacts with 95% ethanolic NH_3 to give the dihydroxide	[35]
13	$(1\text{-}C_{10}H_7)_3SbI_2$ probably II	useful for extraction of Cl^- from H_2O into $CHCl_3$	[34]

References:

[1] A. Michaelis, A. Reese (Liebigs Ann. Chem. **233** [1886] 39/60).
[2] A. Michaelis, L. Weitz (Ber. Deut. Chem. Ges. **20** [1887] 52).
[3] K. R. Bhaskar, S. N. Bhat, S. Singh, C. N. R. Rao (J. Inorg. Nucl. Chem. **28** [1966] 1915/25).
[4] S. N. Bhattacharya, M. Singh (Indian J. Chem. A **18** [1979] 515/6).
[5] G. A. Razuvaev, M. A. Shubenko (Dokl. Akad. Nauk SSSR **67** [1949] 1049/52).
[6] G. A. Razuvaev, M. A. Shubenko (Zh. Obshch. Khim. **21** [1951] 1974/9; J.Gen. Chem. [USSR] **21** [1951] 2193/9).
[7] A. D. Beveridge, G. S. Harris, F. Inglis (J. Chem. Soc. A **1966** 520/8).
[8] F. Feigl, D. Goldstein (Mikrochim. Acta **1966** 1/3).
[9] G. O. Doak, G. G. Long, L. D. Freedman (J. Organometal. Chem. **4** [1965] 82/91).
[10] F. W. Parrett (Spectrochim. Acta A **26** [1970] 1271/4).

[11] B. A. Nevett, A. Perry (Spectrochim. Acta A **33** [1977] 755/60).
[12] J. Havranek, A. Lycka (Sb. Ved. Pr. Vys. Sk. Chemickotechnol. Pardubice **43** [1980] 123/7).
[13] G. G. Long, J. G. Stevens, R. J. Tullbane, L. H. Bowen (J. Am. Chem. Soc. **92** [1970] 4230/5).
[14] J. G. Stevens, S. L. Ruby (Phys. Letters A **32** [1970] 91/2).
[15] J. N. R. Ruddick, J. R. Sams, J. C. Scott (Inorg. Chem. **13** [1974] 1503/7).
[16] G. M. Bancroft, V. G. K. Das, T. K. Sham, M. G. Clark (J. Chem. Soc. Dalton Trans. **1976** 643/54).
[17] L. H. Bowen, G. G. Long (Inorg. Chem. **15** [1976] 1039/44).
[18] L. M. Kataeva, Yu. V. Rydvanskii, N. I. Trofimova (Zh. Fiz. Khim. **50** [1976] 814/5; Russ. J. Phys. Chem. **50** [1976] 486/7).
[19] N. K. Parab, D. M. Desai (Current Sci. [India] **26** [1957] 389).
[20] N. K. Parab, D. M. Desai (J. Indian Chem. Soc. **35** [1958] 573/5).

[21] R. Sahai, P. C. Pande, V. Singh (Indian J. Chem. A **18** [1979] 217/20).
[22] J. Kawasaki (Inorg. Nucl. Chem. Letters **5** [1969] 805/10).
[23] C. G. Moreland, M. H. O'Brien, C. E. Douthit, G. G. Long (Inorg. Chem. **7** [1968] 834/6).
[24] J. Havranek, J. Mleziva (Angew. Makromol. Chem. **84** [1980] 105/7).
[25] J. Havranek (Sb. Dokl. 1st Nats. Konf. Mladite Nauchni Rab. Spets. Neft Khim., Burgas, Bulg., 1976 [1977], pp. 152/9; C.A. **93** [1980] No. 187219).
[26] H. Chermette, C. Martelet, D. Sandino, J. Tousset (Anal. Chem. **44** [1972] 857/60).
[27] H. Chermette, C. Martelet, D. Sandino, J. Tousset (J. Inorg. Nucl. Chem. **34** [1972] 1627/38).
[28] J. Dahlmann, K. Winsel (J. Prakt. Chem. **319** [1977] 201).
[29] L. Kaufmann (Ger. 606777 [1934]; C.A. **1935** 3783).
[30] J. I. Harris, S. T. Bowden, W. J. Jones (J. Chem. Soc. **1947** 1568/71).

[31] C. Löloff (Ber. Deut. Chem. Ges. **30** [1897] 2834/43).
[32] J. M. Keck, G. Klar (Z. Naturforsch. **27 b** [1972] 591/5).
[33] A. Michaelis, U. Genzken (Liebigs Ann. Chem. **242** [1887] 164/188).
[34] N. Benmalek, H. Chermette, C. Martelet, D. Sandino, J. Tousset (J. Inorg. Nucl. Chem. **36** [1974] 1365/8).
[35] D. E. Worrall (J. Am. Chem. Soc. **52** [1930] 2046/50).
[36] A. Michaelis, U. Genzken (Ber. Deut. Chem. Ges. **17** [1884] 924/5).
[37] G. L. Kuykendall, J. L. Mills (J. Organometal. Chem. **118** [1976] 123/8).
[38] G. S. Zhdanov, Z. P. Razmanova (Dokl. Akad. Nauk SSSR **72** [1950] 1055/7).

[39] V. P. Glushkova, T. V. Talalaeva, Z. P. Razmanova, G. S. Zhdanov, K. A. Kocheshkov (Sb. Statei Obshch. Khim. Akad. Nauk SSSR **2** [1953] 992/6).
[40] A. E. Goddard, V. E. Yarsley (J. Chem. Soc. **1928** 719/23).
[41] L. Kaufmann (U.S. 1917207 [1933]; C.A. **1933** 4630).
[42] N. I. Trofimova, V. E. Zhuravlev, E. N. Sinotova, N. E. Shchepina, M. V. Moshkovskaya (Tr. Estestvennonauchn. Inst. Permsk. Gas. Univ. **13** [1975] 187/93 from Ref. Zh. Khim. B **22** [1976] 652).

2.5.1.1.5 Triorganoantimony Bis(pseudohalides) $R_3Sb(N_3)_2$, $R_3Sb(NCO)_2$, and $R_3Sb(NCS)_2$

General Remarks. The compounds were found to dissolve as monomers in C_6H_6 [2, 3, 16] and CH_3NO_2 [2]. The molecular nature in solution is also confirmed by electrical conductivity measurements in CH_3CN [2, 3] and CH_3NO_2 [2].

$(CH_3)_3Sb(N_3)_2$

Trimethylantimony diazide is obtained by reacting $(CH_3)_3SbCl_2$ with an excess (1:4) of NaN_3 in 1,2-dichloroethane for 24 h at room temperature [1] or in C_6H_6 [2]. The latter solvent is also used for reacting $(CH_3)_3SbCl_2$ with a solution of 20% HN_3 and an excess of NaN_3 [3]. Concentration of the filtrates in vacuum gives the compound as white crystals in yields of about 80% [1]. Recrystallized from C_6H_6 [1, 2] or C_6H_6/HN_3 [3] the substance melts at 91 °C [1] or 91.5 to 92.5 °C [2, 3].

IR spectra [1 to 3] and Raman spectra [3] were measured. Because the substance is very hygroscopic [2] the first measurements [1] are questioned in [2]. The following IR vibrations (in cm^{-1}) are given: 2064vs ($\nu_{as}N_3$) in $CHCl_3$ and 353m (νSbN) in C_6H_6; 2080vs ($\nu_{as}N_3$), 1285m (ν_sN_3), 685m (δN_3), and 576m (ν_{as}SbC) in Nujol. The Raman bands (in cm^{-1}) are in C_6H_6: 576m (ν_{as}SbC), 529s,p (ν_sSbC), 355m,p (νSbN), and in the solid state: 2080w ($\nu_{as}N_3$), 1290w (ν_sN_3), 656w (δN_3), 220m (δSbC out-of-plane), 150m (δSbC in-plane), and 100vs (lattice mode?) [3], see also [2]. The ^1H NMR spectrum shows a singlet at $\delta=1.94$ (in CH_2Cl_2) [1] or 1.97 (in $CDCl_3$) ppm [3]. The ^{121}Sb Mössbauer spectrum at 4.2 K vs. InSb gives values of $\delta=3.06$, $e^2qQ=-21.22$, and $\Gamma=3.54$ mm/s [4]. The compound reacts with $SbCl_5$ in CH_2Cl_2 at 0 °C to give 37% $[(CH_3)_3SbN_3][SbCl_5N_3]$ [5].

$(C_6H_5)_3Sb(N_3)_2$

The compound is obtained by reacting $(C_6H_5)_3SbCl_2$ with NaN_3 (ratio 1:4) in C_6H_6 [2] or from the same reactants with addition of a 20% HN_3/C_6H_6 solution [3]. After concentration of the filtrate, white crystals deposit; they melt at 103.5 to 104.5 °C [2] or 104.5 to 105.5 °C [3]. A compound with a melting point of 138 °C, prepared by reaction of $(C_6H_5)_3SbO$ and HN_3 in ether/water [6], is probably a partially hydrolyzed substance [2].

The following IR vibrations (in cm^{-1}) are given: 2072 ($\nu_{as}N_3$) in $CHCl_3$; 355s (νSbN), 298s (ν_{as}SbC), and 265m in C_6H_6; and 2080vs ($\nu_{as}N_3$), 1268m (ν_sN_3), 648m (δN_3), 458s; 230w (ν_sSbC) in Nujol. The Raman bands (in cm^{-1}) are in C_6H_6: 360m,p (νSbN), 290w (ν_{as}SbC), and 230s,p (ν_sSbC); and in the solid state: 2080vw ($\nu_{as}N_3$), 1270w (ν_sN_3), 650w (δN_3), 455vw, 265w, 205w, and 160w [3], see also [2]. In the ^1H NMR spectrum (in $CDCl_3$) signals are observed at $\delta=7.64$ (t, H-3,4,5) and 8.26 (q, H-2,6) ppm [3]. The compound is extremely sensitive towards hydrolysis [2, 3].

$(CH_3)_3Sb(NCO)_2$

$(CH_3)_3SbCl_2$ and AgOCN are reacted in ether for 24 h. The filtrate of the mixture is concentrated in a vacuum and the remaining residue recrystallized from a $CHCl_3$/petroleum ether mixture. The compound melts at 53 to 54 °C [2, 3].

The following IR vibrations (in cm^{-1}) are given: 2200vs (ν_{as}NCO) in $CHCl_3$, 325m (νSbN) in C_6H_6; 2200vs (ν_{as}NCO), 1365w (ν_sNCO), 620s (δNCO), 588m (ν_{as}SbC) in Nujol. The Raman bands (in cm^{-1}) for the solid state are: 1364w (ν_sNCO), 588 (ν_{as}SbC), 542vs (ν_sSbC), 320m (νSbN), 201m (δSbC out-of-plane), 149m (δSbC in-plane), 100vs (lattice mode?) [3], see also [2]. A ^{1}H NMR spectrum in $CDCl_3$ shows a singlet at $\delta = 1.97$ ppm [3]. The compound is extremely sensitive towards hydrolysis [2, 3].

Other $R_3Sb(NCO)_2$ Compounds (R = alkyl)

The compounds summarized in Table 23 are prepared by the following methods:

Method I: The corresponding R_3SbO and $(NH_2)_2CO$ are heated without a solvent at ca. 130 °C [7 to 9]. The yields are 85 to 95% [8].

Method II: The corresponding R_3SbO is reacted with HOCN in an inert solvent, e.g. C_6H_6, at >60 °C [7, 10].

Method III: The corresponding R_3SbCl_2 is reacted with NaOCN in refluxing CH_3CN [7, 8, 11]. The yields are excellent [8].

Method IV: The corresponding R_3SbCl_2 reacts with $(NH_2)_2CO$ at 140 °C within 2 h [7, 8, 11]. The yields are 20 to 30% [8].

Table 23
Trialkylantimony Diisocyanates $R_3Sb(NCO)_2$.

No.	R	methods of preparation [Ref.]	properties and remarks [Ref.]
1	C_2H_5	I, II, III [7]	–
2	C_4H_9	I [7 to 9], II [10], III [7], IV [11]	b.p. 122 °C/0.2 Torr [10, 11] IR: 2173.8 cm^{-1} [8] distillation with dec. at 190 °C [7, 8]; gives carbamates with alcohols and phenols [8] useful as fungicide, herbicide, insecticide, and as a catalyst for polyurethane foam [9, 10]
3	i-C_4H_9	I [7, 9], II [10], III [11], 4 [7, 8]	b.p. 122 to 124 °C [7, 8, 10, 11] IR: 2173.8 cm^{-1} [8] $n_D^{20} = 1.5128$ [7, 8, 11] same uses as above
4	C_8H_{17}	IV [11]	viscous oil [11]

$(C_6H_5)_3Sb(NCO)_2$

Triphenylantimony diisocyanate is prepared by shaking $(C_6H_5)_3SbCl_2$ with AgOCN in ether for 14 to 24 h. Concentration of the filtrate of the mixture gives about 70% of white crystals [2, 3, 12].

The compound melts at 109 to 110 °C [2], 113 to 114 °C [3], or 111 to 112 °C [12] recrystallized from $CHCl_3$/petroleum ether [3, 12]. The following IR vibrations (in cm^{-1}) were found: 2208vs (ν_{as}NCO) in $CHCl_3$; 355m (νSbN), 298s (ν_{as}SbC), 265m in C_6H_6; 2210vs (ν_{as}NCO), 633m (δNCO), 458s in Nujol. The Raman bands (in cm^{-1}) for the solid state are: 1370 (ν_sNCO), 638vw (δNCO), 455vw; 338m (νSbN), 290w (ν_{as}SbC), 268w; 228s (ν_sSbC), 210w,

References on p. 106

170m [3], see also [2]. Two ^1H NMR signals are observed in $CDCl_3$ at $\delta=7.70$ (t, H-3,4,5) and 8.25 (q, H-2,6) ppm [3].

The compound crystallizes in the monoclinic space group $P2_1/n-C^5_{2h}$ (No. 14) with unit cell dimensions a = 11.920(3), b = 12.693(8), and c = 12.270(13) Å, $\beta=90.22(1)°$; Z = 4; d_m = 1.565 and d_c = 1.572 g/cm³. The structure was refined to R = 0.039 for 1643 reflections. The structure of the molecule with the important bond distances and angles is shown in **Fig. 8**. Antimony has a trigonal bipyramidal configuration with the nearly linear NCO groups in axial positions. Phenyl rings I and III are rotated through 88° and 70°, respectively, in the same sense, while ring II is rotated 45° in the opposite direction from the plane through C(11), C(21), and C(31). Both NCO groups are inclined in the direction of phenyl ring II. The shortest intermolecular contacts occur between the terminal oxygen atoms and neighboring phenyl rings. O(1)···C(33') = 3.351 Å, and O(2)···C(15') = 3.346 Å. These and all other intermolecular contacts correspond to, or are greater than, van der Waals interactions [13].

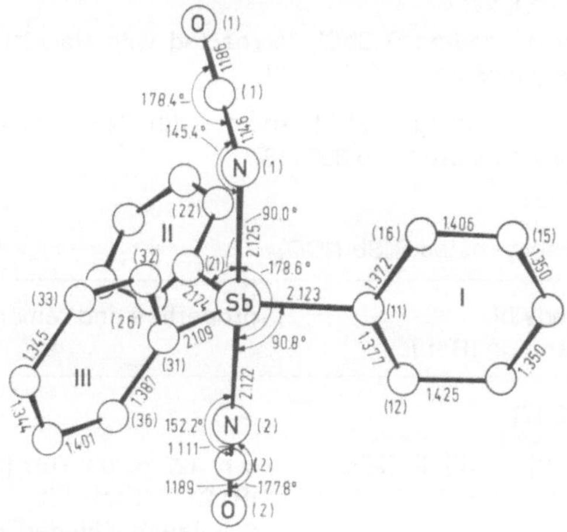

Fig. 8. Molecular structure of $(C_6H_5)_3Sb(NCO)_2$ [13].

Intramolecular non-bonded distances <3.5 Å:

Sb···C(12)	3.054	N(2)···C(11)	3.024
Sb···C(16)	3.041	N(2)···C(12)	2.992
Sb···C(22)	3.066	N(2)···C(21)	2.983
Sb···C(26)	3.048	N(2)···C(26)	3.169
Sb···C(32)	3.048	N(2)···C(31)	2.983
Sb···C(36)	3.058	N(2)···C(36)	2.998
N(1)···C(11)	3.018	C(1)···C(21)	3.405
N(1)···C(16)	2.967	C(1)···C(22)	3.202
N(1)···C(21)	3.004	C(1)···Sb	3.136
N(1)···C(22)	3.181	C(2)···C(21)	3.497
N(1)···C(31)	2.987	C(2)···C(26)	3.379
N(1)···C(32)	3.001	C(2)···Sb	3.148

$(CH_3)_3Sb(NCS)_2$

Trimethylantimony diisothiocyanate is obtained by reacting $(CH_3)_3SbCl_2$ with KSCN, heated in C_2H_5OH [15], or with NaSCN or AgSCN in CH_3OH or $(CH_3)_2CO$. The resulting

References on p. 106

compound melts at 166 to 167 °C [2]. Freshly prepared $(SCN)_2$ (from AgSCN and Br_2 in C_6H_6) reacts with $Sb(CH_3)_3$ in C_6H_6 at 0 °C to form the compound with a melting point of 168 °C [16].

IR [2, 16 to 18] and Raman spectra [17, 18] were measured and assigned; from these data an Sb-N bond (isothiocyanate) is concluded. The following low frequency IR (Nujol) and Raman (solid) vibrations (in cm^{-1}) are observed [17]:

IR	Raman	assignment
584 s	585 m	ν_{as}SbC
	535 vs	ν_sSbC
492 s	492 w	δ_{as}NCS
482 m	483 vw	δ_sNCS
268 vs		ν_{as}SbN
	194 m	ν_sSbN
180 s		δSbC out-of-plane
152 m		δSbC in-plane

The compound dissolves easily in organic solvents and hot water [15]. It slowly decomposes at room temperature and also under N_2 [16].

$(C_6H_5)_3Sb(NCS)_2$

$Sb(C_6H_5)_3$ and freshly prepared $(SCN)_2$ (from $Pb(SCN)_2$ and Br_2 in ether or AgSCN and Br_2 in C_6H_6) react in ether (5 h) [19] or in C_6H_6 at 0 °C [16] to form the compound. In the first case, the substance is precipitated by adding petroleum ether to the mixture. The same reactants in CH_3CN at −5 °C for one hour give a 76% yield of the compound after evaporation of the solvent [20]. From the stibine and $Hg(SCN)_2$ in refluxing $(CH_3)_2CO$, one obtains 70 to 85% of the substance after evaporation of the solvent from the filtrate [21]. $(C_6H_5)_3SbCl_2$ and NaSCN or AgSCN in CH_3OH or $(CH_3)_2CO$ [2] as well as $(C_6H_5)_3SbCl_2$ and $Pb(SCN)_2$ form the title compound, after 8 h shaking in C_6H_6 [19].

Melting points of the compound are given as 104 °C from petroleum ether [20, 21], 104 to 105 °C [2], 105 to 106 °C from C_2H_5OH [19], and 108 °C [16]. IR [2, 16 to 18] and Raman spectra [17, 18] are discussed in several publications; based upon an analysis of these data, an Sb-N bond is concluded. Low frequency IR bands (in cm^{-1}) in Nujol: 496w (δ_{as}NCS), 480m (δ_sNCS), 455vs, 294vs, 254m, 230w; 187m (δSbC and δSbN), and 160w. The Raman bands (in cm^{-1}) for the solid are: 496s (δ_{as}NCS), 486m (δ_sNCS), 455w, 400vw, 264m, and 219vs [17]. The ^{121}Sb Mössbauer spectrum (77K source; 9K absorber vs. $Ca^{121}SnO_3$) gives the following data (in mm/s): $\delta = 5.6 \pm 0.1$, $e^2qQ = -20.4 \pm 0.7$, and $\Gamma = 2.6$. In accordance with the electric field gradient model, a trigonal bipyramidal geometry of the molecule is established [22]. Calculation of the quadrupole coupling constant is in line with this geometry [23].

The compound slowly decomposes at room temperature and also under N_2 [16]. It reacts with boiling water to give $(C_6H_5)_3Sb(OH)NCS$ [19]. With $Pb_2(C_6H_5)_6$ in a ratio of 1:1 at 6 h in refluxing $CHCl_3$, 78% $Pb(C_6H_5)_4$, 65% $(C_6H_5)_2Pb(NCS)_2$, and 54% $Sb(C_6H_5)_3$ are obtained after work up [24].

$(4\text{-}CH_3C_6H_4)_3Sb(NCS)_2$

$Sb(C_6H_4CH_3\text{-}4)_3$ and $Hg(SCN)_2$ react in refluxing $(CH_3)_2CO$. After filtering the $Hg_2(SCN)_2$ precipitate, the filtrate is concentrated to yield 70 to 85% of the compound with a melting

References on p. 106

point of 148 °C [21]. The corresponding stibine reacts with $(SCN)_2$ in CH_3CN, 1 h at −5 °C, in a similar yield to the title compound. It is isolated by driving off the solvent and by recrystallization from petroleum ether [20].

References:

[1] A. Schmidt (Chem. Ber. **101** [1968] 3976/80).
[2] R. G. Goel, D. R. Ridley (Inorg. Nucl. Chem. Letters **7** [1971] 21/3).
[3] R. G. Goel, D. R. Ridley (Inorg. Chem. **13** [1974] 1252/5).
[4] K. Dehnicke, K. Fleck, K. Schmidt, J. Pebler (Z. Anorg. Allgem. Chem. **451** [1979] 109/14).
[5] A. Schmidt (Chem. Ber. **101** [1968] 4015/21).
[6] J. S. Thayer (Organometal. Chem. Rev. **1** [1966] 157/78).
[7] W. Stamm (Trans. N.Y. Acad. Sci. [2] **28** [1966] 396/401).
[8] W. Stamm (J. Org. Chem. **30** [1965] 693/5).
[9] Stauffer Chemical Co. (Neth. 6411266 [1965]; C.A. **63** [1965] 11613).
[10] W. Stamm, Stauffer Chemical Co. (U.S. 3417115 [1965/68]; C.A. **70** [1969] No. 37925).
[11] W. Stamm, Stauffer Chemical Co. (Ger. 1229529 [1966]; C.A. **66** [1967] No. 28896).
[12] F. Challenger, V. K. Wilson (J. Chem. Soc. **1927** 209/13).
[13] G. Ferguson, R. G. Goel, D. R. Ridley (J. Chem. Soc. Dalton Trans. **1975** 1288/90).
[14] D. R. Ridley (Diss. Guelph Univ., Canada, 1973; Diss. Abstr. Intern. B **34** [1974] 5374; C.A. **81** [1974] No. 55362).
[15] A. Hantzsch, H. Hibbert (Ber. Deut. Chem. Ges. **40** [1907] 1508/19).
[16] T. Wizemann, H. Müller, D. Seybold, K. Dehnicke (J. Organometal. Chem. **20** [1969] 211/7).
[17] R. G. Goel, E. Maslowsky Jr., C. V. Senoff (Inorg. Chem. **10** [1971] 2572/7).
[18] R. G. Goel, E. Maslowsky Jr., C. V. Senoff (Inorg. Nucl. Chem. Letters **6** [1970] 833/5).
[19] F. Challenger, A. L. Smith, F. J. Paton (J. Chem. Soc. **123** [1923] 1046/54).
[20] S. N. Bhattacharya, M. Singh (Indian J. Chem. A **16** [1978] 778).
[21] S. N. Bhattacharya, M. Singh (Indian J. Chem. A **18** [1979] 515/6).
[22] J. N. R. Ruddick, J. R. Sams, J. C. James (Inorg. Chem. **13** [1974] 1503/7).
[23] G. M. Bancroft, V. G. K. Das, T. K. Sham, M. G. Clark (J. Chem. Soc. Dalton Trans. **1976** 643/54).
[24] S. N. Bhattacharya, A. K. Saxena (Indian J. Chem. A **17** [1979] 307/9).

2.5.1.1.6 Triorganoantimony Compounds R_3SbX_2 with X or X_2 = Anions of Inorganic Acids

General Remark. The compounds R_3SbX_2 with X = bivalent substituent (e. g. S_2O_4, SO_4, SeO_4, CrO_4, CO_3, and derivatives) are polymeric and covalently bonded in all known cases.

$(CH_3)_3Sb(ClO_4)_2$

From the reaction of $(CH_3)_3SbBr_2$ and $AgClO_4$ in anhydrous CH_3OH a white solid was obtained which immediately and violently explodes [1]. Its Raman spectrum in aqueous solution shows the following vibrations: 3034 s, br (dp), 2939 vs (p), 1408 vw, br, 1249 m (p), 1236 m, sh (dp), 1089 w, br, 932 vs (p), 830 vw, 623 m (dp), 585 s (dp), 537 vs (dp), 461 m (dp), and 168 s (dp) cm^{-1}. From the assignment of these data, a planar $[(CH_3)_3Sb]^{2+}$ cation is concluded [2]. This cation is probably hydrated with two H_2O ligands, forming a trigonal bipyramidal geometry [3].

$[R_3SbD_2][ClO_4]_2$

The known complexes of this type are summarized in Table 24. They are prepared from the corresponding R_3SbCl_2 with stoichiometric amounts of ligand D and $AgClO_4$ and reacted for two hours in C_6H_6; the precipitate is extracted with C_2H_5OH. Upon concentration of the latter, the compounds are obtained [3]. See Table 24 for the physical data.

Table 24
Physical Properties of $[R_3SbD_2][ClO_4]_2$ Complexes [3].

No.	D ligand	m.p. in °C	IR (in Nujol) νXO	νSbO	νClO$_4$ (X=S, P, As), $\tilde{\nu}$ in cm^{-1}	^{1}H NMR (in CH_3NO_2) δ in ppm
	with R=CH_3					
1	$(CH_3)_2SO$	109 to 128 (dec.)	939 s	454 m	1085 vs, 623 s	2.31 (CH_3Sb), 3.06 (CH_3S)
2	$(CD_3)_2SO$		950 m	435 m	1100 vs, 625 m	
3	$(C_6H_5)_3PO$	114 to 119 (dec.)	1127 s		1090 vs, 625 m	2.20 (CH_3Sb)
4	$(C_6H_5)_3AsO$	122 to 126 (dec.)	800 m	405 m	1090 vs, 625 m	1.96 (CH_3Sb)
	with R=C_6H_5					
5	$(CH_3)_2SO$	207 to 208	875 s	476 m	1095 vs, 625 s	2.85 (CH_3S)
6	$(CD_3)_2SO$	204	870 m	461 m	1100 vs, 625 s	
7	$(C_6H_5)_2SO$	197 to 199	877 s		1090 vs, 625 m	
8	$(C_6H_5)_3PO$	257	1125 s		1090 vs, 625 m	
9	$(C_6H_5)_3AsO$	190	802 m	410 m	1090 vs, 625 m	

The molar conductivity of ca. 10^{-3} M solutions in CH_3NO_2 ranged from 150.0 to 183.0 cm^2 · Ω^{-1} · mol^{-1} for these compounds. These values and additional IR measurements of CH_3NO_2 solutions indicate that the complexes behave as 1:2 electrolytes with no dissociation of the D ligand in the solvent. The complexes must be considered as potentially explosive [3]!

$(C_6F_5)_3Sb(ClO_4)_2$

$(C_6F_5)_3SbCl_2$ and $AgClO_4$ are reacted for two hours in C_6H_6. Upon concentration of the filtrate and cooling, 48% yield of a white crystalline solid with a melting point of 95 °C is obtained. The conductivities of 5×10^{-4} M solutions in $(CH_3)_2CO$ or CH_3NO_2 are Λ_M=260.43 and 178.35 cm^2 · Ω^{-1} · mol^{-1}, respectively. This shows that the compound is completely dissociated in solution. From an IR spectrum of the solid it is concluded that the compound exists as a covalent pentacoordinate molecule. With aqueous $(CH_3)_2CO$, the compound forms $[(C_6F_5)_3SbClO_4]_2O$ in a yield of 56% [4].

$(CH_3)_3Sb(NO_3)_2$

Trimethylantimony dinitrate is obtained by reacting $Sb(CH_3)_3$ with dilute HNO_3. NO is evolved, and colorless crystals are formed upon cooling the mixture. $(CH_3)_3SbI_2$ [5] or $(CH_3)_3SbBr_2$ react with $AgNO_3$ in hot water and the compound remains after concentration of the filtrate [6]. The substance is formed similarly from CH_3OH solution [1]. From $(CH_3)_3SbO \cdot H_2O$ and stoichiometric amounts of 60% aqueous HNO_3 in $(CH_3)_2CO$, the compound is isolated by evaporation of the solvent and recrystallized from CH_3OH [7].

The crystals melt at 149 °C (from CH_3OH) [7] or 149 to 150 °C (from C_2H_5OH) [8]. IR spectra are published and discussed in several publications [1, 2, 6, 7, 9, 10]. From these

References on p. 112

measurements it was first assumed [6] that the compound has an ionic structure in the solid state. This assumption was called in question by other authors [1, 7, 9, 10]. The IR vibrations listed in Table 25 show that the compound is molecular in the solid or in inert solvents with a trigonal bipyramidal environment of the Sb atom with the NO_3 substituents in axial and the CH_3 groups in equatorial positions [7]. A Raman spectrum of the solid [9, 10] leads to the same conclusion. In contrast to the behavior in inert solvents, the substance dissociates in H_2O. Beside a nitrate ion, a solvated $[(CH_3)_3Sb]^+$ cation, probably with trigonal bipyramidal geometry, is suggested from the Raman measurements of aqueous solutions [2] (see Table 25). Force constants for ν_sSbC were calculated as k=2.55 [2] and 2.44 mdyn/Å [9, 10]. A ^1H NMR spectrum of the compound dissolved in $CDCl_3$ shows a singlet at δ=2.17 ppm [8].

The compound dissolves as a monomer in C_6H_6 [7]. The substance reacts with KBr in aqueous solution to form $(CH_3)_3SbBr_2$ [7]. Its reaction with $(CH_3)_3SbCl_2$ in $C_6H_5NO_2$ at 32 °C leads to an equilibrium with $2(CH_3)_3Sb(Cl)NO_3$. An equilibrium constant of 20 was calculated from ^1H NMR data [11].

Table 25
IR and Raman Vibrations of $(CH_3)_3Sb(NO_3)_2$.

IR (in cm^{-1}) [7] mull	$CHCl_3$ or $CHBr_3$	assignment	Raman (in cm^{-1}) [2] H_2O	assignment
3040 w	3030 w		3034 s, br, dp	$\nu_{as}CH_3$
2940 vw	2947 vw	ν_sCH_3	2940 vs, p	ν_sCH_3
2797 vw	2817 vw			
2016 vw				
1933 vw				
1812 vw				
1761 vw	1773 vw			
1672 vw				
1545 sh, s; 1527 s	1536 s	$\nu_{as}NO_2$		
1404 w	1403 w	$\delta_{as}CH_3$	1409 w, br	$\delta_{as}CH_3$
			1386 w, sh	νNO_2
1290 s; 1274 s	1282 s	ν_sNO_2		
1242 w	1244 w		1249 m, p	δ_sCH_3
1229 w	1230 w	δ_sCH_3	1237 m, sh, dp	
968 s; 959 s	954	νNO	1048 s, p	νNO
864 s	861 m	ϱCH_3		
793 w	794 w	δNO_3		
729 m	725 w			
707 w	702 vw		716 w, br	δNO_2
			675 vvw	ϱCH_3
580 m	573 w	$\nu_{as}SbC$	582 s, dp	$\nu_{as}SbC$
529 vw			536 vs, p	ν_sSbC
521 vw				
508 vw				
			166 s, dp	δSbC in-plane

References on p. 112

$(C_2H_5)_3Sb(NO_3)_2$

$Sb(C_2H_5)_3$ [12] or $(C_2H_5)_3SbO$ and dilute aqueous HNO_3 form the compound, which is isolated by concentration of the solution [12 to 14]. It is also obtained by reaction of $(C_2H_5)_3SbI_2$ with $AgNO_3$ in dilute HNO_3 [13, 14].

Recrystallization from H_2O gives large rhombohedral crystals [12 to 14] which melt at 62.5 °C [12]. The crystals are easily soluble in H_2O, less soluble in C_2H_5OH, and hardly soluble in ether [12]. Concentrated H_2SO_4 gives $(C_2H_5)_3SbSO_4$, and concentrated HCl the dichloride. The compound is suitable for the preparation of $(C_2H_5)_3SbX_2$ (X=halide) [12].

$(C_5H_{11})_3Sb(NO_3)_2$

The compound is obtained as colorless crystals from the reaction of $(C_5H_{11})_3SbX_2$ (X=Cl, Br, I) with $AgNO_3$ in C_2H_5OH. It has a melting point of ca. 20 °C and is soluble in aqueous C_2H_5OH, insoluble in H_2O and ether [15].

$(C_6H_5)_3Sb(NO_3)_2$

Triphenylantimony dinitrate is prepared by treating $Sb(C_6H_5)_3$ with hot fuming HNO_3. Upon cooling, colorless plates crystallize which are then recrystallized from C_2H_5OH [16, 17]. Similarly, the compound is formed from $(C_6H_5)_3Sb(OH)_2$ with hot HNO_3 [16]. An excess of liquid N_2O_4 and $(C_6H_5)_3SbCl_2$ give, after cooling to 0 °C and concentration of the mixture, the title compound [18]. The usual way for the preparation of dinitrates, the reaction of $(C_6H_5)_3SbCl_2$ with $AgNO_3$ in C_2H_5OH, leads only to a mixture of the compound with $(C_6H_5)_3Sb(OH)NO_3$ [19].

Melting points of the compound are given as 156 °C (from C_2H_5OH) [16], 143 to 145 °C (from HNO_3) [17], or 144 to 146 °C [18]. IR spectra [9, 10, 17, 18] and Raman spectra [9, 10] were measured and assigned. The following IR vibrations (in cm^{-1}) were found (in Nujol): 1069 m, 1060 m, 789 m, 730 s, 708 w, 684 s, 612 w, 458 s, 450 s, 395 w, 300 s, 292 s, 270 m [18]; 460 vs, 295 vs; 275 vs (ν_{as}SbO), 242 m, 225 w, 214 m; 198 m, 180 w (δSbO and δSbC), 158 w, and 140 m [9, 10]. Raman vibrations (in cm^{-1}) are for the solid: 272 s (ν_sSbO) and 216 vs [9, 10]. From these data a molecular structure with trigonal bipyramidal geometry is supposed [9, 10, 17, 18]. The same is concluded from the ^{121}Sb Mössbauer resonances of the compound (source 77 K, absorber 9 K vs. $Ca^{121}SnO_3$): $\delta=-5.7\pm0.1$, $e^2qQ=-21.3\pm1.0$, and $\Gamma=3.0$ mm/s [20]. A quadrupole coupling constant was calculated on the basis of the molecule model as $e^2qQ=-22.8$ mm/s [21].

The compound is soluble in C_2H_5OH and hot concentrated HNO_3, but is insoluble in H_2O [16]. It reacts with wet C_2H_5OH or wet CH_3CN to give $[(C_6H_5)_3SbNO_3]_2O$ [18].

$(C_6F_5)_3Sb(NO_3)_2$

$(C_6F_5)_3SbCl_2$ and $AgNO_3$ react in C_6H_6 for 2 h at room temperature. The filtrate of the mixture is concentrated, and upon cooling, the compound is obtained in 40% yield. It dissolves undissociated in C_6H_6. Conductivity of a 5×10^{-4} M solution in this solvent is $\Lambda_M=0.09\cdot cm^2\cdot\Omega^{-1}\cdot mol^{-1}$. The compound reacts with H_2O to form $[(C_6F_5)_3SbNO_3]_2O$ [4].

$(4\text{-}CH_3OC_6H_4)_3Sb(NO_3)_2$ and **$(4\text{-}C_2H_5OC_6H_4)_3Sb(NO_3)_2$**

Both compounds are synthesized by reacting the corresponding dibromide with $AgNO_3$ in C_2H_5OH. Concentration of the filtrates gives the compounds. $(4\text{-}CH_3OC_6H_4)_3Sb(NO_3)_2$ is purified by crystallization from $CHCl_3$/ether to yield short needles, m.p. 217 °C with

References on p. 112

decomposition; $(4\text{-}C_2H_5OC_6H_4)_3Sb(NO_3)_2$ has a melting point of 151 to 152 °C and decomposes at 170 °C [22].

$(3\text{-}NO_2\text{-}4\text{-}CH_3C_6H_3)_3Sb(NO_3)_2$

$Sb(C_6H_4CH_3\text{-}4)_3$ is treated with a calculated amount of cold fuming HNO_3. Pouring the mixture in cold H_2O gives a precipitate which was recrystallized several times from C_2H_5OH. It melts at 182 °C [23].

The compound reacts with H_3PO_4 in absolute C_2H_5OH to give the corresponding triarylantimony oxide. With NH_4Cl, NH_3, and zinc dust, refluxed for 8 h in 80% C_2H_5OH, it is reduced to $Sb(C_6H_3CH_3\text{-}4\text{-}NH_2\text{-}3)_3$; with PBr_5, an Sb-C cleavage occurs to give 4-bromo-2-nitrotoluene [23].

$(2,4\text{-}(CH_3)_2\text{-}5\text{-}NO_2C_6H_2)_3Sb(NO_3)_2$

$Sb(C_6H_3(CH_3)_2\text{-}2,4)_3$ is heated on a water bath for 20 min with HNO_3 (density 1.16). Pouring the mixture into H_2O gives a precipitate which was recrystallized six times from C_2H_5OH. The pure substance melts at 175 °C [24].

The compound reacts like the previous one with H_3PO_4, zinc dust, or PBr_5 to give analogous products. In addition to these reactions, it may be reduced by H_2S in an ethanolic NH_3 solution to $Sb(C_6H_2NO_2\text{-}5\text{-}(CH_3)_2\text{-}2,4)_3$ [24].

$(C_6H_5)_3SbS_2O_4$

$Sb(C_6H_5)_3$ dissolves in liquid SO_2 to yield a yellow-colored solution containing the adduct $(C_6H_5)_3Sb \cdot SO_2$. Upon standing for eight months the title compound precipitates [25].

The compound decomposes upon heating to yield $C_6H_5SbSO_2$ and $(C_6H_5)_2SO_2$. With HBr in C_2H_5OH, $(C_6H_5)_2S_2$, and with HCl in the same solvent, $(C_6H_5)_3SbCl_2$ is formed [25].

$(CH_3)_3SbSO_4$

$(CH_3)_3SbX_2$ (X=Br, I) and Ag_2SO_4 are reacted in hot aqueous solution. Concentrating the filtrate gives the compound as a crystalline crust [5, 6]. IR spectra of the compound were recorded and discussed [1, 6]. The following vibrations are found (in cm^{-1}, as mull or in KBr): 3050 w, 2950 w (νCH), 1415 m (δ_{as}CH), 1285 s ($\nu_{as}SO_2$), 1230 m (δ_sCH), 1145 s (ν_sSO_2), 950 s ($\nu_{as}SO_2$), 860 s (ϱCH_3), 825 s (ν_sSO_2), 650 s (ϱSO_4), 600 s (δSO_2), 495 m (ϱSO_4), 428 w (δSO_2), and 250 s (lattice mode). From these vibrations, a covalent polymeric structure with bridging sulfate groups is suggested. In KBr an anion exchange is observed within a few hours [1].

$(C_2H_5)_3SbSO_4$

$(C_2H_5)_3SbI_2$ reacts with Ag_2SO_4 [13, 14] and $(C_2H_5)_3SbS$ reacts with $CuSO_4$ [12] in aqueous solution to form the compound, which is isolated by evaporation of the filtrate as a gummy mass [13, 14] or as small white crystals [12]. $(C_2H_5)_3SbBr_2$ and Ag_2SO_4 in 70% C_2H_5OH give the compound in 45% yield [6].

The compound is very soluble in H_2O, soluble in C_2H_5OH, but hardly soluble in ether [12]. It is hygroscopic and reacts with $BaBr_2$ in water to form $(C_2H_5)_3SbBr_2$ [13, 14].

$(C_5H_{11})_3SbSO_4$

$(C_5H_{11})_3SbX_2$ (X=Cl, Br, or I) and Ag_2SO_4 in C_2H_5OH give the compound as an oily residue after evaporating the solvent from the filtrate [15].

References on p. 112

$(C_6H_5)_3SbSO_4$

A preparation of the compound from $Sb(C_6H_5)_3$ and concentrated H_2SO_4 [24] was called in question subsequent to a reinvestigation [17]. The best method to prepare the substance is to react $Sb(C_6H_5)_3$, dissolved in 1,2-dichloroethane, with SO_3/N_2 under cooling (ice/NaCl). Addition of a 20-fold amount of ether gives a precipitate in a yield of 89%, which melts at 300 to 310 °C [27]. From $(C_6H_5)_3SbCl_2$ and concentrated H_2SO_4 the compound is also obtained with evolution of HCl. The resulting solid is treated with ether until it is white. No suitable solvent for recrystallization was found [17].

IR bands associated with the sulfate group are (as a mull): 1282 s, 1140 s, 950 to 850 s, and 623 m (br) cm^{-1} [17]. Considering the IR spectra [17, 27], the high melting point, and the slight solubility even in polar solvents, a polymeric structure is possible [17].

The compound reacts with aqueous NaOH [24] or hot H_2O to form $(C_6H_5)_3Sb(OH)_2$. With hot aqueous acetic acid, $(C_6H_5)_3Sb(O_2CCH_3)_2$ is formed [27].

$(CH_3)_3SbSeO_4$ and **$(C_6H_5)_3SbSeO_4$**

Both compounds are prepared from the appropriate organoantimony dihalide R_3SbX_2 (X=Cl, Br) with Ag_2SeO_4 in H_2O ($R=CH_3$) or CH_3OH ($R=C_6H_5$). The $(CH_3)_3Sb$ derivative decomposes at 380 °C without melting and shows IR vibrations for the selenate moiety (as a mull) at 955 s, 908 s, 780 to 720 s, 485 s, 438 m, 388 s, and 350 sh cm^{-1}. The $(C_6H_5)_3Sb$ derivative melts at 308 to 309 °C with decomposition and has IR bands for the selenate group (as a mull) at 950 s, 900 s, 802 s, 760 to 740 vs, 475 m, 445 s, 388 s, and 365 m cm^{-1}. The compounds have a nonionic polymeric structure [28].

$(CH_3)_3SbCrO_4$

$(CH_3)_3SbBr_2$ and Ag_2CrO_4 react in H_2O. Concentration of the filtrate gives the compound. From its IR spectra, a covalently bonded polymeric structure is concluded. The following vibrations (in cm^{-1}, as a mull) were observed: 3040, 2940 ($\nu_{as}CH$), 1460 ($\delta_{as}CH_3$), 1230 (δ_sCH_3), 852 (ϱCH_3), 575 ($\nu_{as}SbC$), 530 w (ν_sSbC or νSbO), and CrO_4 vibrations at 964 s, 940 s, 838 s, 700 s, 420 m, 390 m, 355 m, and 313 m [1].

$(C_6H_5)_3SbCrO_4$

$(C_6H_5)_3SbX_2$ (X=Cl, Br) and Ag_2CrO_4 give the compound in nitrobenzene. It was precipitated by adding excess petroleum ether. The compound decomposes without melting at 180 °C. IR vibrations of the nonionic polymeric substance (as a mull) are given as 960 s, 940 s, 830 sh, 780 vs, 420 w, 385 m, and 325 sh cm^{-1} for the CrO_4 part of the molecule [28].

$(CH_3)_3SbCO_3$

$(CH_3)_3SbBr_2$ and Ag_2CO_3 form the compound in liquid SO_2. An IR spectrum is given and discussed. From the following vibrations a polymeric structure with CO_3-bridged planar $Sb(CH_3)_3$ units is concluded. IR (in cm^{-1}, as a mull): 2950 m, 2880 sh, 1730 s, 1395 w, 1280 s, 1225 m, 1115 s, 1100 s, 875 s, 790 s, 740 s, 632 s, 575 s, 525 w, 510 w, 450 m, 375 m, and 250 s [1].

$(C_2H_5)_3SbCO_3$

The compound is mentioned as a syrupy mass, resulting from the reaction of $(C_2H_5)_3SbI_2$ with Ag_2CO_3 in aqueous solution [13, 14].

References on p. 112

Other R_3SbX_2 Compounds (X_2 = carbonate and derivatives)

The compounds summarized in Table 26 are described in the patent literature as four-membered heterocylic derivatives of carbonic acid with the general Formula I (X=O or S, Y=O, S or NR, Z=O or S, NR), but they are probably polymeric. They are prepared from the reactants shown in Table 26 in ether for 2 h at 20 °C and are claimed to be useful as biocides or as cellular plastic blowing agents [29].

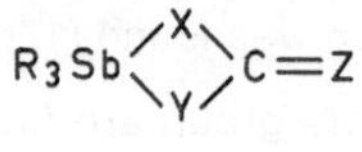

I

Table 26
R_3SbX_2 Compounds with X_2 = Carbonate and Derivatives [29].

No.	compound	preparation from	refractive index n_D
1	$(C_4H_9)_3SbCO_3$	$(C_4H_9)_3SbO + CO_2$	1.5141 (20 °C)
2	$(i\text{-}C_4H_9)_3SbCO_3$	$(i\text{-}C_4H_9)_3SbO + CO_2$	1.4896 (27 °C)
3	$(C_4H_9)_3SbCOS_2$	$(C_4H_9)_3SbO + CS_2$	1.5430 (26 °C)
4	$(i\text{-}C_4H_9)_3SbCOS_2$	$(i\text{-}C_4H_9)_3SbO + CS_2$	1.5318 (27 °C)
5	$(C_4H_9)_3SbCS_3$	$(C_4H_9)_3SbS + CS_2$	1.5630 (23 °C)
6	$(C_4H_9)_3SbCO_2N(C_2H_5)$	$(C_4H_9)_3SbO + C_2H_5N{=}C{=}O$	
7	$(C_4H_9)_3SbCO_2N(C_8H_{17})$	$(C_4H_9)_3SbO + C_8H_{17}N{=}C{=}O$	
8	$(C_4H_9)_3SbCO_2NC_6H_5$	$(C_4H_9)_3SbO + C_6H_5N{=}C{=}O$	
9	$(C_4H_9)_3SbCO_2NC_6H_4Cl\text{-}3$	$(C_4H_9)_3SbO + 3\text{-}ClC_6H_4N{=}C{=}O$	
10	$(i\text{-}C_4H_9)_3SbCO_2NC_6H_5$	$(i\text{-}C_4H_9)_3SbO + C_6H_5N{=}C{=}O$	

References:

[1] H. C. Clark, R. G. Goel (Inorg. Chem. **5** [1966] 998/1003).
[2] A. J. Downs, I. A. Steer (J. Organometal. Chem. **8** [1967] P21/P24).
[3] R. G. Goel, H. S. Prasad (J. Organometal. Chem. **59** [1973] 253/7).
[4] A. Otero, P. Royo (J. Organometal. Chem. **154** [1978] 13/9).
[5] H. Landolt (J. Prakt. Chem. **84** [1861] 328/39).
[6] G. G. Long, G. O. Doak, L. D. Freedman (J. Am. Chem. Soc. **86** [1964] 209/13).
[7] M. Shindo, R. Okawara (J. Organometal. Chem. **5** [1966] 537/44).
[8] R. L. McKenney, H. H. Sisler (Inorg. Chem. **6** [1967] 1178/82).
[9] R. G. Goel, E. Maslowsky Jr., C. V. Senoff (Inorg. Chem. **10** [1971] 2572/7).
[10] R. G. Goel, E. Maslowsky Jr., C. V. Senoff (Inorg. Nucl. Chem. Letters **6** [1970] 833/5).
[11] C. G. Moreland, R. J. Beam (Inorg. Chem. **11** [1972] 3112/4).
[12] C. Löwig, E. Schweizer (Liebigs Ann. Chem. **75** [1850] 315/55).
[13] W. Merck (J. Prakt. Chem. **66** [1855] 56/72).
[14] W. Merck (Liebigs Ann. Chem. **97** [1856] 329/33).
[15] F. Berlé (J. Prakt. Chem. **65** [1855] 385/418).
[16] A. Michaelis, A. Reese (Liebigs Ann. Chem. **233** [1886] 39/60).
[17] G. O. Doak, G. G. Long, L. D. Freedman (J. Organometal. Chem. **4** [1965] 82/91).
[18] G. C. Tranter, C. C. Addison, D. B. Sowerby (J. Organometal. Chem. **12** [1968] 369/76).

[19] G. T. Morgan, F. M. G. Micklethwait, G. S. Whitby (Proc. Chem. Soc. **25** [1909] 302; J. Chem. Soc. **97** [1910] 34/6).
[20] J. N. R. Ruddick, J. R. Sams, J. C. Scott (Inorg. Chem. **13** [1974] 1503/7).
[21] G. M. Bancroft, V. G. K. Das, T. K. Sham, M. G. Clark (J. Chem. Soc. Dalton Trans. **1976** 643/54).
[22] C. Löloff (Ber. Deut. Chem. Ges. **30** [1897] 2834/43).
[23] A. E. Goddard, V. E. Yarsley (J. Chem. Soc. **1928** 719/23).
[24] A. E. Goddard (J. Chem. Soc. **123** [1923] 2315/23).
[25] S. I. A. El Sheikh, B. C. Smith (Chem. Commun. **1968** 1474).
[26] P. May (Proc. Chem. Soc. **26** [1910] 218; J. Chem. Soc. **97** [1910] 1956/60).
[27] M. Becke-Goehring, H. Thielemann (Z. Anorg. Allgem. Chem. **308** [1961] 33/51).
[28] R.G. Goel, P. N. Joshi, D. R. Ridley, R. E. Beaumont (Can. J. Chem. **47** [1969] 1423/7).
[29] A. W. Breindel, S. Herbstman, Stauffer Chemical Co. (U.S. 3317575 [1963/67]; C.A. **67** [1967] No. 32780).

2.5.1.1.7 Triorganoantimony Oxides and Dihydroxides

General Remarks. Differing from the usual arrangement in this book, triorganoantimony oxides and dihydroxides are included in one chapter. This results from the fact that the compounds are not always properly characterized in the original literature. Authors often do not differentiate between the fully hydrated $R_3SbO \cdot H_2O = R_3Sb(OH)_2$, the partially hydrated $R_3SbO \cdot nH_2O$ ($n<1$), and the water-free R_3SbO. If in the original literature the formulation $R_3Sb(OH)_2$ is used, these compounds are described in addition to the corresponding triorganoantimony oxides.

The trialkylantimony oxides are extremely sensitive to moisture, which makes the handling under ordinary atmospheric conditions impossible [16].

2.5.1.1.7.1 R_3SbO and $R_3Sb(OH)_2$ Compounds with R = Alkyl

$(CH_3)_3SbO$

A compound of this composition, verified by physical measurements (see below) is obtained by drying $(CH_3)_3Sb(OH)_2$ in a high vacuum at 110 °C [1, 11] or as a crystalline colorless solid by sublimation of the crude product [3] from the reaction of $(CH_3)_3SbBr_2$ with Ag_2O in H_2O [2, 3] at low pressures. The melting point is given as 95 to 97 °C [3]. A compound with a melting point of 180 °C is obtained by reacting $Sb(CH_3)_3$ with $t\text{-}C_4H_9OOH$ in a molar ratio of 1:1 in C_6H_6. Beside $t\text{-}C_4H_9OH$, the title compound is recovered from the solution in 82% yield. The product of a reaction in aqueous solutions of $(CH_3)_3SbSO_4$ with $Ba(OH)_2$ [17] or of $(CH_3)_3SbBr_2$ with Ag_2O [2] is probably $(CH_3)_3Sb(OH)_2$ [6]. The same is presumed for the product obtained by evaporation of an ethereal solution of $Sb(CH_3)_3$ in the air [5].

The infrared and Raman spectra of the compound were recorded and assigned. The data are shown in Table 27 [3]. A normal coordinate analysis was carried out assuming C_{3v} symmetry. Some of the calculated force constants are: 3.43 (νSbO), 2.46 (νSbC), and 4.81 (νCH) mdyn/Å, 0.67 (δCSbO), 0.71 (δCSbC), 0.60 (δSbCH), and 0.55 (δHCH) mdyn · Å. The calculated values are in good agreement with the experimental ones [3].

A ^{121}Sb Mössbauer spectrum (at 4.2 K vs. $Ba^{121}SnO_3$) gives values of $\delta = -4.14(5)$, $e^2qQ = -11.4(6)$, and $\Gamma = 2.8$ mm/s with an asymmetry parameter of $\eta = 0$. From these data a polymeric structure with oxygen-linked trigonal bipyramidal units (CH_3 groups in equatorial positions) is concluded [1]. It was not possible to grow single crystals for an X-ray study [3].

Table 27
IR and Raman Vibrations (in cm^{-1}) of $(CH_3)_3SbO$ [3].

IR solid	mull	Raman solid (relative intensity)	assignment
3016 m	3007 m	3012 (16)	$\nu_{as}CH_3$
2931 m	2920 m	2920 (76)	ν_sCH_3
		2798 (2)	
1561 m		1553 vw	
1421 m		1415 vw	$\delta_{as}CH_3$
1310 m	1313 w		
1231 m	1230 m	1231 (5)	δ_sCH_3
1199 w	1199 w	1197 (18)	δ_sCH_3
		1191 (11)	δ_sCH_3
		1051 vw	
821 s	800 s	810 vw	ϱCH_3
		733 vw	
708 s	713 s	711 vw	νSbO
645 vs	644 vs		νSbO
		574 (14)	$\nu_{as}SbC_3$
553 s	551 s	554 (34)	$\nu_{as}SbC_3$
526 s	524 s	517 (100)	ν_sSbC_3
485 m			ν_sSbC_3
		455 vw	ν_sSbC_3
415 m			
		307 vw	
253 s		250 vw	$\delta_{as}SbC_3$
	222 s	227 vw	$\delta_{as}SbC_3$
	184 w	187 (7)	$\delta CSbO$
	173 m	170 (24)	δ_sSbC_3
	105 m		
	99 w		lattice modes
		74 (5)	lattice modes
		41 vw	lattice modes

The compound is hygroscopic [11] and thus is dissolved in H_2O as $(CH_3)_3Sb(OH)_2$ [7], which is a weak base with a dissociation constant of $K = 1.38 \times 10^{-5}$ [2]. It reacts in aqueous solution with HX (X=Cl, Br, I) or H_2S to form $(CH_3)_3SbX_2$ or $(CH_3)_3SbS$ [5]. The latter compound is also obtained in CH_3OH [8]. With t-C_4H_9OOH it forms $(CH_3)_3Sb(OOC_4H_9\text{-t})_2$ [9].

The compound may be used as a catalyst for the polycondensation of ethylene glycol with dimethyl terephthalate [10].

$(CH_3)_3Sb(OH)_2$

$(CH_3)_3SbCl_2$, dissolved in hot H_2O, is treated with ion exchange material M 500 [1], or Amberlite 1R4B [11]. The eluate is evaporated to dryness in a vacuum, and the residue is recrystallized from $(CH_3)_2CO$. The yield of the compound is 85% [11]. The substance could be also obtained by reacting $(CH_3)_3Sb(OH)Br$ with Ag_2O, heated for one hour. From the filtrate of the mixture, small colorless needles crystallize upon concentration [12].

References on p. 118

In an IR spectrum in KBr the $\nu_{as}SbC$ vibration is observed at 566 cm^{-1} [11]. From the ^{121}Sb Mössbauer spectrum (at 4.2 K vs. $Ba^{121}SnO_3$) with $\delta = -4.52(5)$, $e^2qQ = -16.7(6)$, $\Gamma = 2.1$ mm/s, and $\eta = 0$ a trigonal bipyramidal structure of the molecule is concluded [1]. A ^1H NMR spectrum in D_2O shows a signal at $\delta = 1.59$ ppm [4].

The compound loses H_2O upon heating to 110 °C under vacuum to form $(CH_3)_3SbO$ [1, 11]. With PCl_5 [12] or concentrated HCl [11], it forms $(CH_3)_3SbCl_2$. With 60% HNO_3 in $(CH_3)_2CO$, depending on the molar ratio of reactants, $(CH_3)_3Sb(NO_3)_2$ or $[(CH_3)_3SbNO_3]_2O$ are formed [6]. $C_6H_5CO_2H$ [6] and the compound react in refluxing C_6H_6 to give $(CH_3)_3Sb(O_2CC_6H_5)_2$. With other organic acids RCO_2H (R = H, CH_3, C_2H_5, C_3H_7, C_4H_9) in excess, the compound was reacted without a solvent to quantitatively give $(CH_3)_3Sb(O_2CR)_2$ [6]. Refluxing with t-C_4H_9OOH in C_6H_6 gives $(CH_3)_3Sb(OOC_4H_9\text{-t})_2$ in a yield of about 70% [13, 14].

The compound is claimed to be useful as a catalyst for the polycondensation of dimethyl terephthalate with ethylene glycol [10] or as a cocatalyst together with Mo or W for the epoxidation of olefins with H_2O_2 [15].

$(CD_3)_3SbO$

The compound is obtained from $(CD_3)_3SbBr_2$ and Ag_2O. Sublimation of the crude product at low pressure gives a colorless substance, melting at 101 to 102 °C.

IR and Raman vibrations are shown in Table 28. A normal coordinate analysis and a potential energy distribution of vibrations are given in the original as well as a figure of a mass spectrum. In this latter the most intense peak is that of the dimer which has lost one CD_3 group [3].

$(C_2H_5)_3SbO$

The reaction of a suspension of HgO in ether with $Sb(C_2H_5)_3$ gives, after shaking for one week and evaporation of the solvent from the filtrate, white crystals of the compound in 70 to 90% yield; m.p. 144 to 146 °C. The compound could not be sublimed in vacuum, since it was found to decompose easily [16]. Former authors performed the same reaction in C_2H_5OH [1, 10], but they did not obtain the compound in a pure state [16]. Another method of preparation is to treat $(C_2H_5)_3SbBr_2$ with an anion exchange resin (Amberlite 1R 4B[OH]) and to heat the resulting dihydroxide for 2 h at 120 °C and 0.1 Torr. Sublimation of the residue gives the compound in a yield of 52%. Reaction of $Sb(C_2H_5)_3$ with t-C_4H_9OOH (molar ratio 1:1) in C_6H_6 gives a white powder, which is thoroughly washed with ice-cold C_5H_{12}, in a 93% yield, m.p. 167 to 168 °C. Addition of more hydroperoxide gives $(C_2H_5)_3$-$Sb(OOC_4H_9\text{-t})_2$ [4]. The compound is claimed to be obtained by evaporation in air of C_2H_5OH from a solution of $Sb(C_2H_5)_3$ in this solvent [17], by reacting $(C_2H_5)_3SbI_2$ with Ag_2O in aqueous solution [19, 20] or by reaction of $(C_2H_5)_3SbSO_4$ with aqueous $Ba(OH)_2$ [17].

In an IR spectrum in CCl_4, νSbO vibrations were observed at 678 and 478 cm^{-1} [16] and compared with those of other R_3MX_2 (R = alkyl, M = P, As, Sb, X_2 = O, S, Se, Te) compounds [21]. The compound dissolves as a monomer in $CHCl_3$, probably as a $CHCl_3$ solvate. In air, the oxide is hydrolyzed back to the dihydroxide [4]. The compound is reduced by K under slight heating to $Sb(C_2H_5)_3$ [17]. It reacts with aqueous solutions of HCl, HBr, HI, or with gaseous HCl to form the corresponding dihalides [17], and with HNO_3 [17, 19, 20] or CH_3CO_2H to yield the corresponding dinitrate or diacetate [19, 20]. Treatment with H_2S gives $(C_2H_5)_3SbS$ after evaporation [17, 19, 20]. It reacts with C_2H_5OH to give $(C_2H_5)_3Sb(OC_2H_5)_2$ [16].

References on p. 118

Table 28
IR and Raman Vibrations (in cm^{-1}) of $(CD_3)_3SbO$ [3].

IR solid	mull	Raman solid (relative intensity)	assignment
2258 m	2250 vw	2254 (9)	$\nu_{as}CD_3$
2150 m	2145 w	2110 (75)	ν_sCD_3
		2018 vw	
1360 m	1342 w		
1161 m	1154 w	1150 vw	$\delta_{as}CD_3$
1041 s	1034 m	1040 vw	$\delta_{as}CD_3$
948 s	950 m	942 (18)	δ_sCD_3
		930 (39)	δ_sCD_3
846 s	846 w		
791 s			
761 vs	769 m		
726 s		728 vw	
628 vs	625 m	625 vw	νSbO
596 vs	594 m		ϱCD_3
546 vs	544 m		$\nu_{as}SbC_3$
528 vs	525 m	525 vw	$\nu_{as}SbC_3$
		505 (36)	$\nu_{as}SbC_3$
481 s	479 m	468 (100)	ν_sSbC_3
454 s			ν_sSbC_3
		425 vw	
408 s			
	258 s		$\delta_{as}SbC_3$
	214 m		$\delta_{as}SbC_3$
	177 m		$\delta CSbO$
	169 m		$\delta CSbO$
	157 s	153 (48)	δ_sSbC_3
		70 vw	lattice mode

$(C_2H_5)_3Sb(OH)_2$

Triethylantimony dihydroxide is obtained by hydrolysis of $(C_2H_5)_3SbBr_2$ on an Amberlite 1R 4B[OH] anion exchange resin. Treatment of the dihydroxide at 120 °C and 0.1 Torr for 2 h gives the oxide. The oxide is hydrolyzed back to the dihydroxide in air. It is a viscous oil [4]. It may also have been formed instead of or together with the oxide in some of the earlier investigations, e.g. [1, 10, 17, 19, 20].

$(C_3H_7)_3SbO$

Tripropylantimony oxide is prepared by shaking $Sb(C_3H_7)_3$ with HgO for one week in ether. Evaporation of the ether gives the compound with a melting point of 168 to 170 °C [16]. The same reaction performed in C_2H_5OH [18] gives a gelatinous mass, described to be the title compound, but probably other products are obtained.

IR vibrations for νSbO are found at 650 and 450 cm^{-1} in CCl_4 [16]. They are compared with those of other R_3MX_2 (R = alkyl, M = P, As, Sb, X_2 = O, S, Se, Te) compounds [21]. The

References on p. 118

compound reacts with H_2S in C_2H_5OH to form $(C_3H_7)_3SbS$. A 1:1 adduct of the compound with Sb_2O_3 is reported to be obtained by air oxidation of $Sb(C_3H_7)_3$ [18].

$(C_4H_9)_3SbO$

$Sb(C_4H_9)_3$ and HgO react in C_2H_5OH [18, 24], or better in ether, when shaken for one week to form the compound, which melts at 177 to 179 °C when isolated after evaporation of the solvent [16].

IR vibrations for νSbO are found at 650 and 450 cm^{-1} in CCl_4 [16]. They are compared with those of other R_3MX_2 (R=alkyl, M=P, As, Sb, X_2=O, S, Se, Te) compounds in [21]. The compound reacts with C_2H_5OH in C_6H_6 to give 44% of $(C_4H_9)_3Sb(OC_2H_5)_2$ [24]. With $(NH_2)_2CO$ at about 130 °C, $(C_4H_9)_3Sb(NCO)_2$ is formed, accompanied by NH_3 evolution [22, 25, 26]. The same compound is formed by reacting the title compound with HNCO in an inert solvent at about 60 °C [27]. $(C_4H_9)_3SbX_2$ compounds with $X_2=OCO_2^{2-}$, $SC(O)S^{2-}$, and $OC(NR)O^{2-}$ (R=C_2H_5, C_8H_{17}, C_6H_5, and 3-ClC_6H_4) are obtained by reacting the compound with CO_2, CS_2, or RNCO [28].

The compound is useful as a catalyst for the copolymerization of organoisocyanates with compounds containing active hydrogen atoms [29, 30], and for the trimerization of isocyanates [23, 24].

Several adducts of the compound are known. $(C_4H_9)_3SbO \cdot Sb_2O_3$ is obtained besides the oxide upon air oxidation of $Sb(C_4H_9)_3$. It is scarcely soluble in $Sb(C_4H_9)_3$, ether, C_6H_6, C_2H_5OH, or H_2O [18]. $(C_4H_9)_3SbO \cdot H_2O_2$ is obtained by reacting the corresponding stibine with 30% H_2O_2 in toluene [31]. The adduct $(C_4H_9)_3SbO \cdot C_6H_5NCO$ is formed when $Sb(C_4H_9)_3$ and phenylisocyanate are allowed to stand for days at room temperature. It is a catalyst for the trimerization of isocyanates [22, 25].

$(C_4H_9)_3Sb(OH)_2$

An aqueous solution of $(C_4H_9)_3SbBr_2$ is passed through an anion exchange resin. The eluate is concentrated to give the compound [32].

$(i\text{-}C_4H_9)_3SbO$

The compound is assumed to be obtained by oxidation of the corresponding stibine with HgO in C_2H_5OH [33].

Heating the compound with $(NH_2)_2CO$ to 125 to 140 °C for one hour [26] or reaction with HNCO in an inert solvent leads to $(i\text{-}C_4H_9)_3Sb(NCO)_2$ [27]. With CO_2, CS_2, or C_6H_5NCO reacted for 2 h at 20 °C, the compound forms the corresponding carbonic acid derivatives $(i\text{-}C_4H_9)_3SbX_2$ with $X_2=OCO_2^{2-}$, $SC(O)S^{2-}$, and $OC(NC_6H_5)O^{2-}$ [28].

Oxidation of $Sb(C_4H_9\text{-}i)_3$ in air gives an insoluble powder, infusible up to 240 °C, which is presumed to be the adduct $(i\text{-}C_4H_9)_3SbO \cdot Sb_2O_3$ [33].

$(C_5H_{11})_3SbO$

A liquid with a boiling point of 109 to 110 °C at 1.5 Torr is obtained if $Sb(C_5H_{11})_3$ is shaken for one week in ether with HgO. Distillation of the filtrate gives the compound [16]. Other previous preparations, for example the oxidation of $Sb(C_5H_{11})_3$ by air without a solvent [18], or in ether [34, 35], or by HgO in C_2H_5OH [18], led to poorly characterized syrupy or resinous masses.

References on p. 118

An IR spectrum of the liquid shows νSbO vibrations at 650 and 450 cm^{-1} [16]. These values are compared with those of other R_3MX_2 (R=alkyl, M=P, As, Sb, X_2=O, S, Se, Te) compounds in [21].

$(C_7H_{15})_3SbO$

$Sb(C_7H_{15})_3$ and HgO refluxed in ether for half an hour are reported to give the compound as a syrupy mass after evaporation of the ether from the filtrate of the reaction mixture. Spontaneous evaporation of an ether solution of the stibine in air left a white residue which is believed to consist of the oxide and the adduct $(C_7H_{15})_3SbO \cdot Sb_2O_3$. The oxide was removed by extracting several times with ether. The adduct is insoluble in ether, C_2H_5OH, and H_2O [36].

$(C_8H_{17})_3SbO$

A potentiometric titration with $HClO_4$ was performed in CH_3NO_2, and an acidity constant of $pK_a = 16.58$ was calculated [37].

$(c\text{-}C_6H_{11})_3SbO$

Upon standing in air, $Sb(C_6H_{11}\text{-}c)_3$ is oxidized with evolution of heat and formation of a surface film. The oxidation product appears to be mixed consisting of $(c\text{-}C_6H_{11})_3SbO$ and $(c\text{-}C_6H_{11})_3SbO \cdot Sb_2O_3$. The title compound can be extracted with ether, from which it crystallizes [38].

$(i\text{-}C_3H_7O_2CCH_2)_3SbO$

The corresponding stibine, dissolved in ether, is treated for 5 d at 20 °C with Ag_2O. Distillation of the filtrate of the mixture gives 55% of the dark colored substance. Its boiling point is 137 to 138 °C at 8.5×10^{-3} Torr; $d_4^{20} = 1.3056$, and $n_D^{20} = 1.4984$. The νC=O vibrations are in the region 1725 to 1723 cm^{-1}, and the ^1H NMR spectrum shows a characteristic CH_2Sb signal in the region $\delta = 2.25$ to 2.50 ppm [39].

$(C_6H_5CH_2)_3SbO$ and $(C_6H_5CH_2)_3Sb(OH)_2$

The dihydroxide is formed by reacting $C_6H_5CH_2MgCl$ with $SbCl_3$ (ratio 4:1) in ether and workup of the mixture with H_2O, ether, and C_2H_5OH [41, 42]. It has a melting point of 161 °C, then solidifies, and remains unchanged to 250 °C [41]. The product is described in [43] as $(C_6H_5CH_2)_3SbO \cdot H_2O$. It is mentioned, without details, that hydrolysis of $(C_6H_5CH_2)_3SbCl_2$ gives the oxide, which melts at 240 °C [40].

The dihydroxide is oxidized by alkaline MnO_4^-, yielding $C_6H_5CO_2H$. Warming with diluted HCl gives $C_6H_5CH_2Cl$ [41]. It reacts with concentrated HCl in $CHCl_3$ [41] or ether [42] to give $(C_6H_5CH_2)_3SbCl_2$. With HNO_3 ($d = 1.5\ g/cm^3$) at −5 to 0 °C $(NO_2C_6H_4CH_2)_3Sb(OH)NO_3$ is probably obtained [41].

References:

[1] J. Pebler, F. Weller, K. Dehnicke (Z. Anorg. Allgem. Chem. **492** [1982] 139/47).
[2] T. M. Lowry, J. H. Simons (Ber. Deut. Chem. Ges. **63** [1930] 1595/1602).
[3] W. Morris, R. A. Zingaro, J. Laane (J. Organometal. Chem. **91** [1975] 295/306).
[4] A. G. Davies, S. C. W. Hook (J. Chem. Soc. C **1971** 1660/5).
[5] H. Landolt (J. Prakt. Chem. **84** [1861] 328/339).
[6] M. Shindo, R. Okawara (J. Organometal. Chem. **5** [1966] 537/44).
[7] J. J. Monagle (J. Org. Chem. **27** [1962] 3851/5).

[8] M. Shindo, Y. Matsumura, R. Okawara (J. Organometal. Chem. **11** [1968] 299/305).
[9] A. Rieche, J. Dahlmann (Ger. [East] 44608 [1961/66]; C.A. **65** [1966] 10623/4).
[10] H. Terada, I. Takeshi, N. Osamu, T. Shunichi (Japan. 68-15999 [1965/68]; C.A. **70** [1969] No. 29594).

[11] G. G. Long, G. O. Doak, L. D. Freedman (J. Am. Chem. Soc. **86** [1964] 209/13).
[12] G. T. Morgan, V. E. Yarsley (Proc. Roy, Soc. [London] A **110** [1926] 534/7).
[13] A. Rieche, J. Dahlmann, D. List (Liebigs Ann. Chem. **678** [1964] 167/82).
[14] A. Rieche, J. Dahlmann (Ger. 1158975 [1961/63]; C.A. **60** [1964] 9313/4).
[15] M. Pralus, J. P. Schirmann, S. Y. Delavarenne (Ger. 2605041 [1976]; C.A. **86** [1977] No. 16526).
[16] G. N. Chremos, R. A. Zingaro (J. Organometal. Chem. **22** [1970] 637/46).
[17] C. Löwig, E. Schweizer (Liebigs Ann. Chem. **75** [1850] 315/55).
[18] W. J. C. Dyke, W. J. Jones (J. Chem. Soc. **1930** 1921/7).
[19] W. Merck (J. Prakt. Chem. **66** [1855] 56/72).
[20] W. Merck (Liebigs Ann. Chem. **97** [1856] 329/33).

[21] G. N. Chremos, R. A. Zingaro (J. Organometal. Chem. **22** [1970] 647/51).
[22] W. Stamm (Trans. N.Y. Acad. Sci **28** [1966] 396/401).
[23] S. Herbstman, Stauffer Chemical Co. (U.S. 3278492 [1964/66]; C.A. **66** [1967] 66142).
[24] S. Herbstman (J. Org. Chem. **30** [1965] 1259/60).
[25] W. Stamm (J. Org. Chem. **30** [1965] 693/5).
[26] Stauffer Chemical Co. (Neth. Appl. 6411266 [1965]; C.A. **63** [1965] 11613/4).
[27] W. Stamm, Stauffer Chemical Co. (U.S. 3417115 [1965/68]; C.A. **70** [1969] No. 37925).
[28] A. W. Breindel, S. Herbstman, Stauffer Chemical Co. (U.S. 3317575 [1963/67]; C.A. **67** [1967] No. 32780).
[29] F. Hostettler, E. F. Cox (Brit. 845827 [1960]; C.A. **1961** 8352).
[30] F. Hostettler, E. F. Cox (U.S. 3235518 [1957/1966]; C.A. **64** [1966] 12902).

[31] H. Rudolph, K. Reinking, Farbenfabriken Bayer A.-G. (Ger. 1268619 [1967/68]; C.A. **69** [1968] No. 77507).
[32] J. L. Zollinger, Minnesota Mining and Manufg. Co. (U.S. 3470176 [1967/69]; C.A. **71** [1969] No. 124513).
[33] M. E. Brinnand, W. J. C. Dyke, W. Haydn Jones, W. J. Jones (J. Chem. Soc. **1932** 1815/19).
[34] F. Berlé (J. Prakt. Chem. **65** [1855] 385/418).
[35] F. Berlé (Liebigs Ann. Chem. **97** [1856] 316/22).
[36] Chao-Lun Tseng, Wen-Yu Shih (J. Chinese Chem. Soc. **4** [1936] 183/6).
[37] V. V. Yakshin, N. A. Lyubosvetova, M. I. Tymonyuk, B. N. Laskorin (Dokl. Akad. Nauk SSSR **245** [1979] 1406/9; Dokl. Chem. Proc. Acad. Sci. **244/249** [1979] 207/10).
[38] Z. M. Manulkin, A. N. Tatarenko, F. Yu. Yusupov (Dokl. Akad. Nauk SSSR **88** [1953] 687/90; C.A. **1954** 2631).
[39] E. A. Besolova, V. L. Foss, I. F. Lutsenko (Zh. Obshch. Khim. **38** [1968] 1574/8; J. Gen. Chem. [USSR] **38** [1968] 1523/6).
[40] G. J. Morgan, F. M. G. Micklethwait (Proc. Chem. Soc. **28** [1912] 68).

[41] F. Challenger, A. T. Peters (J. Chem. Soc. **1929** 2610/21).
[42] J. P. Tsukervanik, D. Smirnov (J. Gen. Chem. [USSR] **7** [1937] 1527/31).
[43] L. Kolditz, M. Gitter, E. Rösel (Z. Anorg. Allgem. Chem. **316** [1962] 270/7).

2.5.1.1.7.2 R_3SbO and $R_3Sb(OH)_2$ Compounds with R = Aryl and 2-Thienyl

2.5.1.1.7.2.1 Triphenylantimony Oxide and Dihydroxide

Incomplete and conflicting data in the literature make it difficult to ascertain the validity of the compounds described therein. In many cases, compositions of $(C_6H_5)_3SbO$, $(C_6H_5)_3SbO \cdot H_2O$, $(C_6H_5)_3Sb(OH)_2$, and structures between monomers or polymers of these compositions are not explicitly differentiated.

$(C_6H_5)_3SbO$

Preparation and Formation. The methods of preparation and formation of the compound are summarized in Table 29.

Table 29
Preparation and Formation of $(C_6H_5)_3SbO$.

reactants, reaction conditions (yield in %)	melting point and remarks	Ref.
$Sb(C_6H_5)_3$ + 30% H_2O_2 in $(CH_3)_2CO$ with cooling, precipitate heated with refluxing C_6H_6, dried at 80 °C (80)	280 °C, white powder, identified as $(-(C_6H_5)_3SbO-)_n$	[1]
+20 to 30% H_2O_2 in $(CH_3)_2CO$ for 1 to 2 h at 20 °C (90)	249 to 251 °C	[6]
+3% H_2O_2 in $(CH_3)_2CO$, dried in air	amorphous, associated form	[7, 30]
+3% H_2O_2 in $(CH_3)_2CO$	$(C_6H_5)_3SbO_{1.4}$, possibly an H_2O_2 adduct	[8]
+30% H_2O_2 in $CH_3C_6H_5$	$(C_6H_5)_3SbO \cdot H_2O_2$	[9]
$(C_6H_5)_3SbCl_2$ in $(CH_3)_2CO$ + ethanolic KOH, concentrating, addition of H_2O, drying the precipitate at 150 °C until no OH bands can be detected by IR	280 °C, identified as $(-(C_6H_5)_3SbO-)_n$	[1]
$(C_6H_5)_3SbBr_2$ treated as in the previous method	217 °C, is probably not the compound	[1]
$(C_6H_5)_4SbOH$ in $CH_3C_6H_4CH_3$-4 heated for 6 to 7 d on 60 to 65 °C, cooling to 0 °C gives crystals which are washed with cold $CH_3C_6H_4CH_3$-4/$(CH_3)_2CO$ and dried (20 to 68)	221.5 to 222 °C, monomeric in refluxing C_6H_6, partially associated in the solid state	[2, also 1, 18 to 22]
$(C_6H_5)_3Sb(OCH_3)_2$ + H_2O/$(CH_3)_2CO$ standing for 24 h (80)	colorless needles, soluble in CCl_4, monomeric in C_6H_6	[2, 20, 26, 80]
$Sb(C_6H_5)_3$ + t-C_4H_9OOH in C_6H_6 at 40 to 60 °C (94)	219 °C	[10]

References on p. 125

Table 29 [continued]

reactants, reaction conditions (yield in %)	melting point and remarks	Ref.
$Sb(C_6H_5)_3 + [(CH_3)_3Si]_2O_2$ in petroleum ether for 1.5 h at −10 °C (77) or 18 h at 25 °C (81)	221 °C	[12]
$Sb(C_6H_5)_3 + NO_2PF_6$ or NO_2BF_4 in CH_2Cl_2 for 5 to 10 min at −78 °C (98)	282.5 °C from petroleum ether/ C_6H_6	[13]
$Sb(C_6H_5)_3 + CH_3SO_2NSO$ in refluxing C_6H_6 for 7 h, then +2N NaOH (38)	221 to 225 °C (=oxidehydrate), recrystallization from $CH_3C_6H_5$	[14]
$Sb(C_6H_5)_3 + SeO_2$ in refluxing C_6H_6 or C_2H_5OH	235 to 237 °C from C_6H_6/petroleum ether, is $(C_6H_5)_3SbO \cdot SeO_2$	[16]
$Sb(C_6H_5)_3 + UO_2(O_2CCH_3)_2$ or UO_2SO_4 or $UO_2(NO_3)_2$ in dioxane/H_2O (6:1), $h\nu \geqq 400$ nm	–	[17]
photolysis of $(C_6H_5)_4SbOOC_4H_9$-t for 100 h in $CH_3C_6H_5$ (80) or	220 °C, $+(C_6H_5)_4SbOH$ (17)	[22]
for 80 h in C_6H_5Cl (86.5), CCl_4 (19.5), or $CHCl_3$ (35.5)	$+(C_6H_5)_4SbCl$ (16, 75, 61% yield, respectively)	[23]
$(C_6H_5)_4SbOOM(C_6H_5)_3$ (M = Si, Ge) in $CH_3C_6H_5$, C_6H_5Cl, or C_9H_{20} for 11 to 12 h at 100 °C (60 to 95)	230 °C	[24]
$(C_6H_5)_3SbCl_2 + H_2O$ or $H_2^{18}O$, refluxing for some days	–	[25]
$(C_6H_5)_3Sb{=}NSO_2C_6H_4CH_3\text{-}4 + H_2O$ in CH_3CN	280 to 285 °C	[27]
$(C_6H_5)_3Sb{=}NSO_2R$ (R = C_6H_5, $C_6H_4CH_3$-4) $+C_6H_5CHO$ in CH_2Cl_2 for 2 h at 20 °C (95 to 98)	220 to 221 °C (from C_6H_6), dimeric in C_6H_6	[28]
$(C_6H_5)_4C_5Sb(C_6H_5)_3 + RC_6H_4OH$ (R = H, NO_2-4) in CCl_4 (25 to 50)	+phenylfulvene derivatives	[29]

The oxidation enthalpy of $Sb(C_6H_5)_3$ with t-C_4H_9OOH at a 1:1 ratio was determined as -240.6 ± 1.7 kJ/mol at 298.15 K. The calculated values for the enthalpy of formation of crystalline or of gaseous $(C_6H_5)_3SbO$ are 135.4 ± 17 kJ/mol and 244.2 ± 21 kJ/mol [11].

Properties. The products obtained by reaction of $Sb(C_6H_5)_3$ with 30% H_2O_2 or from $(C_6H_5)_3SbCl_2$ with ethanolic KOH give the IR (in Nujol) and Raman (solid) bands (in cm^{-1}) listed on p. 122 [1].

X-ray powder photographs and differential thermoanalyses were performed with these products. The slight solubility of the compound in CCl_4, $CHCl_3$, or C_6H_6 (solubility 5 to 8 g/L) precludes a characterization by NMR. A mass spectrum was measured by direct insertion of the solid at a temperature of 100 to 150 °C. The observed fragments are:

References on p. 125

$[(C_6H_5)_5Sb_2O_2]^+$, $[(C_6H_5)_2Sb_2O_3]^+$, $[(C_6H_5)_2Sb_2O_2]^+$, $[(C_6H_5)_3SbO]^+$, $[(C_6H_5)_nSb]^+$ (n=1 to 3), $[C_6H_5SbO]^+$, $[(C_6H_5)_2]^+$, $[C_6H_5OH]^+$, $[C_6H_5]^+$, and $[C_4H_3]^+$ [1].

IR	Raman (relative intensity)	assignment
744 s		ν_{as}SbO
726 s	737 (4)	ν_4, δCH out-of-plane
694 s	696 (4)	ν_8, out-of-plane ring deformation
669 s, br		ν_sSbO
665 vw, sh	664 (12), 656 (100)	ν_2, ring breathing
624 w, 617 w	615 (28)	ν_{18}, in-plane ring deformation
464 m, 450 s		ν_{19}, δCH perpendicular
400 vw		ν_{20}, ring deformation

The compound prepared in [2] from $(C_6H_5)_3Sb(OCH_3)_2$ shows ^{1}H NMR signals in CCl_4 at δ=7.28 and 7.58 ppm, and in the IR spectrum νSbO bands at 664 and 478 cm^{-1} [2].

IR spectra were taken in both CCl_4 (4000 to 1400 cm^{-1}) and CS_2 (1400 to 600 cm^{-1}) solutions (in cm^{-1}): 3020 w, 1485 w, 1435 m, 1335 w, 1308 w, 1188 w, 1070 m, 1020 w, 998 m, 730 s, 692 s, 664 s, and 655 to 646 br [2]. The values obtained from a KBr pellet are (in cm^{-1}): 3020 m, 1670 w, 1490 m, 1435 s, 1335 w, 1305 m, 1180 m, 1165 w, 1062 s, 1021 m, 998 m, 738 s, 728 s, 694 s, 664 s, 655 s, and 650 s [2], see also [31]. A UV spectrum in i-C_3H_7OH shows absorptions at λ_{max}(ε in $L\cdot cm^{-1}\cdot mol^{-1}$)=224 (22000), 257 (1670), 263 (1800), and 270 (1320) nm [25]. A ^{121}Sb Mössbauer spectrum (4 K vs. InSb) of $(C_6H_5)_3SbO$ gives an isomer shift of δ=5.27 mm/s, with $e^2qQ=-10.6$, Γ=2.8 mm/s, and η=0 [3]. Binding energies were determined from photoelectron spectra; Sb 3d(3/2)=538.8, Sb 3d(5/2)=529.4, O 1s=531.0 eV [4] or ca. 530.6 eV [5], and Sb 4d(5/2)=34.86 eV. The energy of dissociation of the Sb=O bond was determined as 440±25 kJ/mol, and the enthalpy of solution of $(C_6H_5)_3SbO$ in C_6H_6 up to 1 mol% is 15.2±0.4 kJ/mol [11].

Reactions. Pyrolysis of $(C_6H_5)_3SbO$ at 230 to 260 °C in an evacuated system gives C_6H_5-C_6H_5 and C_6H_6 [2]. The autoxidation is discussed, and a mechanism is proposed. The photochemical oxidation was also attempted [19]. Triphenylantimony oxide reacts with concentrated HCl in $(CH_3)_2CO$ to form $(C_6H_5)_3SbCl_2$ [20]. With SF_4 in CH_2Cl_2 at −78 °C it gives $(C_6H_5)_3SbF_2$ in a yield of 94% [32]. The compound reacts with $COCl_2$ or $SOCl_2$ in CH_3CN to form $(C_6H_5)_3SbCl_2$ [34]. With CH_3OH and a molecular sieve, stirred for 12 h, it gives $(C_6H_5)_3Sb(OCH_3)_2$ in a yield of 25%. With acetylacetone under the same conditions it forms 85% $(C_6H_5)_3Sb(OH)$acac, whereas the decomposition product $(C_6H_5)_3Sb(O_2CCH_3)_2$ is formed under reflux [33]. 2,3-Dimethyl-2,3-butanediol or 1,2-diphenyl-1,2-ethanediol heated with the substance shortly to 120 °C form the corresponding heterocycles I (R, R′=CH_3 or R=H, R′=C_6H_5). With 2,3-dimethyl-2,3-butanediol, heated for a longer period, or with 2,3-butanediol compounds of type II (R=CH_3 or H) are formed [8]. Many organic acids RCO_2H (R=H, CH_3, CH_2Cl, CCl_3, C_2H_5, CH_2=CH, CH_2=C(CH_3), C_6H_5, 2-ClC_6H_4, 4-$NH_2C_6H_4$ [7], or CF_3 [32]) were reacted in CH_3OH with the compound to give the corresponding $(C_6H_5)_3Sb(O_2CR)_2$ in yields of about 40 to 80%. The compounds $(C_6H_5)_3SbX_2$ with X=CH_2=$CHCO_2$ [35, 36], 4-$HOC_6H_4CO_2$, 2-$CH_3O_2CC_6Y_4CO_2$ (Y=H, Cl, Br), CH_3CO_2(Y)C=C(Y)CO_2 (Y=H, Cl, Br) [37], or 8-quinolinolate [35] were prepared directly from $(C_6H_5)_3SbO$ and organic acids or phenols with evolution of H_2O. Triphenylantimony oxide refluxed with $Pb_2(C_6H_5)_6$ for 8 h in $CHCl_3$ gives $Pb(C_6H_5)_4$ (81%), $(C_6H_5)_2PbO$ (60%), and $Sb(C_6H_5)_3$ (40%) [38]. The compound reacts with $P(C_6H_5)_3$ and catalytic amounts of $[(C_2H_5)_2NC(S)S]_2MoO$ in C_6H_6 to form 86% $Sb(C_6H_5)_3$ and $(C_6H_5)_3PO$ [39]. $(C_6H_5)_3SbO$, suspended in $HOCH_2CH_2NHCH_3$ and heated at 100 °C, gives $(CH_3HNCH_2CH_2OSbOH(C_6H_5)_2)_2O$ in 97% yield [79]. Reaction of the compound with

References on p. 125

$(C_6H_5)_3SbCl_2$, refluxed for 10 min in C_6H_6, gives 96% $[(C_6H_5)_3SbCl]_2O$ [2, 20]. The compound was refluxed for 4 d in para-xylene or was allowed to react in dioxane at room temperature with exposure to air; a precipitate (45% yield) consisting of $[(C_6H_5)_2SbO]_2O$ and $[(C_6H_5)_3SbOC_6H_5]_2O$ is formed [2].

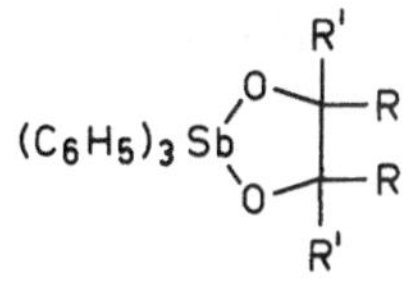

I

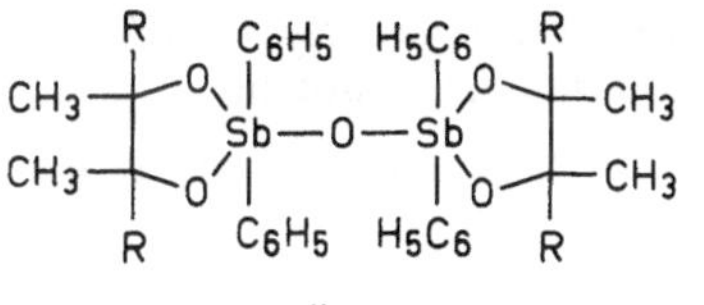

II

The compound forms a 1:1 adduct with iodine, $(C_6H_5)_3SbO \cdot I_2$, which was isolated. It is identified by a charge transfer absorption at 365 nm in CH_2Cl_2. The IR spectrum in Nujol shows the following vibrations (in cm^{-1}): 910 br, 736 vs, 730 vs, 692 vs; 668 w, 660 w, 653 w, 575 br, 550 sh (νSbO), 618 w; 475 sh (δSbO), 458 s, and 445 sh. An equilibrium exists between $(C_6H_5)_3SbO$, I_2, and $(C_6H_5)_3SbO \cdot I_2$ in CH_2Cl_2 solution. From equilibrium constants the enthalpy of formation is determined as $\Delta H° = -8.16$ kJ/mol, and $\Delta S° = 41.5$ $J \cdot mol^{-1} \cdot K^{-1}$. The free energy is $\Delta G = -20.55$ kJ/mol at 25 °C [40].

Triphenylstibine oxide also gives a 1:1 adduct with H_2O_2, $(C_6H_5)_3SbO \cdot H_2O_2$. Infrared data (in cm^{-1}) for the adduct are: 736 vs, 730 vs, 692 vs; 670 w, 665 w, 653 sh (νSbO), 620 w, 495 s (δSbO), 458 s, and 445 s [40].

A partially hydrated form of the compound, $(C_6H_5)_3SbO \cdot 0.3H_2O$, is mentioned [41].

Uses. The activity as a catalyst for the preparation of carbodiimides from isocyanates is proved in [6]. The catalytic activity of the compound for the formation of polyester is claimed in [42, 43]. The compound is useful as a catalyst or cocatalyst for the ring-opening polymerization of alkene oxides [44, 45] as well as for the reaction of epoxides with CO_2 to give cyclic carbonates [46, 47]. The compound was tested as a retarding agent in the burning of epoxy resins [48]. It passivates metals which poison cracking catalysts [49, 50]. It is an antioxidant for polypropylene, nylon-6, and polystyrene, and shows a synergistic effect with phenolic antioxidants [51]. The H_2O_2 adduct of the compound is useful as a polymerization catalyst [52].

$(C_6H_5)_3Sb(OH)_2$

A compound of this composition with a melting point of about 210 °C (compare $(C_6H_5)_3SbO$) is obtained instead of the oxide, if preparations or the workup are carried out in H_2O or H_2O-containing solvents. The methods of preparation and formation are summarized in Table 30.

Table 30
Preparation and Formation of $(C_6H_5)_3Sb(OH)_2$.

reactants, reaction conditions (yield in %) [Ref.]	melting point and other properties [Ref.]
$Sb(C_6H_5)_3 + SeO_2$ in C_6H_6, one week at 20 °C [15, 53]; precipitate dissolved in glacial CH_3CO_2H and poured into H_2O [53]	212 °C [53]

References on p. 125

Table 30 [continued]

reactants, reaction conditions (yield in %) [Ref.]	melting point and other properties [Ref.]
$Sb(C_6H_5)_3+C_6H_5S(O)SC_6H_5$ in CH_3OH/C_6H_6 for 39 h at 70 °C, concentration, and addition of C_6H_{14} (92)	208 to 210 °C from C_6H_6/petroleum ether (1:4) as prisms [54]
$Sb(C_6H_5)_3+FeCl_3$ in $CHCl_3$ for 6 h at 100 °C, then hydrolysis (95)	210 °C [55]
$Sb(C_6H_5)_3$ + 3% H_2O_2 and diluted KOH, or $+Na_2O_2$	212 °C, crystals [56]
$Sb(C_6H_5)_3+4\text{-}IC_6H_4OCH_3$ in CH_3OH, irradiated for 40 h with UV light (43)	205 to 206 °C [57]
$Sb(C_6H_5)_3$ + excess MnO_4^- + diluted H_2SO_4, warmed on a water bath, precipitate extracted with C_2H_5OH (small quantity)	210 °C [58]
$(C_6H_5)_3SbCl_2$ in C_6H_6 [64], C_2H_5OH [58], or $(CH_3)_2CO$ [1] + aqueous NaOH [64, 58], set aside for some days [58], or + ethanolic KOH for 1 h at 30 °C, concentration, and addition of H_2O [1], see also [59]	213 °C [64], 210 °C, beautiful leaflets, washed with H_2O [58]
$(C_6H_5)_3SbCl_2$ + aqueous $(H_2NCH_2)_2$	208 °C [61]
$(C_6H_5)_3SbBr_2$ + ethanolic KOH under boiling, precipitate dissolved in glacial CH_3CO_2H, addition of H_2O [59, 60]	212 °C [59], 214 to 215 °C from C_6H_6/ether [60]
$(C_6H_5)_3SbBr_2+(H_2NCH_2)_2$ in H_2O	210 °C [61]
$(C_6H_5)_3SbI_2$ + ethanolic KOH, refluxed, concentrated, precipitate dissolved in glacial CH_3CO_2H, addition of H_2O	206 to 208 °C, light powder, insoluble in ether [62]
$(C_6H_5)_3SbSO_4$ in hot $CH_3CO_2H+H_2O$	212 °C [63]
$(C_6H_5)_3SbSO_4$ + warm aqueous NaOH [58]	–
$(C_6H_5)_3Sb(O_2CCH_3)_2$ in CH_2Cl_2, shaken with H_2O, evaporation of organic layer [15]	–
$RSO_2(C_6H_5CO)NSb(C_6H_5)_3Cl$ ($R=C_6H_5$, $4\text{-}CH_3C_6H_4$) in aqueous C_2H_5OH, refluxed for 2 h (98)	211 to 212 °C [65]
$Sb(C_6H_5)_5$ in C_2H_5OH for 40 h at 100 °C (71)	206 °C [66]

No dipole moment could be measured [60]. The solubility at 25 °C is 3 mol/L in C_6H_6, and 0.3 mol/L in CCl_4. The distribution coefficient for C_6H_6/H_2O and for CCl_4/H_2O is 10^4 [67]. A mass spectrum (70 eV) shows the following fragments: $[M-nOH]^+$ (n=0 to 2), $[M-O]^+$, $[M-3OH-nC_6H_5]^+$ (n=0 to 3), $[(C_6H_5)_2Sb-2H]^+$, $[C_6H_5SbOH]^+$, $[SbOH]^+$, $[SbO]^+$, $[(C_6H_5)_2]^+$, $[(C_6H_5)_2-2H]^+$, $[C_6H_5OH]^+$, $[C_6H_5H]^+$, and $[C_6H_5]^+$ [15]. Triphenylantimony dihydroxide loses H_2O when dried in a vacuum [1, 66] or when heated to the melting point (214 to 215 °C) [60] to give the oxide [1, 66]. It reacts with ethanolic HCl to form $(C_6H_5)_3SbCl_2$ [54, 66]. The reaction with HX (X=halogen) and HNO_3 to form the corresponding dihalides and dinitrate is mentioned without details [59]. With concentrated H_2SO_4 a solution of $(C_6H_5)_3SbSO_4$ in H_2SO_4 is obtained [58]. With stoichiometric amounts of H_2S, $(C_6H_5)_3SbS$ may be formed [68]; but with an excess in the presence of $N(C_2H_5)_3$ in C_2H_5OH, the compound is reduced to $Sb(C_6H_5)_3$ [59]. Dissolved in hot formic acid [69] or acetic acid [30], the corresponding triphenylantimony dicarboxylates crystallize upon cooling. With HN_3 the compound reacts in ether to give $(C_6H_5)_3Sb(N_3)_2$ [70]. Organic peroxides like t-C_4H_9OOH or $C_6H_5(CH_3)_2COOH$ react with the compound in refluxing C_6H_6 to give the corresponding $(C_6H_5)_3Sb(OOCR_3)_2$ compounds in good yields [71 to 73]. The compound was irradiated with neutrons (n,γ-reaction), and a method for separation of the recoil products is described [74].

The title compound may be used as an extracting reagent for phosphate, acetate, propionate, and benzoate from H_2O at low pH into $CHCl_3$ solutions [75]. It is also useful as an additive to epoxy resins [76], or halogen-containing polyester resins [77], and to impart fire resistance. It is used as a catalyst instead of Sb_2O_3 for the polycondensation of terephthalate with ethylene glycol. These polymers have less coloration than usual [78].

References:

[1] D. L. Venezky, C. W. Sink, B. A. Nevett, W. F. Fortescue (J. Organometal. Chem. **35** [1972] 131/42).

[2] W. E. McEwen, G. H. Briles, D. N. Schulz (Phosphorus **2** [1972] 147/53).

[3] L. H. Bowen, G. G. Long (Inorg. Chem. **17** [1978] 551/4).

[4] T. Birchall, J. A. Connor, I. H. Hillier (J. Chem. Soc. Dalton Trans. **1975** 2003/6).

[5] S. Hoste, D. F. Van de Vondel, G. P. Van der Kelen (J. Electron Spectrosc. Relat. Phenom. **17** [1979] 191/5).

[6] J. J. Monagle (J. Org. Chem. **27** [1962] 3851/5).

[7] J. Havranek, J. Mleziva, A. Lycka (J. Organometal. Chem. **157** [1978] 163/6).

[8] F. Nerdel, J. Buddrus, K. Höher (Chem. Ber. **97** [1964] 124/31).

[9] H. Rudolph, K. Reinking, Farbenfabriken Bayer A.-G. (Ger. 1268619 [1967/68]; C.A. **69** [1968] No. 77507).

[10] G. A. Razuvaev, T. G. Brilkina, E. V. Krasil'nikova, T. I. Zinov'eva, A. I. Filimonov (J. Organometal. Chem. **40** [1972] 151/7).

[11] V. G. Tsvetkov, Yu. A. Aleksandrov, V. N. Glushakova, N. A. Skorodumova, G. M. Kol'yakova (Zh. Obshch. Khim. **50** [1980] 256/8; J. Gen. Chem. [USSR] **50** [1980] 198/201).

[12] D. Brandes, A. Blaschette (J. Organometal. Chem. **73** [1974] 217/27).

[13] G. A. Olah, B. G. B. Gupta, S. C. Narang (J. Am. Chem. Soc. **101** [1979] 5317/22).

[14] A. Senning (Acta Chem. Scand. **19** [1965] 1755/9).

[15] C. Glidewell (J. Organometal. Chem. **116** [1976] 199/209).

[16] S. I. A. El Sheikh, M. S. Patel, B. C. Smith, C. B. Waller (J. Chem. Soc. Dalton Trans. **1977** 641/4).

[17] A. S. Brar, S. S. Sandhu, A. S. Sarpal (Indian J. Chem. A **18** [1979] 19/22).

[18] W. E. McEwen, F. L. Chupka Jr. (Phosphorus **1** [1972] 277/82).

[19] D. N. Schulz (Diss. Univ. Massachusetts, Amherst, Mass., 1971; Diss. Abstr. Intern. B **32** [1971] 1464).

[20] G. H. Briles, W. E. McEwen (Tetrahedron Letters **1966** 5299/302).

[21] F. L. Chupka Jr. (Diss. Univ. Massachusetts, Amherst, Mass., 1969; Diss. Abstr. Intern. B **30** [1969] 117).

[22] G. A. Razuvaev, T. I. Zinov'eva, T. G. Brilkina (Izv. Akad. Nauk SSSR Ser. Khim. **1969** 2007/13; Bull. Acad. Sci. USSR Div. Chem. Sci. **1969** 1855/9).

[23] G. A. Razuvaev, T. I. Zinov'eva, T. G. Brilkina, E. P. Silkovskaya (Dokl. Akad. Nauk SSSR **193** [1970] 355/8; Dokl. Chem. Proc. Acad. Sci. USSR **190/195** [1970] 497/9).

[24] G. A. Razuvaev, T. I. Zinov'eva, T. G. Brilkina (Dokl. Akad. Nauk SSSR **188** [1969] 830/2; Dokl. Chem. Proc. Acad. Sci. USSR **184/189** [1969] 805/7).
[25] J. Bernstein, M. Halmann, S. Pinchas, D. Samuel (J. Chem. Soc. **1964** 821/4).
[26] G. H. Briles, W. E. McEwen (Tetrahedron Letters **1966** 5191/6).
[27] G. Wittig, D. Hellwinkel (Chem. Ber. **97** [1964] 789/93).
[28] A. M. Pinchuk, Z. I. Kuplennik, Zh. N. Belaya (Zh. Obshch. Khim. **46** [1976] 2242/6; J. Gen. Chem. [USSR] **46** [1976] 2155/8).
[29] B. H. Freeman, D. Lloyd, M. I. C. Singer (Tetrahedron **28** [1972] 343/52).
[30] H. Schmidt (Liebigs Ann. Chem. **429** [1922] 123/52).

[31] K. A. Jensen, P. H. Nielsen (Acta Chem. Scand. **17** [1963] 1875/85).
[32] L. M. Yagupol'skii, N. V. Kondratenko, V. I. Popov (Zh. Obshch. Khim. **46** [1976] 620/3; J. Gen. Chem. [USSR] **46** [1976] 618/21).
[33] R. G. Goel, D. R. Ridley (J. Organometal. Chem. **182** [1979] 207/12).
[34] R. Appel, W. Heinzelmann, Badische Anilin- & Soda-Fabrik A.-G. (Ger. 1192205 [1962/ 65]; C.A. **63** [1965] 8405).
[35] J. R. Leebrick, M & T Chemicals Inc. (U.S. 3287210 [1962/66]; C.A. **66** [1967] No. 85070).
[36] J. Musher, K. Su (U.S. 3939190 [1972/76]; C.A. **84** [1976] No. 181136).
[37] M. M. Y. Chang, K. Su, J. I. Musher (Israel J. Chem. **12** [1974] 967/70).
[38] S. N. Bhattacharya, A. K. Saxena (Indian J. Chem. A **17** [1979] 307/9).
[39] L. Xiyan, S. Junhui, T. Xiaochun (Synthesis **1982** 185/6).
[40] J. F. C. Boodts, W. A. Bueno (J. Chem. Soc. Faraday Trans. I **76** [1980] 1689/93).
[41] R. L. McKenney, H. H. Sisler (Inorg. Chem. **6** [1967] 1178/82).
[42] O. K. Carlson, J. A. Price (U.S. 3415787 [1966/68]; C.A. **70** [1969] No. 29547).
[43] S. B. Maerov (J. Polym. Sci. Polym. Chem. Ed. **17** [1979] 4033/40).
[44] R. Nomura, H. Hisada, A. Ninagawa, H. Matsuda (Makromol. Chem. Rapid Commun. **1** [1980] 135/8).
[45] R. Nomura, Y. Shiomura, A. Ninagawa, H. Matsuda (Makromol. Chem. **184** [1983] 1163/9).
[46] R. Nomura, A. Ninagawa, H. Matsuda (J. Org. Chem. **45** [1980] 3735/8).
[47] H. Matsuda (Japan. 80122776 [1979/80]; C.A. **94** [1981] No. 139779).
[48] J. Havranek, J. Mleziva (Angew. Makromol. Chem. **84** [1980] 105/17).
[49] D. L. McKay (U.S. 4257876 [1978/81]; C.A. **94** [1981] No. 211339).
[50] D. L. McKay (Eur. 51689 [1980/82]; C.A. **97** [1982] No. 95317).

[51] T. Ohseki, M. Watanabe, Mitsubishi Rayon Co., Ltd. (Japan. 71-22104 [1967/71]; C.A. **76** [1972] No. 154856).
[52] H. Schnell, H. Rudolph, K. Reinking, Farbenfabriken Bayer A.-G. (Ger. 1270816 [1967/68]; C.A. **69** [1968] No. 36585).
[53] N. N. Mel'nikov, M. S. Rokitskaya (Zh. Obshch. Khim. **8** [1938] 834/8).
[54] J. F. Carson, F. F. Wong (J. Org. Chem. **26** [1961] 1467/70).
[55] Z. M. Manulkin, A. N. Tatarenko (Zh. Obshch. Khim. **21** [1951] 93/8; J. Gen. Chem. [USSR] **21** [1951] 103/7).
[56] L. Kaufmann (Ger. 360973; C.A. **1924** 841).
[57] M. A. Shubenko (Sb. Stat. Obshch. Khim. Akad. Nauk SSSR **2** [1953] 1043/5).
[58] P. May (Proc. Chem. Soc. **26** [1910] 218; J. Chem. Soc. **97** [1910] 1956/60).
[59] A. Michaelis, A. Reese (Liebigs Ann. Chem. **233** [1886] 39/60).
[60] K. A. Jensen (Z. Anorg. Allgem. Chem. **250** [1943] 257/67).

[61] W. J. Lile, R. J. Menzies (J. Chem. Soc. **1950** 617/21).
[62] G. A. Razuvaev, M. A. Shubenko (Zh. Obshch. Khim. **21** [1951] 1974/9; J. Gen. Chem. [USSR] **21** [1951] 2193/9).

[63] M. Becke-Goehring, H. Thielemann (Z. Anorg. Allgem. Chem. **308** [1961] 33/51).
[64] A. N. Nesmeyanov, A. E. Borisov (Izv. Akad. Nauk SSSR Ser. Khim. **1969** 939/40; Bull. Acad. Sci. [USSR] Div. Chem. Sci. **1969** 853/5).
[65] Z. I. Kuplennik, A. M. Pinchuk (Zh. Obshch. Khim. **49** [1979] 155/60; J. Gen. Chem. [USSR] **49** [1979] 135/9).
[66] G. A. Razuvaev, N. A. Osanova, N. P. Shulaev, B. M. Tsigin (Zh. Obshch. Khim. **30** [1960] 3234/7; J. Gen. Chem. [USSR] **30** [1960] 3203/5).
[67] M. Benmalek, H. Chermette, C. Martelet, D. Sandino, J. Tousset (J. Organometal. Chem. **67** [1974] 53/9).
[68] L. Kaufmann (Ger. 223694 [1908] 3122; C.A. **1910** 3122; Brit. 18896 [1909]; C.A. **1911** 2905).
[69] G. O. Doak, G. G. Long, L. D. Freedman (J. Organometal. Chem. **4** [1965] 82/91).
[70] J. S. Thayer (Organometal. Chem. Rev. **1** [1966] 157/78).

[71] A. Rieche, J. Dahlmann, D. List (Liebigs Ann. Chem. **678** [1964] 167/82).
[72] A. Rieche, J. Dahlmann (Ger. 1158975 [1961/63]; C.A. **60** [1964] 9313).
[73] A. Rieche, J. Dahlmann (Ger. [East] 44608 [1961/66]; C.A. **65** [1966] 10623).
[74] G. Grossmann, G. Krabbes, G. Tschernko (Isotopenpraxis **4** [1968] 307/10).
[75] G. K. Schweitzer, S. W. McCarty (J. Inorg. Nucl. Chem. **27** [1965] 191/9).
[76] H. Watanabe, T. Kawashima, T. Suzuki, N. Ashikari (Japan. 74-02038 [1970/74]; C.A. **81** [1974] No. 92643).
[77] B. O. Schoepfle, B. S. Marks, P. Robitschek, Hooker Chemical Corp. (U.S. 2913428 [1959]; C.A. **1960** 5162).
[78] H. Tereda, T. Imaida, O. Nakagawa, S. Takashima, Mitsubishi Rayon Co., Ltd. (Japan. 68-15999 [1965/68]; C.A. **70** [1969] No. 29594).
[79] R. Nomura, M. Kori, H. Matsuda (Chem. Letters **1985** 579/80).
[80] W. E. McEwen, G. H. Briles, B. E. Giddings (J. Am. Chem. Soc. **91** [1969] 7079/84).

2.5.1.1.7.2.2 Other Triarylantimony Oxides and Dihydroxides, and Tris(2-thienyl)antimony Oxide

$(4\text{-}CH_3OC_6H_4)_3SbO$

This compound is prepared from $(4\text{-}CH_3OC_6H_4)_3SbBr_2$ in $CHCl_3$ and ethanolic NaOH. The precipitate is extracted with hot C_6H_6, and the solvent evaporated. The melting point of the compound is 191 °C, recrystallized from C_2H_5OH. It is easily soluble in C_6H_6 and $CHCl_3$, soluble in ether, and slightly soluble in C_2H_5OH. Reaction with HX (X=halogen) gives the corresponding dihalides [1].

$(3\text{-}NO_2C_6H_4)_3Sb(OH)_2$

$Sb(C_6H_5)_3$ [3] or $(C_6H_5)_3Sb(NO_3)OH$ [2] reacted with the nitrating acid HNO_3:H_2SO_4 (ca. 4:1) at 40 to 50 °C for two hours gives a yellow precipitate when poured into ice-water. It is recrystallized from glacial acetic acid [2, 3], m.p. 190 to 191 °C [3], and 170 to 191 °C, possibly due to the presence of other isomers [2].

It forms light yellow leaflets [2, 3] and is soluble to some extent in C_2H_5OH or ether, insoluble in H_2O, and almost insoluble in C_6H_6 or light petroleum ether [3]. The compound is reduced with Zn in ethanolic NH_4Cl to $Sb(C_6H_4NH_2\text{-}3)_3$. The compound reacts with Br_2 in H_2O, or PBr_5 in $CHCl_3$, to form $1\text{-}Br\text{-}3\text{-}NO_2\text{-}C_6H_4$ [2].

$(4\text{-}CH_3CONHC_6H_4)_3SbO \cdot 4H_2O$

The corresponding stibine is oxidized in CH_3OH with 3% H_2O_2. Addition of HCl and ice-cold aqueous NaOH gives a colorless substance, melting at 200 °C if recrystallized from aqueous CH_3OH. This compound is easily soluble in hot, anhydrous CH_3OH, from which crystallizes another modification, which melts at 250 °C. It gives with alcoholic HCl a chloride, which reverts to the hydroxide of m.p. 200 °C upon addition of methanolic NaOH [4].

$(3\text{-}N_3C_6H_4)_3SbO$

The compound may be used for the photochemical preparation of printing plates [5].

$(2\text{-}CH_3C_6H_4)_3SbO$ and $(2\text{-}CH_3C_6H_4)_3Sb(OH)_2$

Reaction of the corresponding triarylantimony dibromide with ethanolic KOH and evaporation of the filtrate gives a powder which melts at 220 °C and which is formulated as the oxide. The solubility resembles that of $(4\text{-}CH_3C_6H_4)_3SbO$ (see below) [6]. The compound, formulated as the dihydroxide, shows the following solubilities: 8×10^{-3} mol/L in C_6H_6, 4×10^{-3} mol/L in CCl_4, and 6×10^{-3} mol/L in $CHCl_3$. The distribution coefficients between an organic solvent and water are given as 40 for C_6H_6/H_2O, 20 for CCl_4/H_2O, and 30 for $CHCl_3/H_2O$ [7].

$(3\text{-}CH_3C_6H_4)_3SbO$

The compound is prepared by reaction of the corresponding triarylantimony dibromide with ethanolic KOH. Evaporation of the filtrate gives a powder which melts at about 185 °C. It is insoluble in H_2O, and slightly soluble in aqueous NaOH, in C_2H_5OH, C_6H_6, $CHCl_3$, and ether [6].

$(4\text{-}CH_3C_6H_4)_3SbO$ and $(4\text{-}CH_3C_6H_4)_3Sb(OH)_2$

$(4\text{-}CH_3C_6H_4)_3SbBr_2$ is treated with ethanolic KOH or NaOH. The resulting mass is washed with warm water to give a white amorphous powder with a melting point of 220 °C [6] or 223.5 °C [9]. It can be recrystallized from C_6H_6 to form small white needles. This compound was formulated as the oxide [6, 9], but it is actually the dihydroxide, which is easily transformed to the oxide upon drying in a desiccator or upon melting. The oxide has a melting point of 270 °C, recrystallized from C_6H_6 [8]. The dihydroxide is also obtained upon reaction of $Sb(C_6H_4CH_3\text{-}4)_3$ with 3% H_2O_2 in the presence of dilute KOH, or with Na_2O_2; m.p. 225 °C [10], see also [8].

The oxide has a dipole moment of $\mu=2.0\pm0.1$ D at 25 °C in C_6H_6, and $\mu=2.3\pm0.1$ D at 40 °C in dioxane. It is nearly insoluble in most organic solvents, and slightly soluble in hot C_6H_6 or dioxane [8]. The solubility of the dihydroxide is 5.5×10^{-2} mol/L in $CHCl_3$, and the partition coefficient for $CHCl_3/H_2O$ is 10^3 [7]. Reaction with HX (X = halogen) gives the corresponding $(4\text{-}CH_3C_6H_4)_3SbX_2$ [6]. The difluoride is also obtained by reacting the oxide with SF_4 in CH_2Cl_2 at −10 °C in a yield of 86%. With CF_3CO_2H in CH_3OH, 30 min at 20 °C, $(4\text{-}CH_3C_6H_4)_3Sb(O_2CCF_3)_2$ is formed in a yield of 86% [12]. Reaction with formic acid or acetic acid is described to give $(4\text{-}CH_3C_6H_4)_3Sb(OH)O_2CR$ (R = H or CH_3) [6]. The compounds were tested as retarding agents for the burning of epoxy resins [13].

$(4\text{-}HOCH_2CH(OH)CH_2C_6H_4)_3Sb(OH)_2$

The compound is obtained by shaking $Sb(C_6H_4CH_2CH{=}CH_2\text{-}4)_3$ with 1 N KOH and 3% H_2O_2 for two hours and filtration. It decomposes above 150 °C [14].

References on p. 130

$(2\text{-}C_6H_5C_6H_4)_3Sb(OH)_2$

The $(2\text{-}C_6H_5C_6H_4)_3SbBr_2 \cdot CHCl_3$ solvate is treated with ethanolic NH_3 to give a powder which melts at 243 to 244 °C [15].

$(4\text{-}C_6H_5C_6H_4)_3Sb(OH)_2$

The corresponding dihalide $(4\text{-}C_6H_5C_6H_4)_3SbX_2$ (X = Cl, Br, I) is hydrolyzed for 1 h with boiling C_2H_5OH containing NH_3. The compound is obtained in quantitative yield. It begins to sinter at 205 °C and melts at 210 to 211 °C. It is sparingly soluble in hot C_2H_5OH, crystallizing out in small flat needles. It is almost insoluble in C_6H_6. Reaction with H_2S in C_2H_5OH gives the corresponding sulfide [16].

$(3\text{-}NO_2\text{-}4\text{-}CH_3C_6H_3)_3SbO$

The corresponding triarylantimony dinitrate is heated for 8 h with H_3PO_4 in C_2H_5OH. The precipitate is washed with hot H_2O, dried, and crystallized from C_2H_5OH. The compound is obtained as a white precipitate, m.p. 225 °C. On further heating, it blackens and becomes pyrophoric [17].

$(2,4\text{-}(CH_3)_2\text{-}5\text{-}NO_2C_6H_2)_3SbO$

The compound is obtained by reacting the corresponding dinitrate with H_3PO_4 in boiling absolute C_2H_5OH for 5 h. The mixture is poured into water, the precipitate dried, and recrystallized from $CHCl_3$ solution with light petroleum ether. It forms a white crystalline powder which sinters slightly at 110 °C and melts at 218 °C. It is also one of the products in the reaction of $Sb(C_6H_3(CH_3)_2\text{-}2,4)_3$ with fuming HNO_3 [18].

The compound is soluble in C_2H_5OH and forms an orange solution in $CHCl_3$. It is reduced to the corresponding stibine by aqueous $Na_2S_2O_3$ in boiling C_2H_5OH. With glacial acetic acid, the diacetate is formed. The compound is brominated by Br_2 in $CHCl_3$/petroleum ether to give $(2,4\text{-}(CH_3)_2\text{-}5\text{-}NO_2\text{-}6\text{-}BrC_6H)_3SbO$ (see below) [18].

$(2,4\text{-}(CH_3)_2\text{-}5\text{-}NO_2\text{-}6\text{-}BrC_6H)_3SbO$

See the preceding compound for its preparation. It is a white crystalline substance which melts at 162 °C. It is reduced by Zn in C_2H_5OH containing aqueous NH_3 and NH_4Cl to give $Sb(C_6H_2NH_2\text{-}5\text{-}(CH_3)_2\text{-}2,4)_3$. Reaction with $Na_2S_2O_3$ in C_2H_5OH/H_2O probably gives $Sb(C_6H_2NO_2\text{-}5\text{-}(CH_3)_2\text{-}2,4)_3$ [18].

$(1\text{-}C_{10}H_7)_3SbO$

Tris(1-naphthyl)antimony oxide is prepared from $(1\text{-}C_{10}H_7)_3SbX_2$ (X = Cl, Br) and ethanolic KOH. It crystallizes along with one molecule C_6H_6 which is lost at 90 °C. A melting point of 219 to 220 °C is given [19]. The compound is also isolated besides $Sb(C_{10}H_7\text{-}1)_3$ in an attempt to prepare $Sb(C_{10}H_7\text{-}1)_5$ from $(1\text{-}C_{10}H_7)_3SbBr_2$ and $LiC_{10}H_7\text{-}1$. The melting point is given as 244.5 to 246.5 °C [11].

$(C_4H_3S)_3SbO$ (C_4H_3S = 2-thienyl)

Tris(2-thienyl)antimony dibromide and hot ethanolic KOH give the compound as a powder which melts with decomposition at 217 °C (corrected). The substance gives Sb_2O_3 after one year in air. A solution in C_6H_6 reacts with diluted HCl or HBr back to the corresponding dihalide. I_2 is formed upon treatment with aqueous HI, and finally SbI_3 is obtained [20].

References on p. 130

References:

[1] C. Löloff (Ber. Deut. Chem. Ges. **30** [1897] 2834/43).
[2] G. T. Morgan, M. G. Micklethwait (Proc. Chem. Soc. **27** [1911] 274; J. Chem. Soc. **99** [1911] 2286/98).
[3] P. May (Proc. Chem. Soc. **26** [1910] 218; J. Chem. Soc. **97** [1910] 1956/60).
[4] H. Schmidt (Liebigs Ann. Chem. **429** [1922] 123/52).
[5] T. Jurre, E. G. Guk, L. A. Busygina, G. A. Artamonova, A. V. El'tsov (3rd Vses. Konf. Besserebryan Neobych. Fot. Protsessam 1980, Vil'nyus 1980, pp. 211/3; C.A. **94** [1981] No. 39492).
[6] A. Michaelis, U. Genzken (Liebigs Ann. Chem. **242** [1887] 164/88).
[7] M. Benmalek, H. Chermette, C. Martelet, D. Sandino, J. Tousset (J. Organometal. Chem. **67** [1974] 53/9).
[8] K. A. Jensen (Z. Anorg. Allgem. Chem. **250** [1943] 268/76).
[9] A. Michaelis, U. Genzken (Ber. Deut. Chem. Ges. **17** [1884] 924/5).
[10] L. Kaufmann (Ger. 360973; C.A. **1924** 841).
[11] A. N. Nesmeyanov, A. E. Borisov, N. V. Novikova (Izv. Akad. Nauk SSSR Ser. Khim. **1964** 1202/9; Bull. Acad. Sci. USSR Div. Chem. Sci. **1964** 1116/21).
[12] V. I. Popov, N. V. Kondratenko (Zh. Obshch. Khim. **46** [1976] 2597/601; J. Gen. Chem. [USSR] **46** [1976] 2477/80).
[13] J. Havranek, J. Mleziva (Angew. Makromol. Chem. **84** [1980] 105/17).
[14] F. Yu. Yusupov, Z. M. Manulkin (Tr. Tashkent. Farm. Inst. **4** [1966] 531/7; C.A. **68** [1968] No. 59676).
[15] D. E. Worrall (J. Am. Chem. Soc. **62** [1940] 2514/5).
[16] D. E. Worrall (J. Am. Chem. Soc. **52** [1930] 2046/50).
[17] A. E. Goddard, V. E. Yarsley (J. Chem. Soc. **1928** 719/23).
[18] A. E. Goddard (J. Chem. Soc. **123** [1923] 2315/23).
[19] K. Matsumiya (Mem. Coll. Sci. Univ. Kyoto **8** [1925] 11/8; C.A. **1925** 1704).
[20] E. Krause, G. Renwanz (Ber. Deut. Chem. Ges. **65** [1932] 777/84).

2.5.1.1.8 Triorganoantimony Peroxides R_3SbO_2 and Bis(hydroperoxides) $R_3Sb(OOH)_2$

$(CH_3)_3SbO_2$

A polymeric substance of this composition is obtained in 92% yield if stoichiometric amounts of $(CH_3)_3Sb(OC_2H_5)_2$ in pentane are reacted with H_2O_2 in ether. After some hours at room temperature the compound precipitates and is filtered and dried. The compound melts at 222 to 223 °C and is sensitive towards hydrolysis [1].

$(C_6H_5)_3SbO_2$

$(C_6H_5)_3SbBr_2$, in C_6H_6, and H_2O_2, in ether, are reacted with stirring at room temperature and with simultaneous passing of NH_3 or amines ($N(CH_3)_3$ or $N(C_2H_5)_3$) through the mixture. The resulting precipitate is washed with H_2O to remove NH_4Br, with C_2H_5OH and $CHCl_3$, and dried. The yield is 98% [1]. The compound is also obtained from $(C_6H_5)_3Sb(OC_2H_5)_2$ and H_2O_2 in ether [1, 2].

It melts at 205 to 210 °C [1]. Another melting point with decomposition of the substance is given as 150 °C [2]. It is only slightly sensitive towards hydrolysis and hardly soluble in nonpolar organic solvents. In glacial acetic acid or boiling C_2H_5OH, heterolytic cleavage of the Sb-O bond occurs [1].

$(CH_3)_3Sb(OOH)_2$

A C_6H_6 solution of $(CH_3)_3Sb(OCH_3)_2$ is dropped into an ice-cold solution of 6% H_2O_2 (3-fold excess) in ether to induce the precipitation of colorless crystals, which are filtered, washed with ether, and dried in vacuum. The yield is 92% of the compound which melts at 60 to 62 °C with decomposition. A recrystallization from boiling $CHCl_3$ is possible with almost no decomposition [3].

$(C_6H_5)_3Sb(OOH)_2$

A C_6H_6 solution of $(C_6H_5)_3Sb(OC_2H_5)_2$ is dropped into a solution of H_2O_2 (5-fold excess) in a 1:1 ether/C_6H_6 mixture with stirring. The precipitating crystals are filtered, washed with ether, and dried in vacuum; the yield of the compound is 85%. This substance, which contains some solvent, begins to darken at 175 °C and melts with decomposition at 210 to 215 °C [3]. A patent describes the preparation of the compound from $(C_6H_5)_3SbCl_2$ and H_2O_2 in benzene [4].

Recrystallization from C_6H_6, $CHCl_3$, or $CHBr_3$ gives various substances which contain 1/3 mol solvent per formula unit [3]. The compound reacts with $Tl(C_2H_5)_3$ (ratio 1:2) in diglyme at room temperature to give $(C_6H_5)_3Sb(OOTl(C_2H_5)_2)_2$ which rearranges to $(C_6H_5)_2(C_6H_5O)$-$Sb(OOTl(C_2H_5)_2)(OTl(C_2H_5)_2)$. With $(C_6H_5)_2SnCl_2$, it forms the polymer $(-(C_6H_5)_2SnOO$-$Sb(C_6H_5)_3OO-)_n$ [5].

References:

[1] A. Rieche, J. Dahlmann, D. List (Liebigs Ann. Chem. **678** [1964] 167/82).
[2] A. Rieche, J. Dahlmann, D. List (Angew. Chem. **73** [1961] 494).
[3] J. Dahlmann, A. Rieche (Chem. Ber. **100** [1967] 1544/9).
[4] A. Rieche, J. Dahlmann (Ger. 1155127 [1960/63]; C.A. **60** [1964] 5554).
[5] V. A. Dodonov, T. I. Starostina, T. G. Brilkina, T. I. Zinov'eva, V. V. Kutyreva (Khim. Elementoorg. Soedin. No. 4 [1976] 69/71; C.A. **88** [1978] No. 23077).

2.5.1.1.9 Triorganoantimony Diolates

2.5.1.1.9.1 $R_3Sb(OR')_2$ Compounds with R' = Alkyl

$(CH_3)_3Sb(OCH_3)_2$

The preparation of the compound is mentioned without details from trimethylantimony dihalides and CH_3OH in $CHCl_3$ or C_6H_6 solution in the presence of NH_3 as the HX acceptor [1].

It reacts, dissolved in C_6H_6, with an excess of H_2O_2 in ether to form 92% of $(CH_3)_3Sb(OOH)_2$ [1]. Reaction with 2-HOC_6H_4CH=NC_6H_4SH-2′ gives I with an octahedral structure [2, 3]. The analogous reaction occurs with 2-HOC_6H_4CH=$NCH_2CH(CH_3)OH$ [4], whereas reaction with $CH_3C(OH)$=$CHC(CH_3)$=NC_6H_4SH-2 leads to the isolation of II [2].

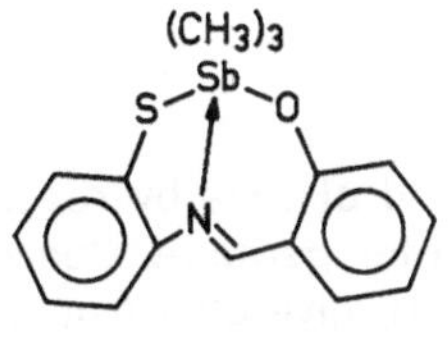

I

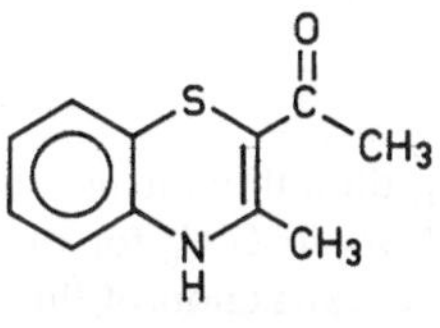

II

$(CH_3)_3Sb(OC_2H_5)_2$

$(CH_3)_3SbBr_2$ and $NaOC_2H_5$ are reacted in C_2H_5OH for 30 min at room temperature. Extracting the mixture with C_6H_6 and evaporation of the solvent gives the hygroscopic compound in a yield of 59%. It boils at 66 to 67 °C/5 Torr [5]. It is also obtained from the dihalide with C_2H_5OH and NH_3 gas in $CHCl_3$ or C_6H_6 [1].

The compound reacts with H_2O_2 in pentane/ether at room temperature to give $(-(CH_3)_3SbO_2-)_n$ [6]. It is reduced to $Sb(CH_3)_3$ by reaction with RSH ($R=C_2H_5$, C_3H_7, C_4H_9, CH_2CH_2OH, C_6H_5) in C_6H_6 for 15 min at room temperature; the thiols give the corresponding disulfides. Thioacetic acid or thiobenzoic acid form $(CH_3)_3Sb(SC(O)R)_2$ ($R=CH_3$ or C_6H_5) in good yields under similar conditions [5]. Reacted with $HXCH_2CO_2H$ (X=O, S) in a 1:1 ratio the substance forms $(CH_3)_3Sb(-XCH_2CO_2-)$. For X=S the structure is not discussed. For X=O, a monomer-dimer equilibrium with chelating or bridging CO_2 groups and hexacoordinated Sb is assumed [7].

$(CH_3)_3Sb(OC_4H_9\text{-}t)_2$

$(CH_3)_3SbCl_2$ and $NaOC_4H_9$-t are reacted for 5 h at 35 °C in ether. Distillation of the filtrate gives 76% of the compound, boiling at 52.5 °C/1 Torr. Its melting point is 22 °C. It decomposes rapidly when distilled at normal pressure [8].

$(C_2H_5)_3Sb(OC_2H_5)_2$

The compound is obtained from $Sb(C_2H_5)_3$ and HgO in C_2H_5OH. Distillation of the filtrate gives the substance which boils at 70 °C/1.5 Torr. Its refractive index is $n_D^{25}=1.3928$ [9]. The compound may be used together with $TiCl_3$ and $Al(C_2H_5)_3$ as a catalyst for the polymerization of propene [10].

$(C_2H_5)_3Sb(OC_4H_9)_2$ and $(C_2H_5)_3Sb(OCH_2CH_2OC_2H_5)_2$

The compounds are mentioned in a patent as catalysts together with $TiCl_3$ and $Al(C_2H_5)_3$ for the preparation of isotactic polypropylene [10].

$(C_3H_7)_3Sb(OCH_2(CF_2)_3CHF_2)_2$

$(C_3H_7)_3Sb{=}NSO_2R$ ($R=C_6H_5$, 4-$CH_3C_6H_4$, and $CHF_2(CF_2)_3CH_2OH$) are reacted in CH_2Cl_2. Upon cooling, RSO_2NH_2 precipitates and is filtered. By concentration of the filtrate, 83% yield of the compound is obtained. It boils at 132 to 135 °C at 0.0931 hPa (0.07 Torr) and dissolves as a monomer in dioxane. Its relative density is $d_4^{20}=1.554$, and its refractive index $n_D^{20}=1.4090$ [25].

$(C_4H_9)_3Sb(OC_2H_5)_2$

Tributylantimony oxide and an excess of C_2H_5OH are refluxed in C_6H_6 on a water separator. Distillation of the mixture gives 44% of the compound as a colorless oil with a boiling point of 104 to 109 °C at 0.1 Torr [11]. The substance is a catalyst for the trimerization of phenylisocyanates [11, 12].

$(C_6H_5)_3Sb(OCH_3)_2$

Triphenylantimony dimethoxide is obtained in a yield of 95% by reacting $NaOCH_3$ in CH_3OH with $(C_6H_5)_3SbBr_2$ in C_6H_6 for one hour at room temperature. Concentration of the mixture in vacuum and extraction of the residue with C_6H_6 give after evaporation the compound with a melting point of 97 to 100 °C [13]. After sublimation at 0.1 Torr, 89% of the

References on p. 135

substance with a melting point of 100 to 102 °C is obtained [13, 14]. The preparation of the compound is also possible from the same educts in methanolic solution [13 to 15]. $(C_6H_5)_3SbO$ and an excess of CH_3OH are stirred together with a molecular sieve (4 Å pore size) for 12 h. Removal of the solvent and sublimation give the compound in a 25% yield, m.p. 100 °C [16]. $(C_6H_5)_4SbOCH_3$ is reacted for five weeks in refluxing absolute CH_3OH. Workup as before gives the compound in a yield of 81%, m.p. 100 to 102 °C [13, 15].

^1H NMR spectra of the compound were recorded in CCl_4 [13, 15]; $\delta = 3.03$(s), 7.3(m), and 7.9(m) ppm [13].

The X-ray crystal structure was determined [17, 18]. The compound crystallizes in the monoclinic space group $P2_1/c-C^5_{2h}$ (No. 14) with a = 11.51(2), b = 9.40(2), c = 7.30(3) Å; β = 101.75°; Z = 4; d_c = 1.505, d_m = 1.50 g/cm³. The structural data are based on 822 independent reflections. Refinements lead to a conventional crystallographic R-factor of 3.4%. The triphenylantimony dimethoxide molecule has a trigonal bipyramidal structure with three carbon atoms of the phenyl substituents and the antimony atom exactly in the equatorial plane. The slight distortion of the O-Sb-O angle to 175.3° probably results from packing effects. The structure and the most important bond distances and angles are given in **Fig. 9** [14].

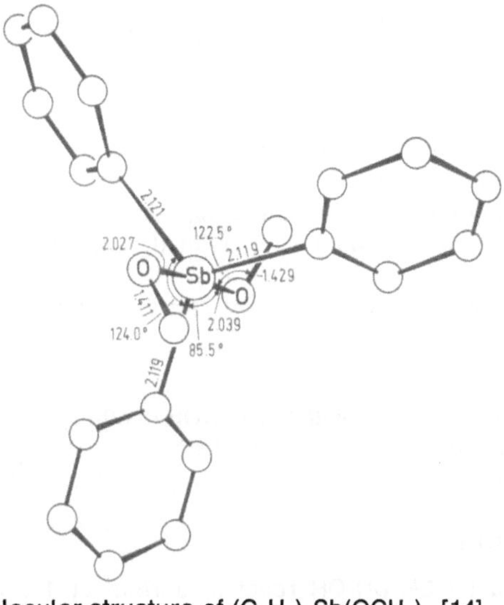

Fig. 9. Molecular structure of $(C_6H_5)_3Sb(OCH_3)_2$ [14].

The compound was found to be very soluble in C_6H_6, CH_3OH, and petroleum ether, but it could not be recrystallized from any of these solvents [13].

Triphenylantimony dimethoxide reacts with H_2O in $(CH_3)_2CO$ to give $(C_6H_5)_3SbO$ in a yield of 80% [13, 15, 18]. With acetylacetone or benzoylacetone, refluxed for two hours in C_6H_6, it gives $(C_6H_5)_3Sb(OCH_3)RCOCHCOR'$ (R = CH_3; R' = CH_3, C_6H_5) in yields of about 65% [19]. Under similar reaction conditions the oximes RR'C=NOH (R = CH_3, R' = C_6H_5, 4-$NO_2C_6H_4$; R = C_6H_5, R' = C_6H_5; R,R' = -$(CH_2)_5$-; R = H, R' = 4-$CH_3OC_6H_4$, 2-furyl) and the compound form the corresponding dioximates $(C_6H_5)_3Sb(ON{=}CRR')_2$ quantitatively [20]. Sodium salts of ortho-substituted aromatic Schiff bases 2-$NaOC_6H_4CR$=NR'-2 (R = R' = CH_3; R = H, R' = CH_3), reacted with the compound in a ratio of 1:1 in C_6H_6 under reflux, give the corresponding $(C_6H_5)_3Sb(OCH_3)OC_6H_4CR$=NR'-2 with an octahedral structure in good yields [21]. Schiff bases of the type 2-HOC_6H_4CR=NR'OH (R = H, R' = -$(CH_2)_2$-, -$(CH_2)_3$-,

References on p. 135

$CH_2CH(CH_3)$; $R=CH_3$, $R'=(CH_2)_2$, $(CH_2)_3$, $CH_2C(H)CH_3$), and 2-$HOC_{10}H_6CH$=NCH_2CH_2OH react with the compound in refluxing C_6H_6 to form III and IV, respectively, in yields of about 80 to 90%, with probably an octahedral structure resulting from the tridendate Schiff base as the ligand [4]. $(CH_3)_3SiN$=$P(C_6H_5)_3$ and the compound, refluxed for two hours in CH_3OH form $(C_6H_5)_3Sb(N$=$P(C_6H_5)_3)_2$ in a yield of 75% [22].

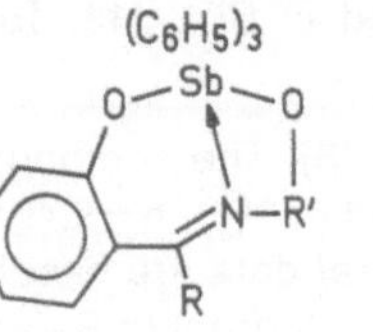

III

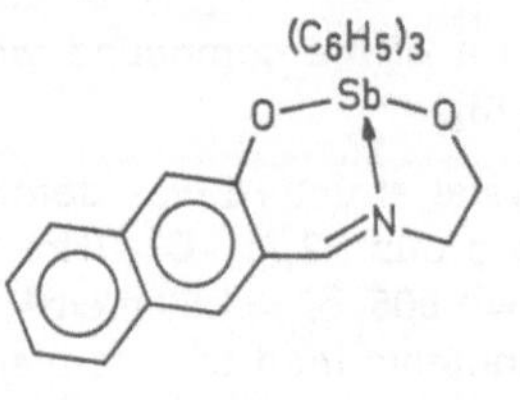

IV

The compound is useful as a fire retardent in halogen-containing polyester resins [23].

$(C_6H_5)_3Sb(OC_2H_5)_2$

The preparation of the compound from $(C_6H_5)_3SbX_2$ (X = halogen) and C_2H_5OH with NH_3 as HX acceptor is mentioned without details [1].

It reacts, dissolved in C_6H_6, with an excess of H_2O_2 in ether to give a precipitate of $(C_6H_5)_3Sb(OOH)_2$ in a yield of 85% [1]. With H_2O_2 in ether/pentane, 92% yield of $(-(C_6H_5)_3SbO_2-)_n$ is obtained. ROOH (R = 1,2,3,4-tetrahydronaphthalen-1-yl or 3,4-dihydro-1H-2-benzopyran-1-yl) react with the compound in ether to form the corresponding $(C_6H_5)_3Sb(OOR)_2$ in yields of about 90%. With $(CH_3)_3COOH$, in pentane in a ratio of 1:1, $(C_6H_5)_3Sb(OC_2H_5)OOC(CH_3)_3$ is formed in a yield of 80% [6]. The compound is useful as a cocatalyst together with $TiCl_3$ and $Al(C_2H_5)_3$ for the polymerization of propene [10].

$(C_6H_5)_3Sb(OC_4H_9)_2$

The reaction of the compound with t-C_4H_9OOH in ether, to form $(C_6H_5)_3Sb(OC_4H_9)OOC_4H_9$-t in a yield of 85% is reported [6]. A patent describes the use of the compound together with $TiCl_3$ and $Al(C_2H_5)_3$ as a catalyst for the polymerization of propene [10].

$(C_6H_5)_3Sb(OCH_2CH_2OH)_2$

$(C_6H_5)_3SbCl_2$ and $LiOCH_2CH_2OH$ react in a ratio of 1:2 to give the compound which is isolated from $CH_3CO_2C_2H_5/C_6H_{14}$. A ^1H NMR spectrum in $CDCl_3$ shows signals at δ = 3.22 (s, OH), 3.68 (m, CH_2), and 7.55 (m, C_6H_5) ppm. The compound was tested as a condensation catalyst for preparing poly(ethylene terephthalate) [27].

$(C_6H_5)_3Sb(OCH_2CH_2OC_2H_5)_2$

A patent describes the compound as useful, together with $TiCl_3$ and $Al(C_2H_5)_3$, for the polymerization of propene [10].

$(C_6H_5)_3Sb(OCH_2(CF_2)_5CHF_2)_2$

This compound is obtained by reacting $(C_6H_5)_3Sb$=NSO_2R (R = C_6H_5, 4-$CH_3C_6H_4$, 4-ClC_6H_4, 4-BrC_6H_4, 4-$NO_2C_6H_4$) with the appropriate alcohol in CH_2Cl_2. Upon cooling, RSO_2NH_2 precipitates and is filtered. The compound is obtained by concentration of the filtrate. The yield is about 90%. It melts at 115 to 116 °C, recrystallized from C_6H_{14}. It is soluble in $CHCl_3$, CH_2Cl_2, and CCl_4, but insoluble in H_2O, and remains unchanged when heated with H_2O [26].

$(C_6H_5)_3Sb(CH_3COCHCONHC_6H_5)_2$

$(C_6H_5)_3SbBr_2$ and the sodium salt of acetoacetanilide in a ratio of 1:3 are reacted for 2 h in refluxing C_6H_6. The filtrate of the mixture is evaporated, and the residue is washed with C_6H_{14} to give 90% yield of the pink compound with a melting point of 173 °C [28].

$(C_6F_5)_3Sb(OCH_3)_2$

Tris(pentafluorophenyl)antimony dichloride and $NaOCH_3$ were reacted in CH_3OH for 4 h at room temperature to give, after evaporation and extraction of the residue with some aqueous CH_3OH, the compound as a residue in a yield of 48% [24]. The compound melts at 60 °C. The conductivity (in $cm^2 \cdot \Omega^{-1} \cdot mol^{-1}$) of 5×10^{-4} M solutions is $\Lambda = 13.63$ in $(CH_3)_2CO$ and 12.56 in CH_3NO_2. The IR spectrum (Nujol or KBr) shows vibrations at 2950(m) cm^{-1} for νCH and at 530 to 500(m) cm^{-1} for νSbO [24].

References:

[1] J. Dahlmann, A. Rieche (Chem. Ber. **100** [1967] 1544/9).

[2] F. Di Bianca, E. Rivarola, A. L. Speck, H. A. Meinema, J. G. Noltes (J. Organometal. Chem. **63** [1973] 293/300).

[3] F. Di Bianca, E. Rivarola (Atti Accad. Sci. Lettere Arti Palermo I [4] **31** [1972] 167/72).

[4] V. K. Jain, R. Bohra, R. C. Mehrotra (Indian J. Chem. A **22** [1983] 445/6).

[5] Y. Matsumura, M. Shindo, R. Okawara (Inorg. Nucl. Chem. Letters **3** [1967] 219/22).

[6] A. Rieche, J. Dahlmann, D. List (Liebigs Ann. Chem. **678** [1964] 167/82).

[7] Y. Matsumura, M. Shindo, R. Okawara (J. Organometal. Chem. **27** [1971] 357/63).

[8] H. Schmidbaur, H. S. Arnold, E. Beinhofer (Chem. Ber. **97** [1964] 449/58).

[9] G. N. Chremos, R. A. Zingaro (J. Organometal. Chem. **22** [1970] 637/46).

[10] S. Yoshida, S. Kitakawa, Mitsubishi Petrochemical Co., Ltd. (Japan. 72-26183 [1969/72]; C.A. **78** [1973] No. 16795).

[11] S. Herbstman (J. Org. Chem. **30** [1965] 1259/60).

[12] S. Herbstman, Stauffer Chemical Co. (U.S. 3278492 [1964/66]; C.A. **66** [1967] No. 66142).

[13] W. E. McEwen, G. H. Briles, B. E. Giddings (J. Am. Chem. Soc. **91** [1969] 7079/84).

[14] K. W. Shen, W. E. McEwen, S. J. La Placa, W. C. Hamilton, A. P. Wolf (J. Am. Chem. Soc. **90** [1968] 1718/23).

[15] G. H. Briles, W. E. McEwen (Tetrahedron Letters **1966** 5191/6).

[16] R. G. Goel, D. R. Ridley (J. Organometal. Chem. **182** [1979] 207/12).

[17] K. W. Shen (Diss. Univ. Massachusetts, Amherst, Mass., 1968; Diss. Abstr. B **29** [1969] 1989).

[18] W. E. McEwen, G. H. Briles, D. N. Schulz (Phosphorus **2** [1972] 147/53).

[19] V. K. Jain, R. Bohra, R. C. Mehrotra (J. Organometal. Chem. **184** [1980] 57/62).

[20] K. Bajpai, R. C. Srivastava (Syn. Reactiv. Inorg. Metal-Org. Chem. **11** [1981] 7/13).

[21] V. K. Jain, R. Bohra, R. C. Mehrotra (Australian J. Chem. **33** [1980] 2749/52).

[22] K. Bajpai, R. C. Srivastava (Syn. Reactiv. Inorg. Metal-Org. Chem. **12** [1982] 47/54).

[23] B. O. Schoepfle, B. S. Marks, P. Robitschek, Hooker Chemical Corp. (U.S. 2913428 [1959]; C.A. **1960** 5162/3).

[24] A. Otero, P. Royo (J. Organometal. Chem. **154** [1978] 13/9).

[25] Z. I. Kuplennik, Zh. N. Belaya, A. M. Pinchuk (Zh. Obshch. Khim. **51** [1981] 2711/5; J. Gen. Chem. [USSR] **51** [1981] 2339/43).

[26] A. M. Pinchuk, Z. I. Kuplennik, Zh. N. Belaya (Zh. Obshch. Khim. **46** [1976] 2242/6; J. Gen. Chem. [USSR] **46** [1976] 2155/8).
[27] S. B. Maerov (J. Polym. Sci. Polym. Chem. Ed. **17** [1979] 4033/40).
[28] S. Gopinathan (Indian J. Chem. A **15** [1977] 660/2).

2.5.1.1.9.2 $R_3Sb(OR')_2$ Compounds with R' = Aryl

The compounds of this type are summarized in Table 31. They can be prepared by the following methods:

Method I: Equivalent amounts of the corresponding triorganoantimony dichloride or dibromide and the appropriate phenol are dissolved in C_6H_6 and then a slight excess (5%) of $N(C_2H_5)_3$ is added. After stirring about 30 min or longer at ambient temperature the $[NH(C_2H_5)_3]X$ (X = Cl, Br) precipitate is filtered. The solvent is removed from the filtrate in vacuum at 60 °C. The resulting residue is recrystallized, depending on its solubility, from CH_3CN, ligroin, C_6H_6, or $(CH_3)_2CO$. The yield is generally 40 to 70% [1].

Method II: The appropriate phenol is added to a solution of $(C_6H_5)_3Sb{=}NSO_2R$ (R = C_6H_5, 4-$CH_3C_6H_4$, 4-ClC_6H_4, 4-BrC_6H_4, 4-$NO_2C_6H_4$) (ratio 2:1) in $CHCl_3$. The mixture is kept for 12 h at 20 °C; the precipitated RSO_2NH_2 is filtered and washed with CH_2Cl_2, and the solvent is removed in vacuum from the filtrate [4].

General Remarks. Alkyl derivatives are more difficult to synthesize by Method I than the aryl derivatives of the same phenolates and some trialkyl compounds cannot be crystallized, although oily products containing impurities may be obtained [1].

The compounds obtained by Method II are described as crystalline substances, soluble in $CHCl_3$, CH_2Cl_2, and CCl_4, but insoluble in H_2O. They remain unchanged when heated with H_2O [4]. The compounds are stable and can be kept in dry air at 20 to 30 °C without any change for several months. Some trialkyl derivatives are hygroscopic [1].

Table 31
$R_3Sb(OR')_2$ Compounds with R' = Aryl
For explanations, abbreviations, and units, see p. X.

No.	compound	method of preparation; properties and remarks	Ref.
1	$(CH_3)_3Sb(OC_6H_5)_2$	I; m.p. 53°	[1]
		^1H NMR ($CDCl_3$): 1.85 (CH_3), 6.65, 6.85, 7.2 (OC_6H_5)	
		^{13}C NMR ($CHCl_3$): 120.79 (C-4), 121.79 (C-2), 130.89 (C-3)	[2]
		IR (mull): 1257 (νC-O)	[1]
2	$(CH_3)_3Sb(OC_6H_4Cl\text{-}2)_2$	I	[1]
		^1H NMR ($CDCl_3$): 1.89 (CH_3), 6.72, 7.04, 7.30 (OC_6H_4)	
3	$(CH_3)_3Sb(OC_6H_4Cl\text{-}3)_2$	I	[1]
		^1H NMR ($CDCl_3$): 1.83 (CH_3), 6.49, 6.54, 6.99, 7.07 (OC_6H_4)	

Table 31 [continued]

No.	compound	method of preparation; properties and remarks	Ref.
4	$(CH_3)_3Sb(OC_6H_4Cl\text{-}4)_2$	I; m.p. 132° ^1H NMR ($CDCl_3$): 1.79 (CH_3), 6.41, 7.02 (OC_6H_4) IR (mull): 1266 (νC-O)	[1]
5	$(CH_3)_3Sb(OC_6H_4Br\text{-}4)_2$	I; m.p. 143° ^1H NMR ($CDCl_3$): 1.79 (CH_3), 6.40, 7.17 (C_6H_4)	[1]
6	$(CH_3)_3Sb(OC_6H_4NO_2\text{-}2)_2$	I; m.p. 115° ^1H NMR ($CDCl_3$): 1.93 (CH_3), 6.72, 7.31 (C_6H_4)	[1]
7	$(CH_3)_3Sb(OC_6H_4NO_2\text{-}3)_2$	I; m.p. 113° ^1H NMR ($CDCl_3$): 1.97 (CH_3), 6.94, 7.30, 7.65, 7.71 (C_6H_4)	[1]
8	$(CH_3)_3Sb(OC_6H_4NO_2\text{-}4)_2$	I ^1H NMR ($CDCl_3$): 1.84 (CH_3), 6.58, 8.10 (C_6H_4) IR (mull): 1293 (νC-O)	[1]
9	$(CH_3)_3Sb(OC_6H_4OCH_3\text{-}2)_2$	I ^1H NMR ($CDCl_3$): 1.78 (CH_3), 1.84 (OCH_3), 6.57, 6.75 (C_6H_4)	[1]
10	$(CH_3)_3Sb(OC_6H_4OCH_3\text{-}4)_2$	I ^1H NMR ($CDCl_3$): 1.15 (OCH_3), 2.08 (CH_3), 7.54, 8.08 (C_6H_4)	[1]
11	$(CH_3)_3Sb(OC_6H_4CH_3\text{-}2)_2$	I; m.p. 80° ^1H NMR ($CDCl_3$): 1.81 (CH_3), 6.55, 6.72, 7.12, 7.24 (C_6H_4)	[1]
12	$(CH_3)_3Sb(OC_6H_4CH_3\text{-}3)_2$	I ^1H NMR ($CDCl_3$): 1.86 (CH_3), 2.26 (3-CH_3), 6.50, 6.65, 7.07 (C_6H_4)	[1]
13	$(CH_3)_3Sb(OC_6H_4CH_3\text{-}4)_2$	I ^1H NMR ($CDCl_3$): 1.78 (CH_3), 2.21 (4-CH_3), 6.56, 6.94 (C_6H_4) IR (mull): 1248 (νC-O)	[1]
14	$(CH_3)_3Sb(OC_6H_4C_2H_5\text{-}4)_2$	I ^1H NMR ($CDCl_3$): 1.84 (CH_3), 1.20, 2.58 (C_2H_5), 6.63, 7.05 (C_6H_4)	[1]
15	$(CH_3)_3Sb(OC_6H_4C_4H_9\text{-t-}4)_2$	I ^1H NMR ($CDCl_3$): 1.86 (CH_3), 6.40, 7.17 (C_6H_4)	[1]
16	$(CH_3)_3Sb(OC_6H_4C_6H_5\text{-}2)_2$	I; m.p. 93°	[1]

References on p. 141

Table 31 [continued]

No.	compound	method of preparation; properties and remarks	Ref.
17	$(CH_3)_3Sb(OC_6H_4C_6H_5\text{-}4)_2$	I; m.p. 161°	[1]
		^{1}H NMR ($CDCl_3$): 1.91 (CH_3)	
18	$(C_2H_5)_3Sb(OC_6H_5)_2$	useful as a catalyst together with $Al(C_2H_5)_3$ and $TiCl_3$ for the polymerization of $CH_3CH{=}CH_2$	[3]
19	$(C_6H_5)_3Sb(OC_6H_5)_2$	I, II (90% yield); m.p. 149° (from CH_3CN), 149 to 150° (from c-C_6H_{12})	[1, 4]
		^{1}H NMR ($CDCl_3$): 6.35, 6.59, 6.85 (OC_6H_5), 7.43, 8.25 (C_6H_5)	[1, 2]
		^{13}C NMR ($CHCl_3$): 136.79, 130.59, 132.69 (C-2,3,4 in C_6H_5), 161.49, 121.29, 129.99, 119.19 (C-1,2,3,4 in OC_6H_5)	[2]
		IR (mull): 1278, 1245 (νC-O)	[1]
		useful as a catalyst together with $TiCl_3$ and $Al(C_2H_5)_3$ for the polymerization of $CH_3CH{=}CH_2$	[3]
20	$(C_6H_5)_3Sb(OC_6H_4Cl\text{-}2)_2$	I; m.p. 153°	[1]
		^{1}H NMR ($CDCl_3$): 6.02, 6.54 (OC_6H_4), 7.48, 8.38 (C_6H_5)	
		IR (mull): 1470, 1434, 452 (C_6H_5), 1272, 1235 (νC-O)	
21	$(C_6H_5)_3Sb(OC_6H_4Cl\text{-}3)_2$	I; m.p. 125°	[1]
		^{1}H NMR ($CDCl_3$): 6.43, 7.20, 7.43 (OC_6H_4), 7.51, 8.17 (C_6H_5)	
		IR (mull): 1470, 1432, 450 (C_6H_5), 1268, 1249 (νC-O)	
22	$(C_6H_5)_3Sb(OC_6H_4Cl\text{-}4)_2$	I, II (93% yield); m.p. 140°, 151 to 152° (from C_6H_6)	[1, 4]
		^{1}H NMR ($CDCl_3$): 6.19, 6.75 (OC_6H_4), 7.49, 8.15 (C_6H_5)	[1]
		^{13}C NMR ($CHCl_3$): 160.09, 122.39, 129.92, 123.79 (C-1,2,3,4 in OC_6H_4)	[2]
		IR (mull): 1474, 1431, 452 (C_6H_5), 1271, 1238 (νC-O)	[1]
23	$(C_6H_5)_3Sb(OC_6Cl_5)_2$	II (90% yield); m.p. 248 to 249° (from CH_2Cl_2)	[4]
		useful as a bactericide	[5]
24	$(C_6H_5)_3Sb(OC_6H_4Br\text{-}4)_2$	I, II (86% yield); m.p. 161°, 168 to 169° (from C_6H_6)	[1, 4]
		^{1}H NMR ($CDCl_3$): 6.14, 6.90, 7.17, 7.24 (OC_6H_4), 7.48, 8.16 (C_6H_5)	[1]
		IR (mull): 1273, 1240 (νC-O)	[1]
25	$(C_6H_5)_3Sb(OC_6H_3Br_2\text{-}2,4)_2$	II (90% yield); m.p. 192 to 193° (from CH_2Cl_2)	[4]
26	$(C_6H_5)_3Sb(OC_6H_4NO_2\text{-}4)_2$	II (80% yield); m.p. 174 to 175° (from C_6H_6)	[4]

References on p. 141

Table 31 [continued]

No.	compound	method of preparation; properties and remarks	Ref.
27	$(C_6H_5)_3Sb(OC_6H_4OCH_3\text{-}4)_2$	I; m.p. 92° ^1H NMR ($CDCl_3$): 3.60 (CH_3), 6.36 (OC_6H_4), 7.49, 8.26 (C_6H_5)	[1]
28	$(C_6H_5)_3Sb(OC_6H_4CH_3\text{-}2)_2$	I; m.p. 200° ^1H NMR ($CDCl_3$): 2.26 (CH_3), 7.04, 7.26, 7.82 (OC_6H_4), 7.42, 8.14 (C_6H_5) IR (mull): 1482, 1431, 456 (C_6H_5), 1273, 1252 (νC-O)	[1]
29	$(C_6H_5)_3Sb(OC_6H_4CH_3\text{-}3)_2$	II; m.p. 110.5° ^1H NMR ($CDCl_3$): 2.00 (CH_3), 6.04, 6.28, 6.65, 7.00 (OC_6H_4), 7.49, 8.27 (C_6H_5) IR (mull): 1481, 1435, 456 (C_6H_5), 1278, 1253 (νC-O)	[1]
30	$(C_6H_5)_3Sb(OC_6H_4CH_3\text{-}4)_2$	I; m.p. 152° ^1H NMR ($CDCl_3$): 2.03 (CH_3), 6.23, 6.65 (OC_6H_4), 7.48, 8.27 (C_6H_5) IR (mull): 1480, 1433, 456 (C_6H_5), 1274, 1232 (νC-O) ^{13}C NMR ($CHCl_3$): 158.99, 120.99, 130.69, 127.99 (C-1,2,3,4 in OC_6H_4)	[1] [2]
31	$(C_6H_5)_3Sb(OC_6H_4C_2H_5\text{-}4)_2$	I ^1H NMR ($CDCl_3$): 1.01, 2.31 (C_2H_5), 6.06, 6.45 (OC_6H_4), 7.45, 8.23 (C_6H_5)	[1]
32	$(C_6H_5)_3Sb(OC_6H_4(C_4H_9\text{-t})\text{-}4)_2$	I; m.p. 120° ^1H NMR ($CDCl_3$): 1.15 (CH_3), 6.28, 6.87 (OC_6H_4), 7.43, 8.26 (C_6H_5) IR (mull): 1274, 1242 (νC-O)	[1]
33	$(C_6H_5)_3Sb(OC_6H_4CHO\text{-}2)_2$	from $(C_6H_5)_3SbBr_2$ and $NaOC_6H_4CHO\text{-}2$ in refluxing C_6H_6; dec. >150°	[6]
34	$(4\text{-}ClC_6H_4)_3Sb(OC_6H_5)_2$	I; m.p. 123° ^1H NMR ($CDCl_3$): 6.33, 6.88 (OC_6H_4), 7.39, 8.00 (C_6H_4) IR (mull): 1241 (νC-O)	[1]
35	$(4\text{-}ClC_6H_4)_3Sb(OC_6H_4Cl\text{-}4)_2$	I; m.p. 151° ^1H NMR ($CDCl_3$): 6.20, 6.83 (OC_6H_4), 7.40, 7.91 (C_6H_4) IR (mull): 1244 (νC-O)	[1]
36	$(4\text{-}ClC_6H_4)_3Sb(OC_6H_4Br\text{-}4)_2$	I ^1H NMR ($CDCl_3$): 7.38, 7.88 (C_6H_4)	[1]

References on p. 141

Table 31 [continued]

No.	compound	method of preparation; properties and remarks	Ref.
37	$(4\text{-}ClC_6H_4)_3Sb(OC_6H_4NO_2\text{-}4)_2$	I; m.p. 185° ^{1}H NMR ($CDCl_3$): 7.30, 7.88 (OC_6H_4), 7.54, 7.97 (C_6H_4) IR (mull): 1290, 1280 (νC-O)	[1]
38	$(4\text{-}ClC_6H_4)_3Sb(OC_6H_4OCH_3\text{-}4)_2$	I; m.p. 123° ^{1}H NMR ($CDCl_3$): 3.63 (CH_3), 6.25, 6.60, 6.79 (OC_6H_4), 7.41, 8.00 (C_6H_4) IR (mull): 1228, 1216 (νC-O)	[1]
39	$(4\text{-}ClC_6H_4)_3Sb(OC_6H_4CH_3\text{-}4)_2$	I; m.p. 167° ^{1}H NMR ($CDCl_3$): 2.11 (CH_3), 6.18, 6.69 (OC_6H_4), 7.36, 7.97 (C_6H_4) IR (mull): 1235 (νC-O)	[1]
40	$(2\text{-}CH_3C_6H_4)_3Sb(OC_6H_5)_2$	I; m.p. 90° ^{1}H NMR ($CDCl_3$): 2.63 (CH_3), 6.02, 6.73 (OC_6H_5), 7.22, 7.33, 7.71 (C_6H_4) IR (mull): 1243 (νC-O)	[1]
41	$(2\text{-}CH_3C_6H_4)_3Sb(OC_6H_4Cl\text{-}4)_2$	I; m.p. 142° ^{1}H NMR ($CDCl_3$): 2.54 (CH_3), 5.91, 6.70 (OC_6H_4), 7.34, 7.61 (C_6H_4) IR (mull): 1263 (νC-O)	[1]
42	$(2\text{-}CH_3C_6H_4)_3Sb(OC_6H_4NO_2\text{-}4)_2$	I; m.p. 199° ^{1}H NMR ($CDCl_3$): 2.22 (CH_3), 5.72, 6.76 (OC_6H_4), 7.26, 7.65 (C_6H_4) IR (mull): 1294 (νC-O)	[1]
43	$(2\text{-}CH_3C_6H_4)_3Sb(OC_6H_4OCH_3\text{-}4)_2$	I; m.p. 113° ^{1}H NMR ($CDCl_3$): 2.59 (CH_3), 3.62 (OCH_3), 5.59, 6.39, 6.70 (OC_6H_4), 6.39, 7.31, 7.71 (C_6H_4) IR (mull): 1226 (νC-O)	[1]
44	$(2\text{-}CH_3C_6H_4)_3Sb(OC_6H_4CH_3\text{-}4)_2$	I; m.p. 113° ^{1}H NMR ($CDCl_3$): 2.07 (4-CH_3), 2.61 (2-CH_3), 5.88, 6.51 (C_6H_4), 7.25, 7.63 (OC_6H_4) IR (mull): 1233 (νC-O)	[1]
45	$(4\text{-}CH_3C_6H_4)_3Sb(OC_6H_5)_2$	I; m.p. 118° ^{1}H NMR ($CDCl_3$): 2.33 (CH_3), 6.31, 6.79 (OC_6H_4), 7.18, 7.91 (C_6H_4) IR (mull): 1243, 1222 (νC-O)	[1]
46	$(4\text{-}CH_3C_6H_4)_3Sb(OC_6H_4Cl\text{-}4)_2$	I; m.p. 200° ^{1}H NMR ($CDCl_3$): 2.35 (CH_3), 6.62, 6.77 (C_6H_4), 7.24, 7.92 (OC_6H_4) IR (mull): 1248 (νC-O)	[1]

Table 31 [continued]

No.	compound	method of preparation; properties and remarks	Ref.
47	$(4\text{-}CH_3C_6H_4)_3Sb(OC_6H_4NO_2\text{-}4)_2$	I; m.p. 200° ^{1}H NMR ($CDCl_3$): 2.38 (CH_3), 6.28 (OC_6H_4), 7.30, 7.89 (C_6H_4) IR (mull): 1261 (νC-O)	[1]
48	$(4\text{-}CH_3C_6H_4)_3Sb(OC_6H_4OCH_3\text{-}4)_2$	I; m.p. 85° ^{1}H NMR ($CDCl_3$): 2.34 (CH_3), 3.62 (OCH_3), 6.34, 6.73 (OC_6H_4), 7.24, 7.98 (C_6H_4) IR (mull): 1227 (νC-O)	[1]
49	$(4\text{-}CH_3C_6H_4)_3Sb(OC_6H_4CH_3\text{-}4)_2$	I; m.p. 170° ^{1}H NMR ($CDCl_3$): 2.07 (CH_3 on OC_6H_4), 2.32 (CH_3 on C_6H_4), 6.22, 6.64 (OC_6H_4), 7.19, 7.95 (C_6H_4) IR (mull): 1243 (νC-O)	[1]
50	$(4\text{-}CH_3C_6H_4)_3Sb(OC_6H_4C_6H_5\text{-}4)_2$	I ^{1}H NMR ($CDCl_3$): 2.35 (CH_3), 7.21, 7.86 (C_6H_4)	[1]

References:

[1] A. Ouchi, M. Nakatani, Y. Takahashi, S. Kitazima, T. Sugihara, M. Matsumoto, T. Uehiro, K. Kitano, K. Kawashima, H. Honda (Sci. Papers Coll. Gen. Educ. Univ. Tokyo **25** [1975] 73/99).
[2] A. Ouchi, T. Uehiro, Y. Yoshino (J. Inorg. Nucl. Chem. **37** [1975] 2347/9).
[3] S. Yoshida, S. Kitakawa, Mitsubishi Petrochemical Co., Ltd. (Japan. 72-26183 [1969/72]; C.A. **78** [1973] No. 16795).
[4] A. M. Pinchuk, Z. I. Kuplennik, Zh. N. Belaya (Zh. Obshch. Khim. **46** [1976] 2242/6; J. Gen. Chem. [USSR] **46** [1976] 2155/8).
[5] J. R. Leebrick; M & T Chemicals Inc. (U.S. 3287210 [1962/66]; C.A. **66** [1967] No. 85070).
[6] S. Gopinathan, C. Gopinathan (Indian J. Chem. A **15** [1977] 660/2).

2.5.1.1.9.3 $R_3Sb(OR')_2$ Compounds with R' = Heterocyclic Group

$(C_6H_5)_3Sb(OC_9H_6N)_2$ (C_9H_6N = 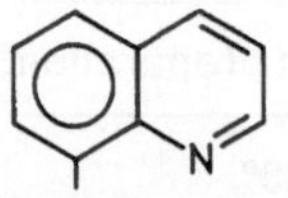)

Sodium 8-quinolinolate is refluxed with $(C_6H_5)_3SbBr_2$ (ratio 3:1) in C_6H_6 for two hours. The solvent from the filtrate is evaporated in vacuum, and the remaining residue is washed with C_6H_{14}. The yield of the compound which melts above 250 °C with decomposition is about 90% [1]. The compound is also obtained by heating $(C_6H_5)_3SbO$ with 8-quinolinol for 15 min at 130 °C [2].

An IR absorption at 1620 cm^{-1} for νCO and UV maxima at λ = 370 and 380 nm indicate strong chelation [1]. The substance is useful as a bactericide and fungicide [2].

$(C_6H_5)_3Sb(OC_8H_7O_3)_2$ ($C_8H_7O_3$ = 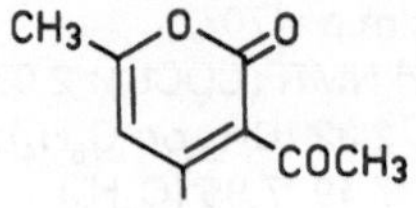)

This substance is prepared analogously to the preceding compound from the sodium salt of 3-acetyl-4-hydroxy-6-methyl-2H-pyran-2-one and $(C_6H_5)_3SbBr_2$ in refluxing C_6H_6. The resulting white powder melts at 155 °C [1].

References:

[1] S. Gopinathan, C. Gopinathan (Indian J. Chem. A **15** [1977] 660/2).
[2] J. R. Leebrick, M & T Chemicals Inc. (U.S. 3287210 [1962/66]; C.A. **66** [1967] No. 85070).

2.5.1.1.9.4 $R_3Sb(-OR'O-)$ Compounds

The compounds described in this section are of the general type shown in Formula I. The triorganoantimony derivatives of carbonic acids (Formula II) are described in Chapter 2.5.1.1.6, pp. 112/3.

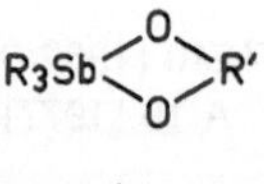

I

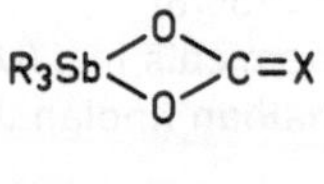

II

The following methods of preparation are used for the compounds listed in Table 32.

Method I: The appropriate R_3SbBr_2 and $CF_3CH(OH)_2$, $CF_3C(OH)_2CF_3$, or $CCl_3CH(OH)_2$ are stirred together in C_6H_6 for 1 h at room temperature in the presence of $N(C_2H_5)_3$. The precipitated $[NH(C_2H_5)_3]Br$ is filtered, and the filtrate is evaporated under vacuum. The residue is recrystallized from CH_3CN [1].

Method II: Compounds of the type $(R_3SbX)_2O$ (X = fluorinated β-diketone) are boiled for several hours or allowed to stand for a few days at room temperature in a moist organic solvent, e.g. CH_3CN. The products are recrystallized from a mixture of dried CH_2Cl_2 and petroleum ether, or form dried CH_3CN [2].

Method III: A C_6H_6 solution of R_3SbBr_2 (R = c-C_6H_{11} or C_6H_5) is added to $NaOCH_3$. The precipitate of NaBr is filtered and $HOCH_2CO_2H$ is added to the filtrate. After evaporation of the solvent under reduced pressure the solid is recrystallized from C_2H_5OH. $(CH_3)_3Sb(-OCH_2CO_2-)$ is prepared similarly, using $(CH_3)_3Sb(OC_2H_5)_2$ [5].

Method IV: To 1,2-dihydroxybenzene, dissolved in C_6H_6 and some $(CH_3)_2CO$, is added Na in a ratio of 1:2. The resulting suspension is given to a solution of $(CH_3)_3SbBr_2$ in $(CH_3)_2CO$. After 20 min at room temperature the mixture is filtered and the filtrate evaporated in vacuum [6].

Method V: $(C_6H_5)_3Sb(OCH_3)_2$ in a slight excess (ca. 20%) and the appropriate Schiff base (2-HOC_6H_4CH=NCH_2CH_2OH, 2-HOC_6H_4CH=$NCH_2CH(CH_3)OH$, 2-HOC_6H_4CH=N-CH_2CH_2OH, 2-$HOC_6H_4C(CH_3)$=NCH_2CH_2OH, 2-$HOC_6H_4C(CH_3)$=$NCH_2CH(CH_3)OH$, 2-$HOC_6H_4C(CH_3)$=$N(CH_2)_3OH$, or 2-$HOC_{10}H_6CH$=NCH_2CH_2OH) are refluxed for 2 to 3 h in C_6H_6. Concentration of the solution in vacuum gives the compounds which are subsequently recrystallized from C_6H_6 or from C_6H_6/C_6H_{14} mixtures [11].

Method VI: The disodium salt of a Schiff base (2-$NaOC_6H_4CH$=NC_6H_4ONa-2′, $NaOC(CH_3)$=CH-$C(CH_3)$=NC_6H_4ONa-2, $NaOC(C_6H_5)$=$CHC(CH_3)$=NC_6H_4ONa-2), prepared from the diol and $NaOCH_3$ in CH_3OH, is added to a stirred suspension of an equimolar amount of R_3SbCl_2 in CH_3OH. The amount of solvent is regulated so that precipitation of NaCl is avoided. Solid complexes deposit from the reaction mixture in 63 to 68% yield [12, 13].

Table 32
Compounds of the Type $R_3Sb(-OR'O-)$.
Further information on numbers preceded by an asterisk is given at the end of the table.
For explanations, abbreviations, and units, see p. X.

No.	R′	method of preparation (yield in %) properties and remarks	Ref.
with R = CH_3			
*1	$CH_2C(O)$	III (quantitative) ^{1}H NMR ($CHCl_3$): 1.59 (CH_3), 4.19 (CH_2)	[5]
2	2-C_6H_4	IV (55) m.p. 113 to 114°, pale yellow transparent crystals (from ether) ^{1}H NMR ($CHCl_3$ or CH_2Cl_2): 1.40 (s, CH_3) from 20° to −70° IR (mull): 564, 550, 530 IR(CH_2Cl_2): 564, 549, 526 dissolves as a monomer in C_6H_6 pseudorotation of the trigonal bipyramidal molecule is assumed decomposes in light and air	[6]
*3	2-C_6H_4CH=NC_6H_4-2′	VI m.p. 188 to 190°, yellowish orange (from c-C_6H_{12}) ^{1}H NMR (CCl_4): 1.15 (s, CH_3), 6.55 to 7.55 (C_6H_4), 8.34 (CH)	[12, 13]
		IR (Nujol): 3070vw, 3010vw, 2920vw (νCH), 1610s, 1600s, 1590s; 1540m, 610m, 605m,	[14, 15]

Table 32 [continued]

No.	R′	method of preparation (yield in %) properties and remarks	Ref.
		595 m (νC–O), 550 m (νSbC_3); 530 vs, 520 vs (νSbC_3), 490 vw, 455 vw, 440 w, 365 mw, 325 vw, 310 vw, 270 vw	
		UV (C_6H_6): λ_{max} (log ε) = 307 (3.90), ca. 350 (sh, 3.70), ca. 370 (sh, 3.74), 424 (4.08)	[12, 13]
		^{121}Sb-γ (4.2 K): δ = −4.41 (vs. $Ca^{121}SnO_3$), $e^2qQ = -20.65$, η = 0.76	[16]
with R = c-C_6H_{11}			
4	$CH_2C(O)$	III (ca. 100)	[5]
		dec. 205 to 206° (from C_2H_5OH)	
		IR (mull): 1580, 1386	
		IR (CH_2Cl_2): 1657, 1340	
		dissolves as a monomer in C_6H_6	
with R = C_6H_5			
*5	$CH(CF_3)$	I (ca. 75)	[1]
		dec. 134°	
		^1H NMR ($CDCl_3$): 5.48 (CH), 7.55, 7.74 (m, C_6H_5)	
		IR (mull): 1110, 1050 (νC–O)	
*6	$CH(CCl_3)$	I (ca. 75)	[1]
		dec. 130°	
		^1H NMR ($CDCl_3$): 5.63 (CH), 7.55, 7.75 (m, C_6H_5)	
		IR (mull): 1105, 1065 (νC–O)	
*7	$C(CF_3)_2$	I (ca. 75)	[1]
		m.p. 122°	
		^1H NMR ($CDCl_3$): 7.60, 7.70 (m, C_6H_5)	
		IR (mull): 1092, 1063 (νC–O)	
8	$C(CF_3)CH_2C(CH_3)O$	II	[2]
		^1H NMR ($CDCl_3$): 2.22 (CH_3), 3.16 (CH_2), 7.46 (H-3,4 in C_6H_5), 7.68 (H-2 in C_6H_5) [2]; 7.36, 7.65 (m, C_6H_5) [1]	[1, 2]
		^{19}F NMR (C_6H_6 vs. c-$CF_2CF_2CCl_2CCl_2$): 37.8	[2]
		IR (mull): 1119, 1089 (νC–O)	[1]
9	$C(CF_3)CH_2C(C_6H_5)O$	II	[2]
		^1H NMR ($CDCl_3$): 3.45 (CH_2), 7.36 (H-3,4 in COC_6H_5), 7.65 (H-2 in COC_6H_5), 7.36, 7.65 (C_6H_5)	
		^{19}F NMR (C_6H_6 vs. c-$CF_2CF_2CCl_2CCl_2$): 37.2	

Table 32 [continued]

No.	R′	method of preparation (yield in %) properties and remarks	Ref.
*10	$C(CH_3)_2C(CH_3)_2$	see further information	[7, 8]
		m.p. 92° (from C_2H_5OH), 93 to 94° (from petroleum ether)	
		^{1}H NMR ($CDCl_3$): 1.28 (s, 12H), 7.0 to 7.3 (m, 9H), 7.8 to 8.02 (m, 6H)	[8]
		^{13}C NMR ($CDCl_3$): 73.35 (CO), 25.68 (CH_3), 128.58 (C-3), 130.06 (C-4), 135.28 (C-2), 140.30 (C-1)	[8]
*11	$CH(C_6H_5)CH(C_6H_5)$	see further information	[7]
		m.p. 168° (dec.), colorless, hexagonal, hard crystals (from $CH_3CO_2C_2H_5$)	
12	$CH_2C(O)$	III (ca. 100)	[5]
		m.p. 179 to 180° (from C_2H_5OH)	
		IR (mull): 1585, 1385 (νC-O)	
		IR ($CHCl_3$): 1675, 1330 (νC-O)	
		dissolves as a monomer in C_6H_6	
13	2-C_6H_4	IV	[6]
		m.p. 141 to 142°, yellow crystals (from ether)	
		dissolves as a monomer in C_6H_6	
*14	2-C_6H_4, 0.5-H_2O solvate	see further information	[9]
		m.p. 149 to 151°	
*15	2-C_6Cl_4	see further information	[10]
		m.p. 185°, colorless crystals (from $CHCl_3$/petroleum ether)	
		IR ($CHCl_3$ or KBr): 1550 (C=C), 740 (CCl)	
		mass spectrum: $[M]^+$, $[C_6H_5Sb]^+$	
*16	2-$C_6H_4CH{=}NCH_2CH_2$	V (74)	[11]
		m.p. 167 to 171° (dec.)	
*17	2-$C_6H_4C(CH_3){=}NCH_2CH_2$	V (80)	[11]
		m.p. 165°	
*18	2-$C_6H_4CH{=}NCH_2CH(CH_3)$	V (97)	[11]
		m.p. 165 to 172° (dec.)	
*19	2-$C_6H_4C(CH_3){=}NCH_2CH(CH_3)$	V (88)	[11]
		m.p. 180 to 185° (dec.)	
*20	$C_{10}H_6CH{=}NCH_2CH_2$ $C_{10}H_6$=naphthalene-2,3-diyl	V (90)	[11]
		m.p. 147 to 160° (dec.)	
21	$C(CH_3){=}CHC(CH_3){=}NC_6H_4$-2	VI	[12, 13]
		m.p. 221 to 222°, 195 to 196°, yellow	
		^{1}H NMR ($CDCl_3$): 1.94 to 2.16 (CH_3), 4.96 (CH), 6.80 to 7.40 (C_6H_4)	[12]

References on p. 153

Table 32 [continued]

No.	R′	method of preparation (yield in %) properties and remarks	Ref.
		IR (Nujol): 1590vs, 1575sh, 1505vs (νCN, CO, CC), 1470vs, 635w, 620m, 570w, 560m, 520s, 465vs, 445vs, 385m, 365s, 330s, 295vs, 280s, 260vs	[15]
		UV (C_6H_6): λ_{max} (log ε) = 320 (3.69), 390 (3.97)	[12, 13]
		dissolves as a monomer in C_6H_6	[15]
		a *mer*-octahedral structure is proposed	[12, 16]
22	$C(C_6H_5)=CHC(CH_3)=NC_6H_4$-2	VI	[12, 13]
		m.p. 217°, orange	
		^1H NMR ($CDCl_3$): 2.28 (CH_3), 5.65 (CH), 6.80 to 7.60 (C_6H_5, C_6H_4)	[12]
		IR (Nujol): 3060m (νCH), 1590s, 1565s, 1500vs (νCN, CO, C=C), 1475vs, 620m, 560s, 550s, 465vs, 445vs, 415s, 400w, 385mw, 355mw, 290vs, 255mw	[15]
		UV (C_6H_6): λ_{max} (log ε) = 340 (3.81), 423 (4.17)	[12, 13]
		dissolves as a monomer in C_6H_6	
		a *mer*-octahedral structure is proposed	
23	2-$C_6H_4CH=NC_6H_4$-2′	VI	[12, 13]
		m.p. 249 to 250°, yellow	
		IR (Nujol): 3050w, 3020vw (νCH), 1610s (νC=N), 1600s, 1550m (νC-O), 615w, 600m, 530s, 490w, 480s, 465mw, 455m, 445ms, 435vw, 365s, 335m, 305ms, 280ms, 260ms	[14, 15]
		UV (C_6H_6): λ_{max} (log ε) = 312 (3.94), ca. 323sh (3.85), ca. 3.57sh (3.76), ca. 372sh (3.81), 423 (4.11)	[12, 13]
		^{121}Sb-γ (4.2 K): δ = −3.97 (vs. $Ba^{121}SnO_3$), e^2qQ = −17.52, η = 0.88	[16]
		^{121}Sb-γ (source ca. 78 K, absorber ca. 9 K): δ = −3.08 ± 0.05 (vs. $Ba^{121}SnO_3$), e^2qQ = −18 ± 0.5, Γ = 2.84 ± 0.08, η = 0.83 ± 0.08	[17]
		a *mer*-octahedral structure is proposed	[12 to 14, 16, 17]
24	2-$C_6H_4C(CH_3)=NC_6H_4$-2′	probably VI	[17]
		^{121}Sb-γ (source ca. 78 K, absorber ca. 9 K): δ = −3.71 ± 0.05 (vs. $Ba^{121}SnO_3$), e^2qQ = −16.7 ± 0.4, Γ = 2.86 ± 0.08, η = 0.93 ± 0.07	
		a *mer*-octahedral structure is proposed	
*25	2-$C_6H_4CH=NCH_2CH_2CH_2$	V (90)	[11]
		m.p. 184 to 185°	
*26	2-$C_6H_4C(CH_3)=NCH_2CH_2CH_2$	V (85)	[11]
		m.p. 195 to 200° (dec.)	

References on p. 153

Table 32 [continued]

No.	R′	method of preparation (yield in %) properties and remarks	Ref.
*27	$C_{28}H_{40}N_2$ Formula IV, p. 151	see further information m.p. 249 to 250° mass spectrum: $[M-nC_6H_5]^+$ (n = 0 to 3), $[M-2C_6H_5-CH_3]^+$, $[Sb(C_6H_5)_n]^+$ (n = 1, 2), $[C_6H_5C_6H_5]^+$, $[C_6H_6]^+$, $[C_6H_5]^+$, $[C_4H_9]^+$	[18]
*28	2-$C_6H_4OC_2H_4OC_2H_4OC_6H_4$-2′	see further information m.p. 202 to 204° (from CH_3CN) IR: 2970, 2940, 2915, 2885, 1590 to 1505, 820, 750	[19]
with R = 4-ClC_6H_4			
*29	$CH(CCl_3)$	I (ca. 75) dec. 145° ^1H NMR ($CDCl_3$): 5.54 (CH), 7.48, 7.64 (q, J = 8, C_6H_4) IR (mull): 1110, 1063 (νC-O)	[1]
*30	$C(CF_3)_2$	I (ca. 75) dec. 160° ^1H NMR ($CDCl_3$): 7.56, 7.64 (q, J = 9, C_6H_4) IR (mull): 1090, 1063 (νC-O)	[1]
*31	$C(CF_3)CH_2C(CH_3)O$	II ^1H NMR ($CDCl_3$): 2.14 (CH_3), 3.06 (CH_2), 7.33 (H-3 in C_6H_4), 7.49 (H-2 in C_6H_4)	[2]
with R = 4-$CH_3C_6H_4$			
*32	$CH(CF_3)$	I (ca. 75) dec. 115° ^1H NMR ($CDCl_3$): 2.44 (CH_3), 5.41 (CH), 7.35, 7.64 (q, J = 8, C_6H_4) IR (mull): 1110, 1070 (νC-O)	[1]
*33	$CH(CCl_3)$	I (ca. 75) dec. 120° ^1H NMR ($CDCl_3$): 2.44 (CH_3), 5.55 (CH), 7.34, 7.63 (q, J = 7, C_6H_4) IR (mull): 1110, 1063 (νC-O)	[1]
*34	$C(CF_3)_2$	I (ca. 75) dec. 132° ^1H NMR ($CDCl_3$): 2.43 (CH_3), 7.31, 7.67 (q, J = 8, C_6H_4) IR (mull): 1110, 1064 (νC-O)	[1]
*35	$C(CF_3)CH_2C(C_6H_5)O$ 0.5-$C_2H_4Cl_2$-1,2 solvate	probably I or II	[4]

References on p. 153

*Further information:

$(CH_3)_3Sb(\text{-}OCH_2CO_2\text{-})$ (Table **32**, No. **1**). Recrystallization from CH_3OH gives two different crystal forms (I = needle-like, II = cubic) with different decomposition points of 224 to 225 °C (I) and 218 to 220 °C (II). The IR spectra of these forms are different in the solid state, but identical in solution (see Table 33). From various physical measurements, such as X-ray powder patterns, molecular weight determinations in different solvents and concentrations, IR spectra, and ^1H NMR spectra at different temperatures, it is concluded that a monomer-dimer equilibrium exists in solution at room temperature. At higher concentrations, the dimer predominates. Below −15 °C, only the dimer species, probably with chelating CO_2 groups, is observed. The structure of form I in the solid may be a repetition of the dimeric moieties [5].

Table 33
IR Vibrations (in cm^{-1}) of $(CH_3)_3Sb(\text{-}OCH_2CO_2\text{-})$ [5].

in mull form I	form II	in $CHCl_3$ or $CHBr_3$	assignment
		1670 vw	νC=O
1603 vs	1623 vs	1590 vs	
1433 m	1437 m	1445 m	δCH_2
1420 s	1408 s	1412 vs	νC-O
1410 s		1333 vw	
1312 s	1307 s	1314 s	CH_2 wagging
1111 vs	1119 vs	1098 vs	νC-O
		1073 w(sh)	
929 w	923 w	917 w	ϱCH_2
865 s	848 vs	852 s	ϱCH_3
844 s		833 s	
723 m	717 m	724 m	δCO_2
587 m	579 vs	587 s	νSbC
556 s		558 s	
549 s		548 (sh)	
526 (sh)	528 w		
512 s	488 vs	515 s	ϱCO_2

$(CH_3)_3Sb(\text{-}2\text{-}OC_6H_4CH{=}NC_6H_4O\text{-}2'\text{-})$ (Table **32**, No. **3**). In $CD_3C_6D_5$ at −65 °C, the ^1H NMR signal for the CH_3 groups is split into two signals at δ = 1.06 and 1.26 ppm in a 1:2 ratio. From this, it is concluded that the CH_3 groups exchange rapidly at room temperature in an octahedral complex [12, 13].

An X-ray crystal structure was determined. The crystals obtained from CH_3OH are monoclinic with a = 10.27, b = 10.28, c = 14.50 Å, and β = 93.45°; space group Ic (standard Cc) − C_s^4 (No. 9); d_m = 1.63, d_c = 1.65 g/cm³. The structure was solved from 1727 independent reflections, R was converged to 0.046. The structure with selected distances and angles is shown in **Fig. 10**. The antimony atom, which is hexacoordinated as a result of Sb-N coordinative bonding, appears to possess a distorted octahedral geometry. The two aromatic rings I and II in the tridentate ligand are slightly twisted from a planar position; the dihedral angle is 11° [13].

References on p. 153

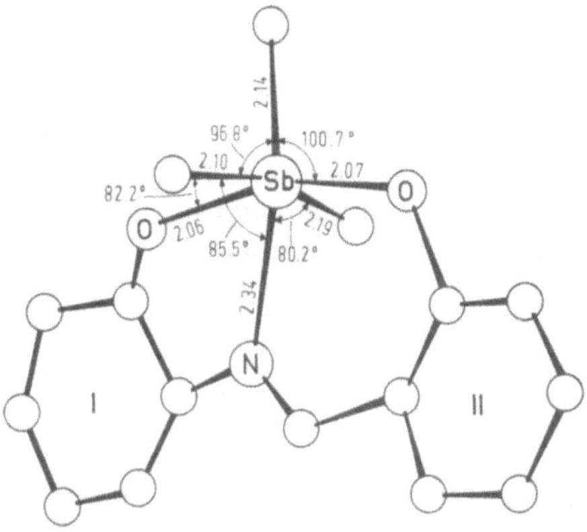

Fig. 10. Molecular structure of $(CH_3)_3Sb(\text{-}2\text{-}OC_6N_4CH{=}NC_6H_4O\text{-}2'\text{-})$ [13].

The compound decomposes upon standing for one month at room temperature [13]. It reacts with R_2SbCl_3 ($R = CH_3$, C_6H_5) in CH_2Cl_2 or c-C_6H_{12} to give $(CH_3)_3SbCl_2$ and $R_2SbCl(\text{-}2\text{-}OC_6N_4CH{=}NC_6H_4O\text{-}2'\text{-})$ [14, 15].

$R_3Sb(\text{-}OR'O\text{-})$ (Table **32**, Nos. **5** to **7**, **29**, **30**, and **32** to **34** with $R = C_6H_5$, $R' = CHCF_3$, $C(CF_3)_2$, $CHCCl_3$; $R = 4\text{-}ClC_6H_4$, $R' = CHCCl_3$, $C(CF_3)_2$; $R = 4\text{-}CH_3C_6H_4$, $R' = CHCF_3$, $CHCCl_3$, and $C(CF_3)_2$). All compounds are white crystalline powders, stable at room temperature for several months, soluble in C_6H_6, $(CH_3)_2CO$, $CHCl_3$, and insoluble in H_2O. They are shown to be monomers in C_6H_6 [1].

$(C_6H_5)_3Sb(\text{-}OC(CH_3)_2C(CH_3)_2O\text{-})$ (Table **32**, No. **10**). $(C_6H_5)_3SbO$ and pinacol are heated together for one to two minutes at 120 to 130 °C. The mixture is dissolved in CH_3OH and, upon standing in a refrigerator, the crude product deposits in 55% yield. Recrystallization from small amounts of C_2H_5OH gives 32% yield of the pure product [7]. The title compound is also obtained by reaction of $Sb(C_6H_5)_3$ with tetramethyl-1,2-dioxetane. The yields are very solvent sensitive. A 1:1 ratio in $CDCl_3$ gives 77% of the title compound and 23% $(CH_3)_2CO$, resulting from catalytic decomposition of the dioxetane. In C_6D_6 the title compound and $(CH_3)_2CO$ are formed in 30 and 70% yield, respectively. With a 3.5-fold excess of dioxetane the conversion of $Sb(C_6H_5)_3$ to the title compound is quantitative in $CHCl_3$ [8].

The compound decomposes upon heating to 240 to 260 °C into $(CH_3)_2CO$ and $Sb(C_6H_5)_3$. Heating in glacial acetic acid gives $(C_6H_5)_3Sb(O_2CCH_3)_2$ and pinacol hexahydrate. It reacts with molar amounts of Br_2 in $CHCl_3$ to form $(C_6H_5)_2SbBr(\text{-}OC(CH_3)_2C(CH_3)_2O\text{-})$ and C_6H_5Br [7].

$(C_6H_5)_3Sb(\text{-}OCH(C_6H_5)CH(C_6H_5)O\text{-})$ (Table **32**, No. **11**). $(C_6H_5)_3SbO$ and $HOCH(C_6H_5)$-$CH(C_6H_5)OH$ are heated in dioxane. Filtration and cooling the filtrate in a refrigerator give the crude product in a yield of 31%. After two recrystallizations from $CH_3CO_2C_2H_5$, 18% yield of the compound is obtained.

The compound decomposes upon heating to 160 °C to form 56% C_6H_5CHO, 22% benzil, and 71% $Sb(C_6H_5)_3$. With Br_2 or 1-Br-2,5-pyrrolidinedione in $CHCl_3$, $(C_6H_5)_3SbBr_2$ is formed. With CH_3CO_2H, $(C_6H_5)_3Sb(O_2CCH_3)_2$ is obtained [7].

$(C_6H_5)_3Sb(\text{-}OC_6H_4O\text{-}2\text{-}) \cdot 0.5H_2O$ (Table **32**, No. **14**). $(C_6H_5)_3SbCl_2$ and 1,2-dihydroxybenzene in a 1:1 molar ratio are dissolved in dry CH_2Cl_2, and dry gaseous NH_3 is passed

References on p. 153

through the solution until no further precipitate is formed. After the NH_4Cl is filtered, the solution is concentrated and cooled to −5 °C whereupon white crystals are obtained. The overall yield of pure material is low due to its ready redox decomposition in solution [9].

The infrared spectrum showed absorptions characteristic of both types of attached groups, but it was not possible to make structurally significant observations because of the general complexity. However, thick concentrated Nujol mulls showed weak bands at 3476 and 3463 cm^{-1} characteristic of coordinated H_2O. The major ions identified in the 70 eV mass spectrum are m/e (% of total ion current): $[(C_6H_5)_3SbO_2C_6H_4]^+$ (1.7), $[Sb(C_6H_5)_3]^+$ and related ions (3.0), and $[SbC_6H_5]^+$ (23.0). Only traces of ions such as $[(C_6H_5)_2SbO_2C_6H_4]^+$, $[C_6H_5SbO_2C_6H]^+$, and $[SbO_2C_6H_4]^+$ were observed [9].

Crystals suitable for X-ray investigation were obtained by slow recrystallization at room temperature from CH_2Cl_2. Crystal data: monoclinic space group $P2_1/c-C^5_{2h}$ (No. 14); a=9.78(1), b=21.10(1), c=19.95(1) Å; β=105.28(5)°; d_c=1.59 g/cm^{-3}; Z=4.

The structure determination shows that the asymmetric unit contains two independent Sb atoms with a water molecule associated only to Sb(1). **Fig. 11** shows the two kinds of molecules, selected bond distances, and angles. Antimony is present in both five- and sixfold coordination. Sb(2) is attached to three carbon and two oxygen atoms in a distorted square pyramidal arrangement. Distortion of this polyhedron arises basically from the presence of the dioxo chelating group. The second polyhedron is a distorted octahedron and is derived from the square pyramidal arrangement mentioned above with the addition of a water molecule weakly coordinated in the sixth position [9].

$(C_6H_5)_3Sb(-OC_6Cl_4O-2)$ (Table **32**, No. **15**). A mixture of equivalent amounts of tetrachloro-1,2-benzoquinone and $Sb(C_6H_5)_3$ in dry C_6H_6 is left for 12 h at room temperature. After evaporation of the solvent in vacuum, the residue is triturated with petroleum ether, and the residue (80% yield) is recrystallized from $CHCl_3$/petroleum ether. A similar yield of the compound is obtained if $(C_6H_5)_3SbS$ and tetrachloro-1,2-benzoquinone are refluxed for 10 h in C_6H_6 and then left for 12 h at room temperature. After evaporation of the solvent in vacuum the residue is washed with C_2H_5OH and with CS_2 to remove sulfur. Recrystallization is as before.

The compound decomposes to $Sb(C_6H_5)_3$ and tetrachloro-1,2-benzoquinone when heated to 250 °C for about 20 min under a pressure of 10 Torr [10].

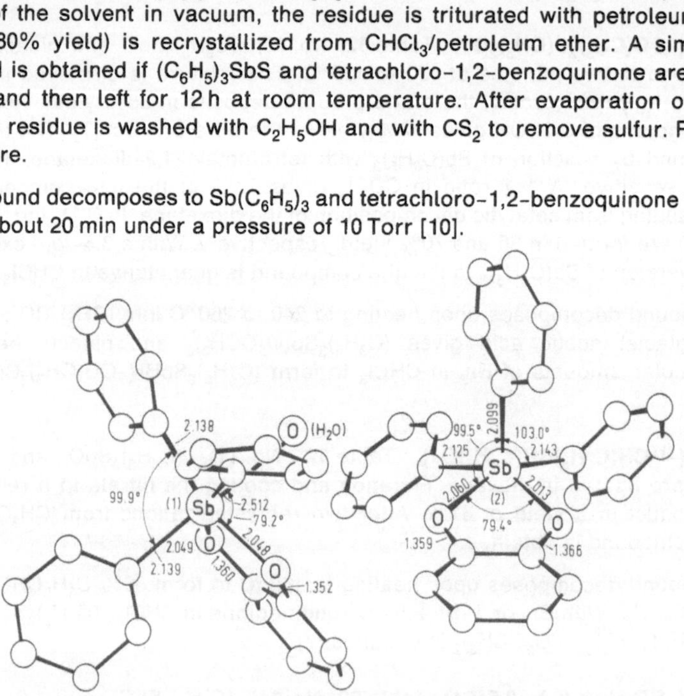

Fig. 11. Molecular structure of $(C_6H_5)_3Sb(-OC_6H_4O-2)_2 \cdot 0.5\ H_2O$ [9].

References on p. 153

$(C_6H_5)_3Sb(-OR'O-)$ (Table **32**, Nos. **16** to **20**, **25**, and **26** with $R'=2\text{-}C_6H_4CH{=}NCH_2CH_2$, $2\text{-}C_6H_4CH{=}NCH_2CH(CH_3)$, $2\text{-}C_6H_4C(CH_3){=}NCH_2CH_2$, $2\text{-}C_6H_4C(CH_3){=}NCH_2CH(CH_3)$, $C_{10}H_6CH{=}NCH_2CH_2$, $2\text{-}C_6H_4CH{=}NCH_2CH_2CH_2$, and $2\text{-}C_6H_4C(CH_3){=}NCH_2CH_2CH_2$). These compounds are yellow crystalline solids soluble in C_6H_6 and $CHCl_3$. They are monomeric in refluxing C_6H_6. The electronic spectra reveal a shift of 20 to 32 nm in the free ligand band at 320 ± 4 nm, indicating chelation of the ligands. The νC=N band for the free Schiff bases (1640 to 1600 cm^{-1}) is shifted by 20 ± 15 cm^{-1} to lower wave numbers except for complex No. 16, Table 32, which appears at a slightly higher wave number. From the spectroscopic data, hexacoordinated Sb is suggested as shown for No. 16 in Formula III [11].

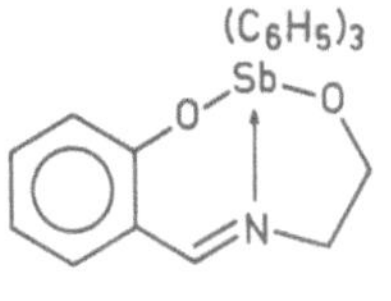

III

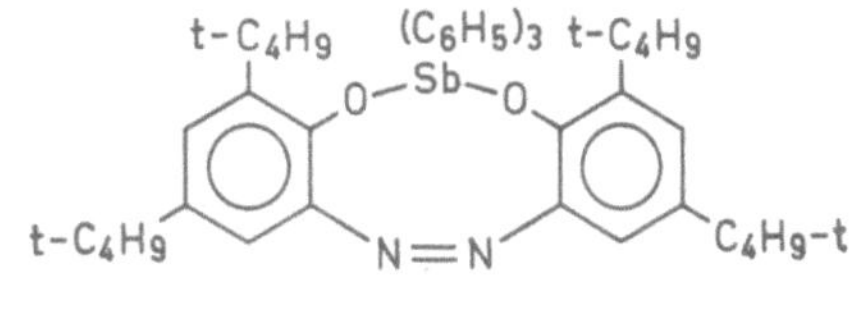

IV

$(C_6H_5)_3Sb\ (C_{28}H_{40}N_2O_2)$ (Formula IV, Table **32**, No. **27**). The compound is obtained by reacting $1\text{-}HO\text{-}2\text{-}NH_2\text{-}4,6\text{-}(t\text{-}C_4H_9)_2C_6H_2$ dissolved in C_2H_5OH and some ethanolic NaOH with a C_6H_6 solution of $Sb(C_6H_5)_3$ in a ratio of 1:1 at temperatures above 30 °C. The solvent is evaporated, and CH_3OH is added. Intensely red crystals precipitate in a yield of 72% [18].

The compound is easily soluble in nonpolar organic solvents. It reacts with mineral acids to form $2\text{-}HO\text{-}3,5\text{-}(t\text{-}C_4H_9)_2C_6H_2N{=}NC_6H_2(C_4H_9\text{-}t)\text{-}3',5'\text{-}OH\text{-}2'$ [18].

$(C_6H_5)_3Sb(-2\text{-}OC_6H_4OCH_2CH_2OCH_2CH_2OC_6H_4O\text{-}2'-)$ (Table **32**, No. **28**). The compound is prepared by reacting $(C_6H_5)_3SbCl_2$ and $(2\text{-}NaOC_6H_4OC_2H_4)_2O$ in dry CH_3CN for 4 h with stirring. Filtration of the precipitated NaCl, concentration, and cooling gives a precipitate in a yield of 75%, which is recrystallized from CH_3CN [19].

The compound crystallizes as a rhombohedral solid in the space group Pbca$-D_{2h}^{15}$ (No. 61) with a=20.134(2), b=18.212(3), and c=16.051(1) Å; Z=8, and $d_c=1.38$ g/cm³. The structure is refined to R=0.075. A view of the molecule and the given bond distances and angles are shown in **Fig. 12**, p. 152. The C-Sb-C angles are given as 109.7(2)°, 109.8(2)°, and 140.4(2)° [19].

$(4\text{-}ClC_6H_4)_3Sb(-OC(CF_3)(CH_2C(CH_3)O)O-)$ (Table **32**, No. **31**). It follows from an X-ray structure determination [2, 3] that the compound crystallizes as a monoclinic solid in the space group $P2_1/c-C_{2h}^5$ (No. 14) with a=11.518(1), b=20.962(2), and c=12.259(1) Å, β=124.08(1)°; Z=4, $d_m=1.68$, $d_c=1.70$ g/cm³. From 4518 independent reflections 2655 were used to calculate the structure. The final R value is 0.059. **Fig. 13**, p. 152, shows the molecular structure together with the most important bond distances and angles. The Sb atom has a distorted octahedral coordination with the three aryl groups in facial positions. The hydrated β-diketone acts as a terdentate ligand [3].

$(4\text{-}CH_3C_6H_4)_3Sb(-OC(CF_3)(CH_2C(C_6H_5)O)O-)\cdot 0.5ClCH_2CH_2Cl$ (Table **32**, No. **35**). The compound crystallizes as a triclinic solid with the space group $P\bar{1}-C_i^1$ (No. 2) with a=11.828(1), b=13.851(1), c=10.516(1) Å; β=113.03(1)°, and γ=84.14(1)°; Z=2; $d_m=1.45$ and $d_c=1.46$ g/cm³. 7059 independent reflections were measured, of which 6159 were used for structure determination. Final refinement, including the H atoms with isotropic thermal factors, gave an R value of 0.028 for all observed reflections. A view of the molecule is shown

References on p. 153

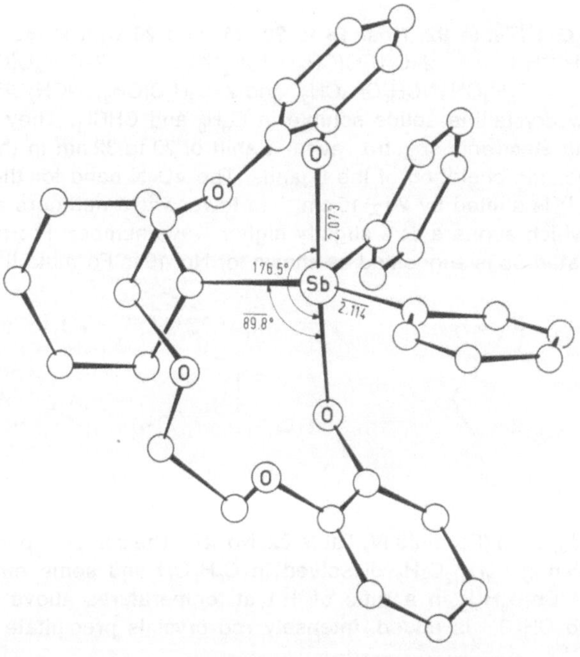

Fig. 12. Molecular structure of $(C_6H_5)_3Sb(-2-OC_6H_4OCH_2CH_2OCH_2CH_2OC_6H_4O-2'-)$ [19].

in **Fig. 14**, p. 153. The main features are, that the compound is monomeric and exhibits a distorted octahedral coordination. The original β-diketone ligand is hydrated at the carbonyl group next to the CF_3 group, to form a terdentate ligand in which three O atoms are bonded to the Sb atom in facial positions [4].

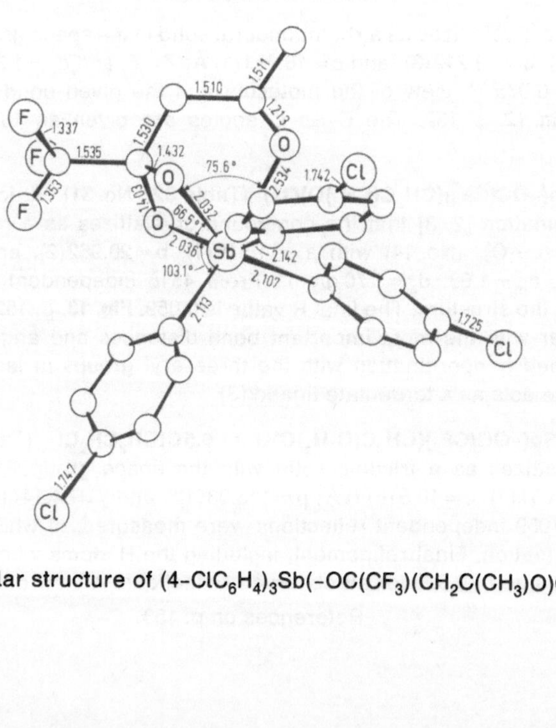

Fig. 13. Molecular structure of $(4-ClC_6H_4)_3Sb(-OC(CF_3)(CH_2C(CH_3)O)O-)$ [3].

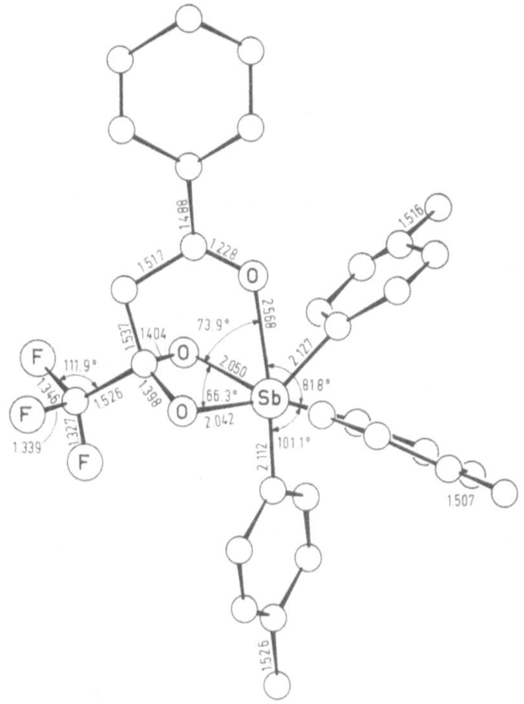

Fig. 14. Molecular structure of $(4\text{-}CH_3C_6H_4)_3Sb(\text{-}OC(CF_3)(CH_2C(C_6H_5)O)O\text{-})$ [4].

References:

[1] A. Ouchi, F. Ebina, T. Uehiro,Y. Yoshino (Bull. Chem. Soc. Japan **51** [1978] 2427/8).
[2] F. Ebina, T. Uehiro, T. Iwamoto, A. Ouchi, Y. Yoshino (J. Chem. Soc. Chem. Commun. **1976** 245/6).
[3] F. Ebina, A. Ouchi, Y. Yoshino, S. Sato, Y. Saito (Acta Cryst. B **33** [1977] 3252).
[4] F. Ebina, A. Ouchi, Y. Yoshino, S. Sato, Y. Saito (Acta Cryst. B **34** [1978] 1512).
[5] Y. Matsumura, M. Shindo, R. Okawara (J. Organometal. Chem. **27** [1971] 357/63).
[6] M. Shindo, R. Okawara (Inorg. Nucl. Chem. Letters **5** [1969] 77/80).
[7] F. Nerdel, J. Buddrus, K.Höher (Chem. Ber. **97** [1964] 124/31).
[8] A. L. Baumstark, M. E. Landis, P. J. Brooks (J. Org. Chem. **44** [1979] 4251/3).
[9] M. Hall, D. B. Sowerby (J. Am. Chem. Soc. **102** [1980] 628/32).
[10] M. M Sidky, M. R. Mahran, W. M. Abdou (Phosphorus Sulfur **15** [1983] 129/35).
[11] V. K. Jain, R. Bohra, R. C. Mehrotra (Indian J. Chem. A **22** [1983] 445/6).
[12] F. Di Bianca, E. Rivarola (Atti Accad. Sci. Lettere Arti Palermo I [4] **31** [1972] 167/72).
[13] F. Di Bianca, E. Rivarola, A. L. Spek, H. A. Meinema, J. G. Noltes (J. Organometal. Chem. **63** [1973] 293/300).
[14] F. Di Bianca, H. A. Meinema, J. G. Noltes, N. Bertazzi, G. C. Stocco, E. Rivarola, R. Barbieri (Atti Accad. Sci. Lettere Arti Palermo I [4] **33** [1974] 173/86).
[15] H. A. Meinema, J. G. Noltes, F. Di Bianca, N. Bertazzi, E. Rivarola, R. Barbieri (J. Organometal. Chem. **107** [1976] 249/55).
[16] N. Bertazzi, F. Di Bianca, T. C. Gibb, N. N. Greenwood, H. A. Meinema, J. G. Noltes (J. Chem. Soc. Dalton Trans. **1977** 957/9).

[17] J. N. R. Ruddick, J. R. Sams (J. Organometal. Chem. **128** [1977] C 41).
[18] G. Bauer, K. Scheffler, H. B. Stegmann (Chem. Ber. **109** [1976] 2231/42).
[19] Yu. A. Sokolova, D. A. D'yachenko, L. O. Atovmyan, N. I. Liptuga, M. O. Lozinskii (Izv. Akad. Nauk SSSR Ser. Khim. **1980** 1446/8).

2.5.1.1.10 Triorganoantimony Bis(organylperoxides) $R_3Sb(OOR')_2$

The compounds of Type $R_3Sb(OOR')_2$ are summarized in Table 34. They are prepared by the following methods:

Method I: The organylhydroperoxide R'OOH is injected through a self-sealing rubber serum cap into a solution of the corresponding stibine SbR_3 in C_6H_6 under nitrogen (molar ratio 3:1) [4] at 5 to 6 °C [5]. The resulting residue is filtered, washed with ether, and dried [4, 5, 10].

Method II: R_3SbX_2 (X=Cl or Br) is added to a suspension of R'OONa with stirring. After 0.5 to 1 h the precipitate is filtered, and the filtrate is evaporated in a vacuum to give the compound.

a. R'OONa is prepared from the corresponding hydroperoxide and $NaNH_2$ in C_5H_{12} or C_6H_6 [2, 4, 6, 7].

b. R'OONa is prepared from the corresponding hydroperoxide and Na in CH_3OH/ether or in C_2H_5OH/ether [2, 6, 7].

Method III: Equivalent amounts of $R_3Sb(OR'')_2$ (R''=alkyl) in ether, and the hydroperoxide R'OOH in the same solvent are mixed slowly with stirring. After some hours at room temperature, the mixture is concentrated in a vacuum to give the compound analytically pure in most cases [6].

Method IV: To a solution of R_3SbX_2 (X=Cl or Br) in $CHCl_3$, C_6H_6, or ether at room temperature, a solution of an equivalent amount of R'OOH in ether or C_6H_6 is added with stirring, and at the same time NH_3 or NH_2CH_3 is passed through the mixture. When the precipitation of NH_4X has ended, the precipitate is filtered, and the filtrate is evaporated in a vacuum. The remaining product is fractionated or recrystallized [2, 6, 7].

Method V: Dry NH_3 is passed through a C_6H_6 solution of R_3SbX_2 (X=Cl or Br) until NH_4Cl is no longer precipitated. The precipitate is filtered, and the solution of $R_3Sb{=}NH$ is treated with the corresponding hydroperoxide in C_6H_6 or ether (molar ratio 1:2). Filtration and concentration of the filtrate gives the compound as a solid [6, 7].

Method VI: R_3SbO or $R_3Sb(OH)_2$ is placed together with R'OOH in C_6H_6 solution in a Soxhlet apparatus filled with anhydrous $CuSO_4$ or Na_2SO_4. Refluxing of the mixture dissolves the compounds. The solution is heated for an additional 2 h. Evaporation of the solvent gives the compounds as residues [2, 6 to 9].

Table 34
Compounds of the Type $R_3Sb(OOR')_2$.
Further information on numbers preceded by an asterisk is given at the end of the table.
For abbreviations, dimensions, and units see p. X.

No.	R	R′	method of preparation (yield in %) properties and remarks [Ref.]
1	CH_3	t-C_4H_9	I (86) [4], IIa (94) [2, 6, 7], IIb (83), V (91) [6, 7], VI (68) [6 to 9] m.p. 79 to 80° [4], 82 to 84° (from C_5H_{12}) [2, 6 to 9] sublimes at 40 to 50°/0.1 [4] ^1H NMR: 1.13 (CH_3), 1.52 (t-C_4H_9) [4] useful as a catalyst for vinyl polymerization [8, 9]
2	CH_3	$C_6H_5(CH_3)_2C$	IV (70) m.p. 47 to 49° (from C_5H_{12}) [2, 6]
3	CH_3	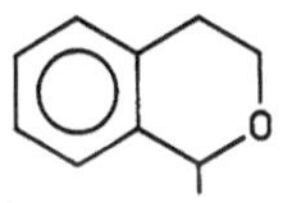	IV (84) m.p. 82° [2, 6]
4	C_2H_5	t-C_4H_9	I (92) yellow liquid ^1H NMR: 1.13 (CH_3), 1.2 to 2.2 (C_2H_5) [4]
5	C_6H_{11}	t-C_4H_9	IV (87) m.p. 128 to 130° (from ether/$(CH_3)_2CO$) [2, 6, 7]
6	$C_6H_5CH_2$	t-C_4H_9	IV (91) liquid [2, 6]
7	$C_6H_5CH_2$	$(C_6H_5)_3C$	IIa (97) m.p. 166 to 168° (from ether) [2, 4, 6, 7]
8	$C_6H_5CH_2$	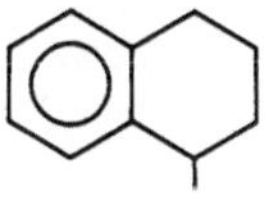	IV (88) oily compound [2, 6]
*9	C_6H_5	t-C_4H_9	I (80 to 82) [5, 10], IIa (77) [6], IIb (85) [2, 6], IV (79), V (75) [6], VI (87) [2, 6] m.p. 99° (from petroleum ether), 101 to 103° (from C_5H_{12}) [2, 6, 9, 10] useful as a catalyst for vinyl polymerization [2, 6, 9] and for bonding polyolefins to metal surfaces [14]
10	C_6H_5	$C_6H_5(CH_3)_2C$	I (71) [5], IV (85) [2, 6, 7], VI (42) [6, 7]
11	C_6H_5		IV (74) [2, 6, 7] m.p. 84 to 85.5° (from ether/petroleum ether) [2, 6, 7]

References on p. 157

Table 34 [continued]

No.	R	R′	method of preparation (yield in %) properties and remarks [Ref.]
12	C_6H_5	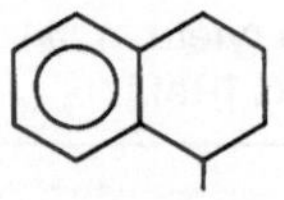	IIa (96), III (92) [6], IV (100) [2, 6] m.p. 186 to 188° (from C_6H_6)
13	C_6H_5	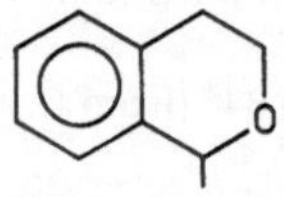	IV (99) [2, 6], III (86) [6] m.p. 178 to 180° [2, 6]
14	C_6H_5	4-$(CH_3)_2CHC_6H_4(CH_3)_2C$	I m.p. 78 to 78.5° [10]
*15	C_6H_5	$Si(C_6H_5)_3$	see further information
*16	C_6H_5	$Tl(C_2H_5)_2$	see further information

* Further information:

$(C_6H_5)_3Sb(OOC_4H_9\text{-t})_2$ (Table **34**, No.**9**). A crystal structure of $(C_6H_5)_3Sb(OOC_4H_9\text{-t})_2$ is described in three publications and compared with those of $[(C_6H_5)_3SbOOC_4H_9\text{-t}]_2O$ [1, 10, 15]. Suitable crystals were obtained by isothermal recrystallization (25 °C) from pentane, hexane, or heptane. Two kinds of crystals are obtained. Those which melt at 101 to 102 °C crystallize as a tetragonal solid in the space group $P\bar{4}2_1c-D_{2d}^4$ (No. 114) with a = 20.513(4) and c = 12.767(4) Å and eight molecules in the unit cell. The structure was solved and refined

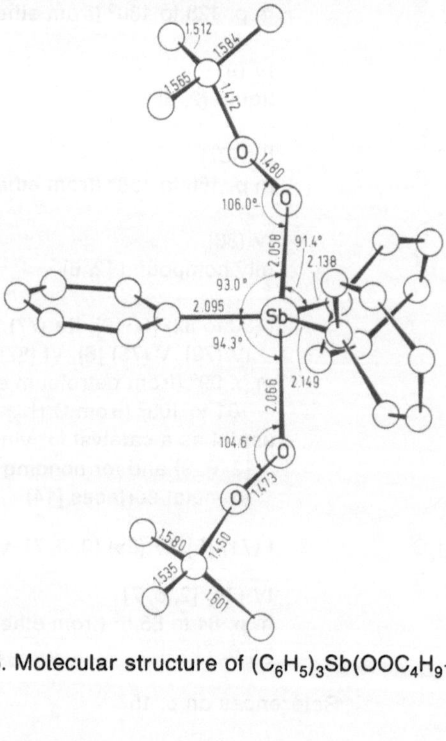

Fig. 15. Molecular structure of $(C_6H_5)_3Sb(OOC_4H_9\text{-t})_2$ [15].

to $R=0.065$ and $R_w=0.045$. The main features of the molecule are that the coordination polyhedron of the antimony atom is a trigonal bipyramid, the oxygen atoms of the organylperoxide groups are located in the axial vertices with an O-Sb-O angle of 172.7°, and the carbon atoms of the phenyl rings are located in the equatorial plane; see **Fig. 15**. The phenyl rings make angles of 56.8°, 4.4°, and 63.5° with the equatorial plane. The intrinsic symmetry of the molecule is close to C_{2v} [15].

$(C_6H_5)_3Sb(OOSi(C_6H_5)_3)_2$ (Table **34**, No. **15**). To $Sb(C_6H_5)_3$, in C_6H_6, is added a solution of t-C_4H_9OOH (molar ratio 1:1) in the same solvent with cooling. After 5 to 10 min a C_6H_6 solution of $(C_6H_5)_3SiOOH$ (molar ratio 1:2) is added and stirred two hours at room temperature. After that the water and the carbinol are distilled azeotropically with C_6H_6 in vacuum. The resulting residue, in a yield of 82%, melts with decomposition at 120 °C. A yield of 74% of the compound is obtained from $(C_6H_5)_3SbCl_2$ and $(C_6H_5)_3SiOOH$ in ether with cooling to 0 °C, in the presence of $N(C_2H_5)_3$ as the HCl acceptor. After concentrating the filtrate of the mixture at reduced pressure the compound precipitates. After filtration it is washed with cold water and dried in a vacuum [5].

The compound, heated in a closed ampule for 30 h at 100 °C, gives oxygen and $(C_6H_5)_3Sb(OSi(C_6H_5)_3)_2$ quantitatively [5].

$(C_6H_5)_3Sb(OOTl(C_2H_5)_2)_2$ (Table **34**, No. **16**). The reaction of $(C_6H_5)_3Sb(OOH)_2$ with $Tl(C_2H_5)_3$ (molar ratio 1:2) in diglyme at room temperature gives the compound, which easily rearranges to $(C_6H_5)_2(C_6H_5O)Sb(OOTl(C_2H_5)_2)OTl(C_2H_5)_2$ [3].

References:

[1] Z. A. Starikova, T. M. Shchegoleva, V. K. Trunov, I. E. Prokovskaya (Kristallografiya **23** [1978] 969/73; Soviet Phys. Cryst. **23** [1978] 547/9).

[2] A. Rieche, J. Dahlmann (Ger. 1155127 [1960/63]; C.A. **60** [1964] 5554).

[3] V. A. Dodonov, T. I. Starostina, T. G. Brilkina, T. I. Zinov'eva, V. V. Kutyreva (Khim. Elementoorg. Soedin. [Gorkiy] **4** [1976] 69/71; C.A. **88** [1978] No. 23077).

[4] A. G. Davies, S. C. W. Hook (J. Chem. Soc. C **1971** 1660/5).

[5] G. A. Razuvaev, T. G. Brilkina, E. V. Krasil'nikova, T. I. Zinov'eva, A. I. Filimonov (J. Organometal. Chem. **40** [1972] 151/7).

[6] A. Rieche, J. Dahlmann, D. List (Liebigs Ann. Chem. **678** [1964] 167/82).

[7] A. Rieche, J. Dahlmann, D. List (Angew. Chem. **73** [1961] 494).

[8] A. Rieche, J. Dahlmann (Ger. 1158975 [1961/63]; C.A. **60** [1964] 9313/4).

[9] A. Rieche, J. Dahlmann (Ger. [East] 44608 [1961/66]; C.A. **65** [1966] 10623/4).

[10] I. E. Pokrovskaya, V. A. Dodonov, Z. A. Starikova, E. N. Kanunnikova, T. M. Shchegoleva, G. P. Lebedeva (Zh. Obshch. Khim. **51** [1981] 1247/53; J. Gen. Chem. [USSR] **51** [1981] 1056/60).

[11] W. Reicherdt, K. Wunsch, J. Dahlmann, L. Stedtler (Brit. 1189629 [1969/70]; C.A. **73** [1970] No. 15710).

[12] W. Reicherdt, K. Wunsch, J. Dahlmann (Fr. 2036424 [1969/70]; C.A. **75** [1971] No. 99083).

[13] W. Reicherdt, K. Wunsch, L. Stedtler, J. Dahlmann (Ger. 1906419 [1970]; C.A. **74** [1971] No. 127790).

[14] W. Reicherdt, K. Wunsch (Ger. [East] 61866 [1967/68]; C.A. **70** [1969] No. 78893).

[15] Z. A. Starikova, T. M. Shchegoleva, V. K. Trunov, I. E. Pokrovskaya, E. N. Kanunnikova (Kristallografiya **24** [1979] 1211/6; Soviet Phys. Cryst. **24** [1979] 694/7).

2.5.1.1.11 Triorganoantimony Dioximates, Dinitroxides, and Disilanolates

The references for the following sections are given on p. 163.

2.5.1.1.11.1 $R_3Sb(ON{=}CR'R'')_2$ Compounds

The compounds are summarized in Table 35. They can be prepared by the following methods:

Method I: R_3SbX_2 (X = Cl or Br) is reacted with the corresponding oxime in the presence of $N(C_2H_5)_3$ in C_6H_6 for one hour at room temperature and then two hours under reflux. Filtration of the $[NH(C_2H_5)_3]X$ precipitate and concentration of the filtrate give the compounds in good yields [1, 2] after recrystallization from petroleum ether/C_6H_6 [1] or petroleum ether/CH_3CN [2]. The yield is generally quantitative [1, 2].

Method II: The corresponding sodium oximate is refluxed with R_3SbBr_2 [3] or R_3SbCl_2 [4] in C_6H_6 for two hours. Evaporation of the filtrate gives a pasty mass as a residue which is distilled at reduced pressure [3] or a solid which is recrystallized from C_6H_6/petroleum ether [4].

Method III: $(C_6H_5)_3Sb(OCH_3)_2$ reacts with the appropriate oxime in refluxing C_6H_6 for 4 h. After removing the volatiles under reduced pressure the compounds are recrystallized from C_6H_6/petroleum ether. The yield is generally quantitative [1].

General Remarks. The compounds described in Table 35 are stable towards air [1, 2] and monomeric in C_6H_6 [1 to 5]. Molar conductance in CH_3CN shows their nonionic character [1 to 4]. They are soluble in common organic solvents [2 to 4] such as $(CH_3)_2CO$, $CHCl_3$, and C_6H_6, but insoluble in H_2O [2]. They are described as stable towards moisture in [1], but as susceptible to hydrolysis in [3, 4].

The triphenyl compounds decompose when heated to 170 to 230 °C in vacuum to $Sb(C_6H_5)_3$, free oxime, and a black residue. Therefore, a purification by distillation is not possible [4].

Table 35
Compounds of the Type $R_3Sb(ON{=}CR'R'')_2$.
For explanations, abbreviations, and units, see p. X.

No.	CR'R'' group	method of preparation (yield in %) properties and remarks [Ref.]
with R = CH_3		
1	$C(CH_3)_2$	I [2], II (81) [3] white crystalline solid [3] m.p. 147° [2], 40°, b.p. 91 to 93°/2.5 ^{1}H NMR (CCl_4): 1.87 (CH_3Sb), 2.07, 2.17 (CH_3C) [3] IR (KBr/Nujol): ca. 1600 (νC=N) insecticidal activity [2]
2	$C(CH_3)C_2H_5$	I [2], II (91) [3] colorless liquid [3] m.p. 220° (dec.) [2]?, b.p. 81°/1.0 n_D^{31} = 1.4879 [3] IR (KBr/Nujol): ca. 1600 (νC=N) insecticidal activity [2]

References on p. 163

Table 35 [continued]

No.	CR′R″ group	method of preparation (yield in %) properties and remarks [Ref.]
3	$C(CH_3)C_3H_7$	II (95) colorless liquid, b.p. 104°/0.5 $n_D^{31}=1.4844$ [3]
4	$C(C_2H_5)_2$	II (95) colorless liquid, b.p. 94°/0.2 $n_D^{31}=1.4810$ ^{1}H NMR ($CDCl_3$): 0.89 to 1.12 (sext, CH_3C), 1.55 (CH_3Sb), 2.02 to 2.38 (octet, CH_2) [3]
5	$C(CH_2)_5$	I m.p. 94° IR (Nujol): 1620w (νCN), 940s (νNO), 575s (νSbC), 450s (νSbO) [1]
6	$C(CH_3)C_6H_5$	I [1], II (76) [3] white solid [3], m.p. 88 to 90° [1], 100 to 101° [3] b.p. 215 to 220°/ 2.5 [3] ^{1}H NMR ($CDCl_3$): 1.73 (s, CH_3Sb), 2.17 (s, CH_3C), 7.20 to 7.93 (m, C_6H_5) [1] ^{1}H NMR (CCl_4): 1.82 (CH_3Sb), 2.25 (CH_3C), 7.25 to 7.80 (C_6H_5) [3] IR (Nujol): 1590w (νCN), 928s (νNO), 550s (νSbC), 430s (νSbO) [1]
7	$C(CH_3)C_6H_4NO_2$-4	I m.p. 112° IR (KBr/Nujol): ca. 1600 (νC=N) [2]
8	$C(NH_2)C_6H_5$	II white, needle-shaped crystals (from C_6H_6), m.p. 110 to 111° ^{1}H NMR ($CDCl_3$): 1.75 (CH_3Sb), 4.88 (NH_2), 7.35 to 7.84 (C_6H_5) IR (Nujol): 3405 (νNH_2), 3300, 3250, 1615, 1600 (νC=N), 275 (Sb-NH_2) IR ($CHCl_3$): 3480 (νNH_2), 3360, 1610 (νC=N) appears to attain an octahedral structure in the solid state by Sb-NH_2 bonding [3]
with R = C_2H_5		
9	$C(CH_3)_2$	II (86) colorless liquid, b.p. 77°/0.08 $n_D^{31}=1.4927$ ^{1}H NMR ($CDCl_3$): 1.72, 1.76 (CH_3), 1.27 to 1.45 (t, CH_3 of C_2H_5), 2.0 to 2.26 (q, CH_2 of C_2H_5) [3]
10	$C(CH_3)C_2H_5$	II (66) colorless liquid, b.p. 104 to 105°/1.0 $n_D^{31}=1.4898$ [3]

References on p. 163

Table 35 [continued]

No.	CR′R″ group	method of preparation (yield in %). properties and remarks [Ref.]
11	$C(CH_3)C_6H_5$	II (88) cream-colored liquid, b.p. 210 to 220°/0.1 $n_D^{31} = 1.5916$ ^1H NMR (CCl_4): 1.40 to 1.65 (t, CH_3 of C_2H_5), 2.17 (CH_3), 2.17 to 2.44 (q, CH_2 of C_2H_5), 7.22 to 7.80 (C_6H_5) [3]
with R = C_6H_5		
12	$CHCH_3$	III m.p. 88 to 90° IR: 1612 (νC=N) [5]
13	$C(CH_3)_2$	I, II (ca. 78), III white crystals, m.p. 116° [4] ^1H NMR ($CDCl_3$): 1.96, 2.25 (CH_3), 7.55 to 8.40 (C_6H_5) [3] IR: 1645 m (νC=N), 1615 m [4]
14	$C(CH_3)C_2H_5$	I, II (ca. 35), III white crystals, m.p. 65 to 66° IR: 1620 w (νC=N) [4]
15	$C(CH_3)C_3H_7$	I, II (ca. 48), III white crystals, m.p. 81 to 82° IR: 1605 w (νC=N) [4]
16	$C(C_2H_5)_2$	I, II (ca. 50), III white crystals, m.p. 64° [4] ^1H NMR ($CDCl_3$): 0.9 to 1.47 (sext, CH_3), 2.06 to 2.88 (octet, CH_2), 7.46 to 8.41 (C_6H_5) [3] IR: 1645 vw, 1615 vw [4]
17	$C(C_2H_5)C_3H_7$	I, II (ca. 54), III white crystals, m.p. 90 to 91° IR: 1620 w (νC=N) [4]
18	$C(CH_2)_5$	II (ca. 100) m.p. 116° IR (Nujol): 1610 w (νCN), 930 s (νNO), 500 s (νSbO) does not react with CH_3OH at 70°, does not insert CS_2 [1]
19	CHC_4H_3O C_4H_3O = 2-furyl	I, III m.p. 168° [1]
20	$CHC_6H_4OCH_3$-4	I, III m.p. 124° [1]
21	$C(CH_3)C_6H_5$	I, III [1], II (ca. 88) [4] white crystals [4], m.p. 146 to 148° [1], 150 to 154° [4] ^1H NMR ($CDCl_3$): 2.52 (CH_3), 7.40 to 8.38 (C_6H_5) [3]
22	$C(CH_3)C_6H_4NO_2$-4	I, III m.p. 194° [1]

References on p. 163

Table 35 [continued]

No.	CR′R″ group	method of preparation (yield in %) properties and remarks [Ref.]
23	$C(C_6H_5)_2$	I, III m.p. 112 to 113° reaction with $C_6H_5TeCl_3$ gives $(C_6H_5)_3SbCl_2$ and $C_6H_5Te(ONC(C_6H_5)_2)Cl_2$ [1]
24	$C(NH_2)C_6H_5$	I, II (ca. 93), III m.p. 143 to 144° [4] ^1H NMR ($CDCl_3$): 5.05 (NH_2), 7.39 to 8.41 (C_6H_5) [3] IR: 1605s (νC=N) [4]
with R = 4-$CH_3C_6H_4$		
25	$C(CH_3)_2$	I m.p. 183° [1]
26	$C(CH_3)C_2H_5$	I m.p. 162° [1]
27	$C(CH_2)_5$	I m.p. 145° reaction with $TeCl_4$ in refluxing C_6H_6 gives $(4\text{-}CH_3C_6H_4)_3SbCl_2$ and an uncharacterized black residue [1]
28	CHC_4H_3O C_4H_3O = 2-furyl	I m.p. 84° [1]
29	$C(CH_3)C_6H_5$	I m.p. 122°, which does not change after stirring with H_2O ^1H NMR ($CDCl_3$): 2.25 (s, CH_3C_6), 2.33 (s, CH_3C), 6.97 to 8.13 (m, C_6H_5) IR (Nujol): 1590w (νCN), 920s (νNO), 480s (νSbO) reaction with $C_6H_5TeCl_3$ in refluxing C_6H_6 gives $C_6H_5Te(ONC(CH_3)C_6H_5)Cl_2$ and $(4\text{-}CH_3C_6H_4)_3SbCl_2$ [1]
30	$C(CH_3)C_6H_4NO_2$-4	I m.p. 85° [1]
31	$C(C_6H_5)_2$	I m.p. 96 to 98° [1]

2.5.1.1.11.2 $R_3Sb(ONR'_2)_2$ Compounds

$(CH_3)_3Sb(ON(CF_3)_2)_2$

$Sb(CH_3)_3$ and $(CF_3)_2NO$ in a molar ratio of 1:2 react at room temperature to give a white solid which converts upon standing to a mixture of a yellow liquid and black and white solids. A needle-shaped white solid in a yield of 28% is obtained from this mixture when it is trapped in a vessel cooled to −10 °C [6, 7].

IR bands are located at 1290s, 1265s, 1215s, 1050m, 965m, 705w, and 545w cm^{-1} [7]. The compound reacts with anhydrous HCl at room temperature to form $(CH_3)_3SbCl_2$ and $(CF_3)_2NOH$ [6, 7].

References on p. 163

$(C_6F_5)_3Sb(ON(CF_3)_2)_2$

$Sb(C_6F_5)_3$ and $(CF_3)_2NO$ in a molar ratio of 1:2 are sealed in an ampule containing some pure $(Cl_2FC)_2$ as solvent. After several minutes a yellow solution is obtained. Removal of the solvent gives a white solid which melts at 143 to 144 °C in a yield of 86.8% [8].

The IR spectrum of the compound shows vibrations at 1643s, 1515s, 1495vs, 1395s, 1290vs (doublet), 1250vs, 1220s, 1205vs, 1192s, 1150vw, 1095vs, 1040vs, 980vs, 965vs, 801m, 735vw, 722vw, 710s, 680vw, 619vw, and 540w cm^{-1} [8].

On heating the compound at 160 °C for 18 h, $Sb(C_6F_5)_3$, $(CF_3)_2NO$, and $CF_3N{=}CF_2$ are formed. The compound is not hydrolyzed in moist air. It reacts with HCl when heated in a closed ampule for 6 days at 100 °C to give $(C_6F_5)_3SbCl_2$ and $(CF_3)_2NOH$ quantitatively [8].

$(C_6H_5)_3Sb(ON(C_6H_5)COC_6H_5)_2$

$(C_6H_5)_3SbBr_2$ and $NaON(C_6H_5)COC_6H_5$ are reacted in a molar ratio of 1:3 in refluxing C_6H_6. Evaporation of the solvent from the filtrate gives a residue (yield about 90%) which is washed with hexane and dried. The buff-colored compound melts at 102 °C [9].

2.5.1.1.11.3 $R_3Sb(OSiR'_3)_2$ Compounds

$(CH_3)_3Sb(OSi(CH_3)_3)_2$

$(CH_3)_3SbCl_2$ and $NaOSi(CH_3)_3$ are reacted in ether for two hours at 35 °C under N_2. After filtration, the filtrate is distilled in a vacuum to yield 70% of a colorless liquid which solidifies at 21 °C [10, 11].

Boiling points are given as 210 °C/725 Torr, 112 °C/30 Torr, 89 °C/10.5 Torr, and 46 °C/1 Torr [10, 11]. An IR spectrum of the liquid shows absorptions at 2957s, 2906m, 1440w, 1405w, 1297w, 1250s, 1259w, 1226w, 969s, 830s, 747s, and 675m cm^{-1}. A ^1H NMR in CCl_4 gives resonances at $\delta = -0.12$ ($SiCH_3$; J(C,H) = 6.60, J(Si,H) = 117.0 Hz), $\delta = 1.48$ (CH_3Sb; J(C,H) = 135 Hz) ppm [11].

The compound is thermally and chemically stable, not sensitive towards oxidation, and only slowly solvolyzed by water and alcohols [11]. It is cleaved by boiling H_2O, by acids, and bases [10]. The presence of HCl gives $(CH_3)_3SbCl_2$ and $(CH_3)_3SiOSi(CH_3)_3$ [11]. Condensation of the substance together with $SOCl_2$ at the temperature of liquid air and repeated thawing give $(CH_3)_3SbCl_2$, $(CH_3)_3SiCl$, and SO_2 quantitatively [10, 11].

$(C_6H_5)_3Sb(OSi(CH_3)_3)_2$

Stoichiometric amounts of $(CH_3)_3SiONa$ and $(C_6H_5)_3SbCl_2$ are reacted in refluxing C_6H_6 for one hour in a N_2 atmosphere. Distillation of the filtrate of the mixture at reduced pressure gives the compound in a yield of 56%, boiling at 161 °C at 1 Torr. The substance melts at 89 °C and decomposes above 280 °C [10, 11].

$(C_6H_5)_3Sb(OSi(C_6H_5)_3)_2$

$(C_6H_5)_3SiOH$ and $(C_6H_5)_3SbCl_2$ are dissolved in ether. NH_3 gas is passed through this solution for 15 min. After storing for 24 h, the precipitate is filtered and carefully washed with H_2O, some $(CH_3)_2CO$, and dried in vacuum. The yield is 85% of the compound, which melts at 293 °C. The compound is obtained quantitatively if $(C_6H_5)_3Sb(OOSi(C_6H_5)_3)_2$ is heated in a closed ampule for 30 h to 100 °C [12]. Thermolysis of $(C_6H_5)_4SbOSi(C_6H_5)_3$ at 250 to 260 °C for one hour gives 50% of the compound (m.p. 310 to 312 °C, from C_6H_6) in addition to 11% C_6H_6, 45% $C_6H_5C_6H_5$, and 45% $Sb(C_6H_5)_3$ [13].

The compound is stable to air [13] and slightly soluble in ether and C_6H_6 [12]. It reacts with ethanolic HCl to give $(C_6H_5)_3SiOH$ and $(C_6H_5)_3SbCl_2$ quantitatively [12, 13].

References:

[1] K. Bajpai, R. C. Srivastava (Syn. Reactiv. Inorg. Metal.-Org. Chem. **11** [1981] 7/13).
[2] K. Bajpai, M. Srivastava, R. C. Srivastava (Indian J. Chem. A **20** [1981] 736/7).
[3] V. K. Jain, R. Bohra, R. C. Mehrotra (Inorg. Chim. Acta **51** [1981] 191/4).
[4] V. K. Jain, R. Bohra, R. C. Mehrotra (J. Indian Chem. Soc. **57** [1980] 408/10).
[5] P. G. Harrison, J. J. Zuckermann (Inorg. Nucl. Chem. Letters **6** [1970] 5/8).
[6] H. G. Ang, W. S. Lien (J. Fluorine Chem. **3** [1973] 235/6).
[7] H. G. Ang, W. S. Lien (J. Fluorine Chem. **15** [1980] 453/70).
[8] H. G. Ang, W. S. Lien (J. Fluorine Chem. **9** [1977] 73/80).
[9] S. Gopinathan, C. Gopinathan (Indian J. Chem. A **15** [1977] 660/2).
[10] H. Schmidbaur, M. Schmidt (Angew. Chem. **73** [1961] 655).

[11] H. Schmidbaur, H. S. Arnold, E. Beinhofer (Chem. Ber. **97** [1964] 449/58).
[12] G. A. Razuvaev, T. G. Brilkina, E. V. Krasil'nikova, T. I. Zinov'eva, A. I. Filimonov (J. Organometal. Chem. **40** [1972] 151/7).
[13] G. A. Razuvaev, N. A. Osanova, T. G. Brilkina, T. I. Zinov'eva, V. V. Sharutin (J. Organometal. Chem. **99** [1975] 93/106).

2.5.1.1.12 Triorganoantimony Dicarboxylates

2.5.1.1.12.1 $R_3Sb(O_2CR')_2$ Compounds with R' = Unsubstituted Alkyl

The compounds are prepared by the following methods:

Method I: The triorganoantimony oxide R_3SbO or dihydroxide $R_3Sb(OH)_2$ is treated with an excess of the corresponding carboxylic acid. Evaporation of the excess in vacuum gives the compound as a residue which is purified by distillation or recrystallization [1].

Method II: Triorganoantimony dihalide R_3SbX_2 (X = Cl, Br), the corresponding carboxylic acid (ratio 1:2), and $N(C_2H_5)_3$ as HX acceptor are dissolved in C_6H_6. Refluxing the mixture for one to three hours, filtering the $[NH(C_2H_5)_3]X$ precipitate, and evaporating the solvent in a vaccum give the compounds [2, 38, 39].

$(CH_3)_3Sb(O_2CH)_2$

Preparation according to Method I [1] or II [2]. The yield is quantitative. The compound melts at 81 °C, recrystallized from petroleum ether [1]. A melting point of 63 to 65 °C (from petroleum ether or petroleum ether/CH_3CN) is given in [2]. The following IR vibrations (in cm^{-1}) were found and assigned [1]:

as mull	in CCl_4 or $CHBr_3$	assignment	as mull	in CCl_4 or $CHBr_3$	assignment
3245vw	3289vw		1377m, 1366w, sh	1362m	δC-H
3012w	3021w	$\nu_{as}CH_3$	1250s, 1231s	1233s	νC-O
2933w	2941vw	ν_sCH_3		1221m	δ_sCH_3
2865w, 2849w	2857w	νC-H	875s, 856s	862s	ϱCH_3
2841w	2703vw		840sh		
			762s	767s	δCO_2
1647s, 1634s	1653s	νC=O	583m	581m	$\nu_{as}SbC_3$
1408w	1404vw	$\delta_{as}CH_3$	535vw		ν_sSbC_3?

From these values a trigonal bipyramidal structure of the molecule is proposed. The compound dissolves as a monomer in C_6H_6 [1].

Antibacterial activities of the compound against Bacillus subtilis and Sorsena lutea were proved [2].

$(CH_3)_3Sb(O_2CCH_3)_2$

The compound is prepared by Method I [1], p. 163, or by treating $(CH_3)_3SbX_2$ (X = halogen) with a suspension of Ag_2O in CH_3CO_2H. Concentration of the filtrate under vacuum and recrystallization from petroleum ether/ether/hexane give the compound, which melts at 79.5 to 80 °C [3] or 80.5 to 81 °C [1].

IR spectra of the solid and of solutions are published and discussed in several publications [1, 3 to 5]. Characteristic bands (in cm^{-1}) for the solid are: 1655, 1648 (νC=O), 1302 (νC-O), 584 (ν_{as}SbC), and in CCl_4: 1658 (νC=O), 1291 (νC-O) [3]. The low frequency IR vibrations in Nujol are: 578vs, 507vs, 279vs, 218m, 212s, 158s, and 156m cm^{-1}. The corresponding Raman lines are: 578m, 538vs, 512w, 275vs, 223m, 168m, and 125m cm^{-1}. Assignments are given in [4, 5]. A ^1H NMR resonance (in $CDCl_3$) is found at 1.85 ppm (CH_3Sb) [3]. ^{121}Sb Mössbauer data (vs. $Ba^{121}SnO_3$ at 8 K, source 80 K) are: $\delta = -5.17 \pm 0.08$, $e^2qQ = -23.3 \pm 0.1$, $\Gamma = 2.51$ mm/s [6]. In connection with these data, an Sb^V orbital population calculation has been performed [7]. All these physical measurements are in agreement with a trigonal bipyramidal structure of the molecule [1, 3 to 7].

The compound dissolves as a monomer in C_6H_6 [1, 3] and shows practically no electric conductivity: $\Lambda \approx 0.15\ cm^2 \cdot \Omega^{-1} \cdot mol^{-1}$ for 10^{-3} to 10^{-4} M solutions in pyridine [3].

The compound may be used as a catalyst for the preparation of colorless polyethylene terephthalates [8]. Concerning the fungitoxicity of the compound in solvents with different concentrations see [9].

$(CH_3)_3Sb(O_2CCD_3)_2$

The deuterated compound is prepared analogously to the previous one by treating trimethylantimony dihalide with Ag_2O in CD_3CO_2H. The melting point is 81 to 81.5 °C, recrystallized from petroleum ether/ether/hexane. IR vibrations (in cm^{-1}) for the solid are: 1655, 1638 (νC=O), 1315 (νC-O), and 578 (ν_{as}SbC); and in CCl_4: 1655 (νC=O), and 1310 (νC-O). The compound dissolves as a monomer in C_6H_6 and has no electric conductivity in pyridine at 25 °C [3].

$(CH_3)_3Sb(O_2CC_2H_5)_2$

The compound is prepared according to Method I, p. 163. The boiling point of the liquid is 110 °C at 6 Torr. The refractive index is $n_D^{25} = 1.4795$. The IR spectrum (liquid film) shows bands at 1647s and 1231s cm^{-1} for νO_2C [1]. It is claimed to be a polymerization catalyst to give colorless polyethylene terephthalates [8].

$(CH_3)_3Sb(O_2CC_3H_7)_2$

The compound is prepared according to Method I, p. 163, and isolated by fractional distillation at 128 °C/2.5 to 3 Torr. IR vibrations (liquid film) are observed at 1658s, 1653s, and 1220s (νO_2C) cm^{-1} [1].

$(CH_3)_3Sb(O_2CCH_2CH(CH_3)_2)_2$

Preparation is performed according to Method II, p. 163. The compound melts at 165 to 168 °C, recrystallized from petroleum ether or CH_3CN/petroleum ether [2].

References on p. 169

$(C_2H_5)_3Sb(O_2CCH_3)_2$

$(C_2H_5)_3SbO$ dissolved in aqueous CH_3CO_2H gives upon evaporation a syrupy mass, which could not be crystallized. It is probably the title compound [10, 11]. It is claimed as a cocatalyst with $TiCl_3$ and $Al(C_2H_5)_3$ for the polymerization of olefins [12].

$(C_3H_7)_3Sb(O_2CCH_3)_2$

Equivalent amounts of $(C_3H_7)_3Sb{=}NSO_2R$ ($R = C_6H_5$ or $4\text{-}CH_3C_6H_4$) and glacial CH_3CO_2H are heated in CH_2Cl_2 at 50 to 60 °C. Upon cooling, the RSO_2NH_2 precipitates (95%) and is filtered; the filtrate is distilled at reduced pressure to give the compound in 30% yield, b.p. 121 to 122 °C at 0.0931 hPa (0.07 Torr) and m.p. 49 °C. The IR spectrum shows absorptions at 1630 (νCO) and 465 (νSbC) cm^{-1} [13].

$(C_4H_9)_3Sb(O_2CCH_3)_2$

The compound is mentioned as a catalyst for the polymerization of organoisocyanates [14].

$(i\text{-}C_4H_9)_3Sb(O_2CCH_3)_2$

$(i\text{-}C_4H_9)_3SbO$ and CH_3CO_2H are refluxed in C_6H_6 on a water separator. Distillation at 144 to 147 °C/3.3 Torr gives a 54% yield of the compound as a colorless oil. It is claimed to be a catalyst for the trimerization of $3\text{-}ClC_6H_4NCO$ [15].

$(C_4H_9)_3Sb(O_2C(CH_2)_{10}CH_3)_2$

The compound is mentioned as a heat stabilizer for halogen-containing resin compositions [16].

$(C_6H_5)_3Sb(O_2CH)_2$

The compound is prepared according to Method I, p. 163 in a yield of 33%, m.p. 157 to 162 °C (from hot HCO_2H) [18]. It is also prepared analogously to Method I in CH_3OH solution and cooling the hot filtered mixture overnight. The resulting crystals (yield 61.2%) melt at 159 to 160 °C, recrystallized from CH_3OH or C_2H_5OH [17].

An IR spectrum shows absorptions (in cm^{-1}) in Nujol at 462 m and 452 m for νSbC, and at 1650 vs, 1610 w, 1367 m, 1339 w, 1333 w, 1230 vs, 844 w, 832 w, 777 m, 772 m (O_2CH), 360 m, and 314 m [18]. A ^1H NMR spectrum in CCl_4 shows resonances at $\delta = 7.45$ (m, H-3,4,5), 7.90 (q, H-2,6), 7.90 (s, O_2CH) ppm [17]. The ^{13}C NMR chemical shift values are: $\delta = 129.48$ (C-3,5; J(C,H) = 163.0 Hz), 131.48 (C-4; J(C,H) = 161.7 Hz), 133.77 (C-2,6; J(C,H) = 165.4 Hz), 136.11 (c-1), and 164.86 (O_2CH; J(C,H) = 211.2 Hz) ppm [19].

The compound is soluble in CH_3OH, C_2H_5OH, $(CH_3)_2CO$, C_6H_6, $CHCl_3$, and CCl_4, and insoluble in petroleum ether [17].

$(C_6H_5)_3Sb(O_2CCH_3)_2$

The compound can be prepared by the methods summarized in Table 36.

An IR spectrum shows the following characteristic absorptions (in cm^{-1}) in Nujol: 1628 s, 1376 m, 1318 s, 1008 w, 930 m, 915 w, 686 m, 672 m, 609 w, and 486 w (O_2CCH_3) [18]. The νC=O and νC–O vibrations appear at 1633 and 1320 cm^{-1} in the solid state, and at 1651 ($CHCl_3$) and 1310 (CCl_4) cm^{-1} in solution [3]. Low frequency IR and Raman spectra were recorded and discussed. IR (Nujol): 461 vs, 306 vs, 288 vs, 240 vs, 237 s, 216 w, 150 m, 136 m, and 126 m

References on p. 169

Table 36
Preparation of $(C_6H_5)_3Sb(O_2CCH_3)_2$.

reactants and reaction conditions, workup (yield)	melting poiont	Ref.
$Sb(C_6H_5)_3$ in $(CH_3)_2CO+3\%$ H_2O_2, precipitate dissolved in hot CH_3CO_2H, cooling	215 °C	[20]
$Sb(C_6H_5)_3$ dissolved in hot $CH_3CO_2H+30\%$ H_2O_2, cooling (85%)	214 to 216 °C	[21]
$Sb(C_6H_5)_3+Hg(O_2CCH_3)_2$ or Cu $(O_2CCH_3)_2$ in refluxing $(CH_3)_2CO$, concentration of the filtrate (70 to 85%)	215 °C	[22]
$Sb(C_6H_5)_3+Pb(O_2CCH_3)_4$ in $CHCl_3$ with catalytic amounts CH_3CO_2H, concentration of the filtrate (83%)	215 °C (from CH_3CO_2H)	[23]
$Sb(C_6H_5)_3+Pb(O_2CCH_3)_4$ in CH_3CO_2H, 10 min reflux, dilution with H_2O (82%)	–	[24]
$Sb(C_6H_5)_3+C_6H_5Pb(O_2CCH_3)_3$ in $CHCl_3$ for 0.5 h, concentration of the filtrate (82%)	–	[23]
$(C_6H_5)_2SbO_2CCH_3+[C_6H_5N_2]O_2CCH_3$ in cold $(CH_3)_2CO$ (26%)	–	[25]
$(C_6H_5)_2SbO_2CCH_3+[C_6H_5N_2]O_2CCH_3$ in boiling $(CH_3)_2CO$, precipitate recrystallized twice from CH_3CO_2H gives the compound as adduct with $2CH_3CO_2H$	204 to 205 °C	[35]
$(C_6H_5ICl)_2SbCl_3+Sb$ powder in various solvents and at different temperatures and heating with CH_3CO_2H, see original	210 °C	[26]
$(C_6H_5)_3SbO$ suspended in $CH_3OH+CH_3CO_2H$ in excess, 10 min reflux, precipitates from hot filtrate upon cooling (37.5%)	213 to 215 °C	[17]
$(C_6H_5)_3SbO_{1.4}+$ hot CH_3CO_2H, cooling	–	[27]
$(C_6H_5)_3SbBr_2+AgO_2CCH_3$ in CH_3OH or C_6H_6, recrystallized from petroleum ether	–	[4]
$(C_6H_5)_3SbOSO_3+$ hot CH_3CO_2H, cooling	215 °C	[28]
$(C_6H_5)_3Sb{=}NSO_2C_6H_4CH_3\text{-}4+H_2O$ in $CH_3CN+CH_3CO_2H$, precipitate dissolved in CH_3CO_2H	215 to 216 °C	[24]
$(C_6H_5)_3Sb(\text{-}OCRR'CRR'O\text{-})$ (R, R'$=CH_3$; R=H, R'$=C_6H_5$), warmed up with CH_3CO_2H, cooling (ca. 80%)	213 °C	[27]
$(C_6H_5)_3SbO$ or $(C_6H_5)_3Sb(OH)(acac)$, 1 h refluxed with $CH_3COCH_2COCH_3$, concentrated filtrate + petroleum ether, recrystallized from C_6H_6	–	[29]

cm^{-1}; Raman: 461m, 305m, 287s, 265vs, 235s, 219vs, and 150m cm^{-1} [4, 5]. A ^1H NMR spectrum shows resonances at $\delta=1.90$ (s, CH_3), 7.40 (m, H-3,4,5), and 7.90 (q, H-2,6) ppm [17]. ^{13}C NMR resonances in $CDCl_3$ are found at $\delta=22.08$ (CH_3; J(C,H)=128.4 Hz), 129.05 (C-3,5; J(C,H)=162.3 Hz), 130.75 (C-4; J(C,H)=161.1 Hz), 133.68 (C-2,6; J(C,H)=165.4 Hz), 138.84 (C-1), and 175.68 (O_2C) ppm [19]. A ^{121}Sb Mössbauer spectrum (source 77 K, absorber 9 K vs. $Ca^{121}SnO_3$) gives an isomer shift of $\delta=-4.8\pm0.1$ with $e^2qQ=-21.8\pm0.5$ and $\Gamma=$

References on p. 169

2.8 mm/s [30]. The coupling constant was calculated as −20.9 mm/s on the basis of a trigonal bipyramidal molecular shape [31]. On the same basis, an orbital population analysis is published in [32]. All these measurements are discussed for a covalently bonded trigonal bipyramidal molecule [4, 5, 18, 19, 30 to 32].

The X-ray crystal structure was determined [34]. Suitable crystals are obtained by recrystallization from toluene. The compound crystallizes as a monoclinic solid with a space group C2/c−C^6_{2h} (No. 15) and a = 12.97(1), b = 10.01(1), c = 15.89(1) Å; β = 107.00°; Z = 4 and d_c = 1.584 g/cm³. Refinement gave a final convergence of R = 4.3% for 1710 reflections. **Fig. 16** shows the trigonal bipyramidal structure of the molecule and the numbering scheme of atoms with the main bond distances and bond angles.

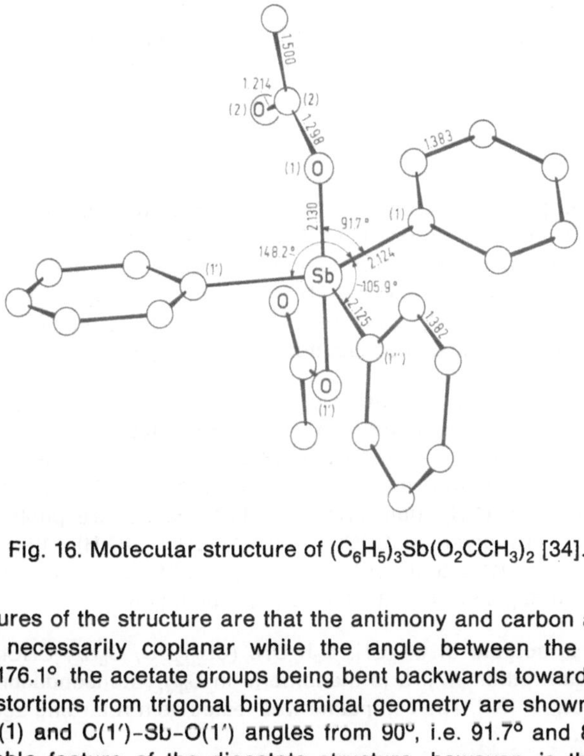

Fig. 16. Molecular structure of $(C_6H_5)_3Sb(O_2CCH_3)_2$ [34].

The main features of the structure are that the antimony and carbon atoms in the equatorial plane are necessarily coplanar while the angle between the two axial groups O(1)-Sb-O(1') is 176.1°, the acetate groups being bent backwards towards the C(1'') phenyl group. Further distortions from trigonal bipyramidal geometry are shown by the deviations of the C(1)-Sb-O(1) and C(1')-Sb-O(1') angles from 90°, i.e. 91.7° and 89.3°, respectively. The most noticeable feature of the diacetate structure, however, is the extent to which the equatorial groups are distorted. The angle between the two symmetry-related phenyl groups is increased to 148.2°, while planarity of the C_3Sb unit is maintained by a decrease in the other two angles to 105.9°. In the acetate structure, both the short Sb···O(2) distance, 2.779(4) Å, compared with a value of 3.6 Å for the sum of the appropriate van der Waals' radii, and the reduction of the Sb-O(1)-C(2) angle to 108.1°, from the ca. 120° value generally found at the oxygen of a covalently bonded acetate group, suggest that intramolecular interactions are important [34].

The compound is soluble in CH_3OH, C_2H_5OH, $(CH_3)_2CO$, C_6H_6, $CHCl_3$, and CCl_4, and insoluble in petroleum ether [17].

The covalent structure is also established in pyridine solution at 25 °C, where practically no electric conductivity was measured. The deuterated derivative $\mathbf{(C_6H_5)_3Sb(O_2CCD_3)_2}$ also

References on p. 169

shows no conductivity [3]. A mass spectrum of triphenylantimony diacetate at 70 eV based on ^{121}Sb shows the following fragments: $[M-nO_2CCH_3]^+$ (n=0 to 2), $[M-O_2CCH_3-nC_6H_5]^+$ (n=2, 3), $[(C_6H_5)_2Sb]^+$, $[(C_6H_5)_2Sb-2H]^+$, $[C_6H_5SbOH]^+$, $[C_6H_5Sb]^+$ (base peak), $[(C_6H_5)_2]^+$, $[(C_6H_5)_2-2H]^+$, $[C_6H_5O_2CCH_3]^+$, $[C_6H_5OH]^+$, $[C_6H_5H]^+$, and $[C_6H_5]^+$ [33].

The compound dissolved in CH_3OH [20] or as a solid reacts with ethanolic [20, 35] or aqueous [23] HCl to give $(C_6H_5)_3SbCl_2$. Dissolved in CH_2Cl_2 and shaken with H_2O the substance forms $(C_6H_5)_3Sb(OH)_2$, isolable by evaporation of the organic phase. A solution of the compound in C_2H_5OH reacts with Na_2S/H_2S to give 62% $(C_6H_5)_3SbS$ [33]. $(C_6H_5)_3Sb(O_2CCH_3)_2$ and LiC_6H_5 in a ratio of 1:10 in ether solution give $Sb(C_6H_5)_5$ in a yield of 80% [24].

The title compound is used as a cocatalyst with $Al(C_2H_5)_3$ and $TiCl_3$ for the preparation of polyolefins [12] or as fire retardant for halogen-containing polyester resins [36]. The title compound renders decomposition products of halogen-containing organic compounds used in electric equipment innocuous [37].

$(C_6H_5)_3Sb(O_2CC_2H_5)_2$

The compound is prepared according to Method II, p. 163 [38, 39] in a yield of about 80% [39]. A yield of only 33% is obtained if $(C_6H_5)_3SbO$, suspended in CH_3OH, is treated with an excess of $C_2H_5CO_2H$. Reflux for 10 min and filtration of the hot mixture gives the compound as a precipitate upon cooling [17]. Treatment of $Sb(C_6H_5)_3$ with $Pb(O_2CC_2H_5)_4$ and catalytic amounts of $C_2H_5CO_2H$ for 0.5 h in $CHCl_3$ gives a precipitate of $Pb(O_2CC_2H_5)_2$. From the filtrate, the compound can be isolated in a 56.2% yield [23].

Melting points of the compound are given as 139 to 140 °C from C_2H_5OH [23], 136 to 138 °C from petroleum ether or petroleum ether/CH_3CN [39], and 134 to 139 °C [17]. IR vibrations were found at 1622s ($\nu_{as}O_2C$) and 1300w (ν_sO_2C) cm^{-1} in KBr/Nujol [39]. The ^1H NMR spectrum in CCl_4 shows resonances at δ=0.9 (t, CH_3), 2.10 (q, CH_2), 7.90 (q, H-2,6 of C_6H_5), 7.40 (H-3,4,5 of C_6H_5) ppm [17]. ^{13}C NMR spectra are published [19, 38], and the following values are given in $CDCl_3$: δ=9.71 (CH_3; J(C,H)=127.9 Hz), 28.86 (CH_2; J(C,H)=127.6 Hz), 129.05 (C-3,5; J(C,H)=162.6 Hz), 130.75 (C-4; J(C,H)=160.0 Hz), 133.77 (C-2,6; J(C,H)=166.0 Hz), 138.94 (C-1), 178.70 (O_2C) ppm [19].

The compound is soluble in CH_3OH, C_2H_5OH, $(CH_3)_2CO$, C_6H_6, $CHCl_3$, and CCl_4, and insoluble in petroleum ether [17]. It is monomeric in C_6H_6, nonconducting in CH_3CN, and stable against hydrolysis [39]. The compound was tested as a retarding agent in the burning of epoxy resins [40].

$(C_6H_5)_3Sb(O_2CC_3H_7\text{-}i)_2$

The compound is prepared according to Method II, p. 163, in a yield of about 80%. It melts at 52 to 54 °C [39]. A melting point of 99.5 to 101 °C (from C_6H_{14}) is observed if the substance is prepared from $Sb(C_6H_5)_3$ and $Pb(O_2CC_3H_7\text{-}i)_4$ or $C_6H_5Pb(O_2CC_3H_7\text{-}i)_3$ in $CHCl_3$ or C_6H_6 with catalytic amounts of i-$C_3H_7CO_2H$. It is isolated in yields of about 45% from the filtrate of the mixture [23].

IR absorptions (in KBr/Nujol) for the O_2C group are observed at 1635s (ν_{as}) and 1280w (ν_s) cm^{-1}. The compound is monomeric in C_6H_6, nonconducting in CH_3CN, and stable against hydrolysis at room temperature [39]. It was tested as a retarding agent in the burning of epoxy resins [40].

$(C_6H_5)_3Sb(O_2C(CH_2)_{16}CH_3)_2$

The substance with a melting point of 50 °C is obtained according to Method II, p. 163 [39].

The IR spectrum shows absorptions for the O_2C group at 1640s (ν_{as}) and 1250m (ν_s) cm^{-1} in KBr/Nujol. The compound is monomeric in C_6H_6, stable against hydrolysis at room temperature, and nonconducting in CH_3CN [39]. It was tested as a retarding agent in the burning of epoxy resins [40].

$(2\text{-}CH_3C_6H_4)_3Sb(O_2CCH_3)_2$

The compound is mentioned without details as the product of a reaction of $Sb(C_6H_4CH_3\text{-}2)_3$ and $Hg(O_2CCH_3)_2$ in C_6H_6. The compound is separated from the byproduct HgO by recrystallization from C_6H_6/C_5H_{12} [41].

$(4\text{-}CH_3C_6H_4)_3Sb(O_2CH)_2$

The title compound is obtained in a yield of about 80% according to Method II, p. 163. Recrystallized from petroleum ether or petroleum ether/CH_3CN, it melts at 106 °C [39].

IR vibrations for the O_2C group are observed at 1650s (ν_{as}), 1250w, and 1230s (ν_s) cm^{-1} in KBr/Nujol. A ^1H NMR spectrum shows resonances at $\delta = 2.35$ (s, CH_3), 3.38 (s, CH), and 7.30 (m, C_6H_5) ppm. From these values, a trigonal bipyramidal geometry with axial formate substituents is concluded. The compound is monomeric in C_6H_6, unaffected by atmospheric moisture, and nonconducting in CH_3CN [39].

$(4\text{-}CH_3C_6H_4)_3Sb(O_2CCH_3)_2$

$Sb(C_6H_4CH_3\text{-}4)_3$ and $Hg(O_2CCH_3)_2$, or $Cu(O_2CCH_3)_2$, are reacted in refluxing $(CH_3)_2CO$. Hg^I- or Cu^I-acetate precipitate. From the filtrate, 70 to 85% of the compound with a melting point of 225 °C is isolated [22]. A melting point of about 165 °C is given in [42].

$(4\text{-}CH_3C_6H_4)_3Sb(O_2CC_3H_7\text{-}i)_2$

The compound is obtained in yields of 70 to 90% by Method II, p. 163, and recrystallization from petroleum ether or petroleum ether/CH_3CN gives a melting point of 64 to 65 °C [39].

IR spectrum (KBr/Nujol, in cm^{-1}): 1630s ($\nu_{as}O_2C$), 1310m, and 1260m (ν_sO_2C). The compound is monomeric in freezing C_6H_6, nonconducting in CH_3CN, and not affected by atmospheric moisture [39].

$(4\text{-}CH_3C_6H_4)_3Sb(O_2C(CH_2)_7CH_3)_2$

See the preceding compound for the preparation and properties, except m.p.: 206 to 208 °C; IR (KBr/Nujol, in cm^{-1}): 1610s ($\nu_{as}O_2C$), 1298s (ν_sO_2C) [39].

$(2,4\text{-}(CH_3)_2\text{-}5\text{-}NO_2C_6H_2)_3Sb(O_2CCH_3)_2$

The compound is prepared by Method I, p. 163. The concentrated mixture is treated with ether to give the product with a melting point of 198 °C [43].

References:

[1] M. Shindo, R. Okawara (J. Organometal. Chem. **5** [1966] 537/44).
[2] K. Bajpai, M. Srivastava, R.C. Srivastava (Indian J. Chem. A **20** [1981] 736/7).
[3] R. G. Goel, D. R. Ridley (J. Organometal. Chem. **38** [1972] 83/9).

[4] R. G. Goel, E. Maslowsky Jr., C. V. Senoff (Inorg. Chem. **10** [1971] 2572/7).
[5] R. G. Goel, E. Maslowsky Jr., C. V. Senoff (Inorg. Nucl. Chem. Letters **6** [1970] 833/5).
[6] R. G. Goel, J. N. R. Ruddick, J. R. Sams (J. Chem. Soc. Dalton Trans. **1975** 67/71).
[7] L. H. Bowen, G. G. Long (Inorg. Chem. **15** [1976] 1039/44).
[8] H. Terada, T. Impaida, O. Nakagawa, S. Takashima (Japan. 69-19270 [1969]; C.A. **71** [1969] No. 125191).
[9] R. E. Burrell, C. T. Corke, R. G. Goel (J. Agr. Food Chem. **31** [1983] 85/8).
[10] W. Merck (J. Prakt. Chem. **66** [1855] 56/72).

[11] W. Merck (Liebigs Ann. Chem. **97** [1856] 329/33).
[12] S. Yoshida, S. Kitakawa, Mitsubishi Petrochemical Co., Ltd. (Japan. 72-26183 [1972]; C.A. **78** [1973] No. 16795).
[13] Z. I. Kuplennik, Zh. N. Belaya, A. M. Pinchuk (Zh. Obshch. Khim **51** [1981] 2711/15; J. Gen. Chem. [USSR] **51** [1981] 2339/43).
[14] S. Herbstman, Stauffer Chemical Co. (U.S. 3278492 [1964/66]; C.A. **66** [1967] No. 66142).
[15] S. Herbstman (J. Org. Chem. **30** [1965] 1259/60).
[16] Nitto Chemical Industry Co., Ltd (Japan. 81-30453 [1981]; C.A. **95** [1981] No. 63222).
[17] J. Havranek, J. Mleziva, A. Lycka (J. Organometal. Chem. **157** [1978] 163/6).
[18] G. O. Doak, G. G. Long, L. D. Freedman (J. Organometal. Chem. 4 [1965] 82/91).
[19] J. Havranek, A. Lycka (Sb. Ved. Pr. Vys. Sk. Chemickotechnol. Pardubice **43** [1980] 123/7).
[20] H. Schmidt (Liebigs Ann. Chem. **429** [1922] 123/52).
[21] T. C. Thepe, R. J. Garascia, M. A. Selvoski, A. N. Patel (Ohio J. Sci. **77** [1977] 134).
[22] S. N.Bhattacharya, M. Singh (Indian J. Chem. A **18** [1979] 515/6).
[23] V. I. Lodochnikova, E. M. Panov, K. A. Kocheshkov (Zh. Obshch. Khim. **34** [1964] 946/9; J. Gen. Chem. [USSR] **34** [1964] 940/2).
[24] G. Wittig, D. Hellwinkel (Chem. Ber. **97** [1964] 789/93).
[25] O. A. Reutov, O. A. Ptitsyna (Dokl. Akad. Nauk SSSR **89** [1953] 877/80).
[26] O. A. Ptitsyna, O. A. Reutov, G. Ertel (Izv. Akad. Nauk SSSR Otd. Khim. Nauk **1961** 265/70; Bull. Acad. Sci. USSR Div. Chem. Sci. **1961** 241/5).
[27] F. Nerdel, J. Buddrus, K. Höher (Chem. Ber. **97** [1964] 124/31).
[28] M. Becke-Goehring, H. Thielemann (Z. Anorg. Allgem. Chem. **308** [1961] 33/51).
[29] R. G. Goel, D. R. Ridley (J. Organometal. Chem. **182** [1979] 207/12).
[30] J. N. R. Ruddick, J. R. Sams, J. C. Scott (Inorg. Chem. **13** [1974] 1503/7).

[31] G. M. Bancroft, V. G. K. Das, T. K. Sham, M. G. Clark (J. Chem. Soc. Dalton Trans. **1976** 643/54).
[32] L. H. Bowen, G. G. Long (Inorg. Chem. **17** [1978] 551/4).
[33] C. Glidewell (J. Organometal. Chem. **116** [1976] 199/209).
[34] D. B. Sowerby (J. Chem. Res. S **1979** 80/1).
[35] A. N. Nesmeyanov, O. A. Reutov, O. A. Ptitsyna, P. A. Tsurkan (Izv. Akad. Nauk SSSR Otd. Khim. Nauk **1958** 1435/44; Bull. Acad. Sci. USSR Div. Chem. Sci. **1958** 1384/92).
[36] B. O. Schoepfle, S. B. Marks, P. Robitschek, Hooker Chemical Corp. (U.S. 2913428 [1959]; C.A. **1960** 5162).
[37] R. L. Jenkins, Monsanto Chemical Co. (U.S. 2566208 [1951]; C.A. **1951** 10442).
[38] A. Ouchi, T. Uehiro, Y. Yoshino (J. Inorg. Nucl. Chem. **37** [1957] 2347/9).
[39] K. Bajpai, R. Singhal, R. C. Srivastava (Indian J. Chem. A **18** [1979] 73/5).
[40] J. Havranek, J. Mleziva (Angew. Makromol. Chem. **84** [1980] 105/17).

[41] G. Deganello, G. Dolcetti, M. Giustiniani, U. Belluco (J. Chem. Soc. A **14** [1969] 2138/40).
[42] L. Kaufmann (Ger. 360973 [1924]; C.A. **1924** 841).
[43] A. E. Goddard (J. Chem. Soc. **123** [1923] 2315/23).

2.5.1.1.12.2 $R_3Sb(O_2CR')_2$ Compounds with R' = Substituted Alkyl

$(CH_3)_3Sb(O_2CR')_2$ (R' = CF_3, CHF_2, CH_2F, CCl_3, $CHCl_2$, CH_2Cl, $CHBr_2$, and CH_2CN)

$(CH_3)_3Sb(O_2CCF_3)_2$ (Table 37, No. 1) is prepared by reaction of $(CH_3)_3SbX_2$ (X = Cl, Br) with stoichiometric amounts of AgO_2CCF_3 in CH_3OH. The other compounds (Table 37, Nos. 2 to 9) are obtained by reacting stoichiometric amounts of $(CH_3)_3SbX_2$ with a freshly prepared aqueous solution of Ag_2O in the appropriate acid. In each case, the product is isolated by concentrating the filtered solution under reduced pressure. $(CH_3)_3Sb(O_2CCCl_3)_2$ (Table 37, No. 4) is also prepared by the reaction of Cl_3CCO_2H with $(CH_3)_3SbS$ in CH_2Cl_2 for 1 h at room temperature. After removal of the solvent in a vacuum, 93% of the substance is obtained. All compounds can be recrystallized from a 4:1 mixture of petroleum ether and hexane [5].

The compounds are white crystalline solids. No. 4 melts at 164 to 166 °C, recrystallized from CH_3OH/petroleum ether [5]. Further melting points, ^1H NMR, IR, and ^{121}Sb Mössbauer data are summarized in Table 37 [1, 2]. The Mössbauer data are discussed and compared with the corresponding values for $(CH_3)_3SnO_2CR'$ compounds [2]. The isomer shift and quad-

Table 37
Melting points, ^1H NMR, IR, and ^{121}Sb Mössbauer Data of $(CH_3)_3Sb(O_2CR')_2$ Compounds.

No.	R'	m.p. [1] (in °C)	^1H NMR [1] in $CDCl_3$ δ (in ppm)	IR (in cm^{-1}) [1] [a] in Nujol / in CCl_4	^{121}Sb Mössbauer (in mm/s) [2] [b] δ [c] / e^2qQ	Γ
1	CF_3	105 to 106	2.08	1725; 1406; 586 1724; 1398	−5.5±0.1 −28.0±0.4	2.71
2	CHF_2	68 to 69	2.03	1692; 1420; 588 1695; 1403	−5.4±0.1 −26.4±0.7	2.67
3	CH_2F	131 to 132	1.98	1688, 1650; 1405; 584 1676 ($CHCl_3$); 1403	−5.3±0.2 −24.4±0.4	2.72
4	CCl_3	138.5 to 139.5	2.10	1699; 1292; 584 1708; 1285	−5.50±0.05 −27.5±0.3	2.53
5	$CHCl_2$	89.5 to 90	2.03	1692, 1680; 1334; 589 1704, 1682; 1318	−5.3±0.2 −25.2±0.5	2.97
6	CH_2Cl	90 to 91	1.95	1669, 1638 sh; 1344; 588 1694, 1666; 1328	−5.27±0.10 −25.1±0.3	3.01
7	$CHBr_2$	102 to 103	2.03	1691, 1678; 1320; 568 1695, 1674; 1310	−5.4±0.1 −26.2±0.1	2.60
8	CH_2Br	86 to 87	1.94	1680, 1668; 1318; 584 1668, 1618; 1318	−5.4±0.1 −26.2±0.1	2.60
9	CH_2CN	120 to 121	1.98	1680 sh, 1667; 1318; 580 1682 ($CHCl_3$); 1318	−5.37±0.06 −25.8±0.9	2.90

[a] Assignment: νC=O; νC-O; $\nu_{as}SbC$. — [b] Sample at 8 K, source at 80 K. — [c] Relative to $Ba^{121}SnO_3$.

References on p. 176

rupole coupling constant for No. 1 are used to obtain the electron populations in the hybrid atomic orbitals used by Sb^V to form bonds [3].

The compounds are stable at room temperature and unaffected by atmospheric moisture. They are soluble in polar as well as nonpolar solvents. Osmometric molecular weight determinations show that they are dissolved as monomers in C_6H_6. The electrical conductances in pyridine at 25 °C and in the concentration range 2.0×10^{-3} to 2.5×10^{-4} M show that the compounds do not ionize in pyridine with the exception of No. 1, where a slight ionization is indicated ($\Lambda=3.08\ cm^2\cdot\Omega^{-1}\cdot mol^{-1}$ at 2×10^{-3} M, $\Lambda=9.20\ cm^2\cdot\Omega^{-1}\cdot mol^{-1}$ at 2.5×10^{-4} M) [1]. Compound No. 4 reacts with $P(C_6H_5)_3$ and cyclopentadiene to give 79% $(CH_3)_3SbCl_2$, 92% $(C_6H_5)_3PO$, and 37% 7,7-dichlorobicyclo[3.2.0]hept-2-en-6-one [5].

The fungitoxicity of compounds Nos. 4 to 6 was tested [4].

$(CH_3)_3Sb(O_2CCH_2C_6H_5)_2$

$(CH_3)_3SbBr_2$ and $C_6H_5CH_2CO_2H$ are reacted in refluxing C_6H_6 in the presence of $N(C_2H_5)_3$. The filtrate of the mixture is concentrated in vacuum, and the residue is recrystallized from petroleum ether or petroleum ether/CH_3CN.

The compound melts at 63 °C. A ^1H NMR spectrum measured at room temperature and at −60 °C shows signals at $\delta=1.73$ (s, CH_3), 3.43 (s, CH_2), and 7.20 (m, C_6H_5) ppm [6].

$(C_4H_9)_3Sb(O_2CCH_2SH)_2$

A heat stable blend is obtained by milling the title compound, dibutyltin maleate, polyvinyl chloride, an acrylonitrile-butadiene-styrene copolymer, and liquid paraffin at 160 °C [7].

$(C_6H_5)_3Sb(O_2CR')_2$ (R′ = CF_3, CHF_2, CH_2F, CCl_3, $CHCl_2$, CH_2Cl, $CHBr_2$, and CH_2Br)

The compounds in Table 38 are prepared by the following methods:

Method I: a. A solution of triphenylantimony dihalide in C_6H_6 reacts with AgO_2CR' in a metathetical reaction. The benzene solution is filtered, and the product is isolated by removal of C_6H_6 under vacuum [1].

b. A benzene solution of triphenylantimony dihalide is added to an aqueous solution of Ag_2O in the desired acid. After filtration, the benzene layer is separated from the aqueous layer. The organic phase is dried with molecular sieves, and C_6H_6 is removed after filtration under vacuum [1].

Method II: a. $(C_6H_5)_3SbBr_2$ reacts with CCl_3CO_2H in the presence of $N(C_2H_5)_3$ in refluxing C_6H_6. Concentration of the filtrate gives a 70 to 90% yield of the product [9].

b. A suspension of $(C_6H_5)_3SbO$ in CH_3OH is treated with an excess of CCl_3CO_2H or $CHCl_2CO_2H$. Boiling the mixture for 10 min and cooling the hot filtrate gives a yield of 75% [10].

Method III: $Sb(C_6H_5)_3$ reacts with $Hg(O_2CCF_3)_2$ in a 1:1 molar ratio in refluxing C_6H_6. The precipitating Hg is separated, and the filtrate is concentrated; 62% yield [8].

General Remarks. All compounds are recrystallized from a 1:4 mixture of light petroleum ether and hexane [1]. They are white crystalline solids, stable at room temperature, and unaffected by atmospheric moisture. They are soluble in polar as well as nonpolar solvents and dissolve in C_6H_6 as monomers [9]. Very low molar conductivities were measured in pyridine at 25 °C (for No. 4 also in CH_3CN) except for No. 1, where a slight ionization is

References on p. 176

indicated ($\Lambda = 3.21\ cm^2 \cdot \Omega^{-1} \cdot mol^{-1}$ at 2×10^{-3} M and $\Lambda = 12.35\ cm^2 \cdot \Omega^{-1} \cdot mol^{-1}$ at 2.5×10^{-4} M) [1].

Compound No. 4 was tested as a retarding agent in the burning of epoxy [11] and polyester [12] resins. The fungitoxicity of No. 4 was tested [4].

Table 38
Compounds of the Type $(C_6H_5)_3Sb(O_2CR')_2$ with R' = Halogenomethyl.
For explanations, abbreviations, and units, see p. X.

No.	R'	method of preparation and physical properties [Ref.]
1	CF_3	I a [1], III [8] m. p. 108 to 109° [1], 106 to 108° (from C_5H_{12}/CH_2Cl_2) [8] IR (Nujol): 1725 (νC=O), 1392 (νC-O) [1] IR (CCl_4): 1734 (νC=O), 1388 (νC-O) [1]
2	CHF_2	I a [1] m. p. 96 to 96.5° [1] IR (Nujol): 1689, 1672 sh (νC=O), 1318 (νC-O) [1] IR (CCl_4): 1709 (νC=O), 1318 (νC-O) [1]
3	CH_2F	I a [1] m. p. 161 to 162° [1] IR (Nujol): 1678, 1656 (νC=O), 1348 (νC-O) [1] IR (CCl_4): 1698 (νC=O), 1338 (νC-O) [1]
4	CCl_3	I b [1], II a [9], II b [10] m. p. 139 to 140° [1], 140° (from petroleum ether or petroleum ether/CH_3CN) [9], 152° with dec. [10] ^{1}H NMR ($CDCl_3$): 7.55 (m, H-3, 4), 7.95 (m, H-2) [10] IR (Nujol): 1715 (νC=O), 1280 (νC-O) [1] IR (CCl_4): 1721 (νC=O), 1279 (νC-O) [1]
5	$CHCl_2$	I b [1] m. p. 147 to 148° [1] ^{13}C NMR ($CDCl_3$): 66.43 (CH; J(C, H) = 180.7), 129.87 (C-3; J(C, H) = 164.2), 132.16 (C-4; J(C, H) = 161.7), 133.77 (C-1,2; J(C, H) = 165.4), 185.84 (CO) [13] IR (Nujol): 1701 sh, 1682 (νC=O), 1304 (νC-O) [1] IR (CCl_4): 1704 sh, 1689 (νC=O), 1302 (νC-O) [1]
6	CH_2Cl	I b [1], II b [10] m. p. 132 to 133° [1], 149 to 150° [10] ^{1}H NMR ($CDCl_3$): 7.55 (m, H-3, 4), 7.95 (m, H-2) [10] ^{13}C NMR ($CDCl_3$): 42.45 (CH_2; J(C, H) = 180.7), 129.58 (C-3; J(C, H) = 163.0), 131.67 (C-4; J(C, H) = 161.1), 133.87 (C-2; J(C, H) = 165.5), 135.63 (C-1), 169.83 (CO) [13] IR (Nujol): 1676, 1654 (νC=O), 1337 (νC-O) [1] IR (CCl_4): 1680 (νC=O), 1328 (νC-O) [1]

References on p. 176

Table 38 [continued]

No.	R′	method of preparation and physical properties [Ref.]
7	$CHBr_2$	Ib [1] m.p. 154 to 155° [1] IR (Nujol): 1684 (νC=O), 1296 (νC–O) [1] IR (CCl_4): 1687 (νC=O), 1298 (νC–O) [1]
8	CH_2Br	Ib [1] m.p. 133 to 134° [1] IR (Nujol): 1673, 1655 (νC=O), 1319 (νC–O) [1] IR (CCl_4): 1678 (νC=O), 1315 (νC–O) [1]

$(C_6H_5)_3Sb(O_2CCH_2OC_6H_5)_2$

$(C_6H_5)_3SbBr_2$ and $C_6H_5OCH_2CO_2H$ (ratio 1:2.5) are reacted in C_6H_6 at room temperature in the presence of $N(C_2H_5)_3$. Evaporation of the solvent from the filtrate and recrystallization from CH_3CN give 60% of a compound which melts at 105 °C. IR absorptions of the CO_2 group are at 1667 (ν_{as}) and 1310 (ν_s) cm^{-1}. A ^1H NMR spectrum (in $CDCl_3$) shows resonances at δ=4.39 (CH_2), 6.81, 6.94, 7.12, 7.51, 7.91 (C_6H_5) ppm [14].

$(C_6H_5)_3Sb(O_2CR')_2$ (R′=$CH_2OC_6H_4Cl$-4, $CH_2OC_6H_4OCH_3$-2, $CH_2OC_6H_4OCH_3$-3, and $CH_2CH_2COCH_3$)

The compounds are prepared in refluxing C_6H_6 from $(C_6H_5)_3SbBr_2$, the corresponding acids, and $N(C_2H_5)_3$. Concentration of the filtrate in vacuum gives 70 to 90% yield of crude products. The compounds are recrystallized from petroleum ether or petroleum ether/CH_3CN mixtures.

Melting points and IR data are as follows [9]:

compound	m.p. in °C	IR (KBr/Nujol) in cm^{-1} $\nu_{as}O_2C$	ν_sO_2C
$(C_6H_5)_3Sb(O_2CCH_2OC_6H_4Cl\text{-}4)_2$	128	1630 s	1320 s
$(C_6H_5)_3Sb(O_2CCH_2OC_6H_4CH_3\text{-}2)_2$	138	1730 s	1300 s
$(C_6H_5)_3Sb(O_2CCH_2OC_6H_4CH_3\text{-}3)_2$	142	1730 s	1300 m
$(C_6H_5)_3Sb(O_2C(CH_2)_2COCH_3)_2$	80	1620 s	1300 m, 1265 m

The compounds are nonconducting in CH_3CN and stable against hydrolysis at room temperature [9]. They were tested as retarding agents in the burning of epoxy resins [11].

$(C_6H_5)_3Sb(O_2CCH_2SR'')_2$ (R″=$CH_2C_6H_5$, CH_3, C_2H_5, C_3H_7, i-C_3H_7, C_4H_9, C_6H_5)

The compounds of this type are prepared from $(C_6H_5)_3SbBr_2$ with the appropriate acid (molar ratio 1:2.5) and in the presence of $N(C_2H_5)_3$ in C_6H_6 at room temperature. The filtrate of the mixture is evaporated, and the resulting residue is recrystallized from CH_3CN. The yield of the compounds is about 60%.

The physical properties are given in Table 39. The compounds and their solutions in $CDCl_3$ are stable to air and moisture [14].

References on p. 176

Table 39
Physical Properties of $(C_6H_5)_3Sb(O_2CCH_2SR'')_2$ Compounds [14].

R''	m.p. in °C	^{1}H NMR ($CDCl_3$), δ in ppm C_6H_5		CH_2	R''			IR (Nujol) in cm^{-1} $\nu_{as}O_2C$	ν_sO_2C
$CH_2C_6H_5$	104	8.09	7.50	2.84	3.34	7.00		1597	1348
CH_3	105	8.07	7.25	2.97	1.88			1635	1315
C_2H_5	109	8.06	7.52	3.05	2.22	1.05		1635	1318
C_3H_7	94	8.07	7.55	2.99	2.14	1.37	0.80	1647	1305
i-C_3H_7	85	8.06	7.55	3.04	2.48	1.03		1630	1313
C_4H_9	73	8.07	7.55	3.01	2.16	1.30	0.83	1630	1300
C_6H_5	140	7.93	7.48	3.47	7.16			1666	1302

$(C_6H_5)_3Sb(O_2C(CH_2)_2C_6H_2(C_4H_9\text{-t})_2\text{-3,5-OH-4})_2$

The compound is described in a patent as a synergistic antioxidant for polyolefins [15].

$(4\text{-}CH_3C_6H_4)_3Sb(O_2CCF_3)_2$

$Sb(C_6H_4CH_3\text{-4})_3$ and $Hg(O_2CCF_3)_2$ are refluxed in C_6H_6 for two hours. The precipitated Hg is separated and the solution is concentrated to yield 58% of the compound, which melts at 204 to 205 °C when recrystallized from C_5H_{12}/CH_2Cl_2. A yield of 86% is obtained from the reaction of $(4\text{-}CH_3C_6H_4)_3SbO$, dissolved in CH_3OH, with CF_3CO_2H after 30 min at 20 °C. The compound is isolated by concentrating the filtrate of the mixture [8].

$(4\text{-}CH_3C_6H_4)_3Sb(O_2CR')_2$ (R' = CH_2Cl, CCl_3, CH_2Br, $CH_2OC_6H_4Cl\text{-4}$, and $CH_2C_6H_5$)

The compounds are prepared by the following method: $(4\text{-}CH_3C_6H_4)_3SbBr_2$, the corresponding substituted acetic acid, and $N(C_2H_5)_3$ in stoichiometric amounts are reacted in C_6H_6, the first two hours at room temperature and then for the same time under reflux. The filtrate of the mixture is concentrated in a vacuum, and the resulting precipitate is recrystallized from petroleum ether or petroleum ether/CH_3CN. The yield is about 70 to 90%.

The melting points and IR vibrations are as follows [9]:

compound	m.p. in °C	IR (KBr/Nujol) in cm^{-1} $\nu_{as}O_2C$	ν_sO_2C
$(4\text{-}CH_3C_6H_4)_3Sb(O_2CCH_2Cl)_2$	196	1630s	1320m
$(4\text{-}CH_3C_6H_4)_3Sb(O_2CCCl_3)_2$	180 (dec.)	1610s	1250m
$(4\text{-}CH_3C_6H_4)_3Sb(O_2CCH_2Br)_2$	120 to 122	1620s	1260m
$(4\text{-}CH_3C_6H_4)_3Sb(O_2CCH_2OC_6H_4Cl\text{-4})_2$	102	1695s	1305w, 1265s
$(4\text{-}CH_3C_6H_4)_3Sb(O_2CCH_2C_6H_5)_2$	110 to 112	1620s	1230s

A ^{1}H NMR spectrum of $(4\text{-}CH_3C_6H_4)_3Sb(O_2CCH_2C_6H_5)_2$ shows resonances at δ = 2.35 (s, CH_3), 3.38 (s, CH_2), and 7.30 (m, C_6H_5, C_6H_4) ppm. The compounds are nonconducting in CH_3CN and are not affected by atmospheric moisture [9].

$(C_{10}H_7)_3Sb(O_2CCH_2C_6H_2(C_4H_9\text{-t})_2\text{-3,5-OH-4})_2$

A compound of this composition is mentioned in a patent as a synergistic antioxidant for polyolefins [15].

References on p. 176

References:

[1] R. G. Goel, D. R. Ridley (J. Organometal. Chem. **38** [1972] 83/9).
[2] H. G. Ang, W. S. Lien (J. Fluorine Chem. **9** [1977] 73/80).
[3] L. H. Bowen, G. G. Long (Inorg. Chem. **15** [1976] 1039/44).
[4] R. E. Burrell, C. T. Corke, R. G. Goel (J. Agr. Food Chem. **31** [1983] 85/8).
[5] T. Okada, R. Okawara (J. Organometal. Chem. **42** [1972] 117/21).
[6] K. Bajpai, M. Srivastava, R. C. Srivastava (Indian J. Chem. Sect. A **20** [1981] 736/7).
[7] F. Kato, M. Yatsu, Katsuta Chemical Industry Co., Ltd. (Japan. 67-2903 [1967]; C.A. **67** [1967] No. 33319).
[8] V. I. Popov, N. V. Kondratenko (Zh. Obshch. Khim **46** [1976] 2597/601; J. Gen. Chem. [USSR] **46** [1976] 2477/80).
[9] K. Bajpai, R. Singhal, R. C. Srivastava (Indian J. Chem. A **18** [1979] 73/5).
[10] J. Havranek, J. Mleziva, A. Lycka (J. Organometal. Chem. **157** [1978] 163/6).
[11] J. Havranek, J. Mleziva (Angew. Makromol. Chem. **84** [1980] 105/17).
[12] J. Havranek, J. Muller, J. Mleziva (Sb. Ved. Pr. Vys. Sk. Chemickotechnol. Pardubice **42** [1980] 123/32).
[13] J. Havranek, A. Lycka (Sb. Ved. Pr. Vys. Sk. Chemickotechnol. Pardubice **43** [1980] 123/7).
[14] A. Ouchi, H. Honda, S. Kitazima (J. Inorg. Nucl. Chem. **37** [1975] 2559/61).
[15] T. Ozeki, M. Watanabe, Mitsubishi Rayon Co., Ltd. (Japan. 72-29573 [1972]; C.A. **78** [1973] No. 30819).

2.5.1.1.12.3 $R_3Sb(O_2CR')_2$ Compounds with R′ = Alkenyl

$(CH_3)_3Sb(O_2CCH{=}CHCH_3)_2$ and $(CH_3)_3Sb(O_2CCH{=}CHC_6H_5)_2$

Both compounds are prepared by reaction of $(CH_3)_3SbBr_2$ with the corresponding acid in C_6H_6 in the presence of $N(C_2H_5)_3$. Heating for 2 h at room temperature with additional 2 h reflux, filtering the $[NH(C_2H_5)_3]Br$, and concentrating the filtrate in a vacuum give the products with melting points of 53 to 60 °C and 120 °C, respectively, after recrystallization from petroleum ether or petroleum ether/CH_3CN [1].

The compounds are stable under aerobic conditions and monomeric in C_6H_6. Molar conductance values indicate their nonionic character [1].

$(C_4H_9)_3Sb(O_2CCH{=}CHCO_2(CH_2)_7CH_3)_2$

The compound is mentioned in a patent as a heat stabilizer for halogen-containing resin compositions [2].

$(C_6H_5)_3Sb(O_2CCH{=}CH_2)_2$

The compound is prepared from $(C_6H_5)_3SbO$, suspended in CH_3OH, and an excess of the corresponding acid. The mixture is boiled for 10 min and filtered hot. Upon cooling overnight, crystals of the compound deposit, which are then recrystallized from CH_3OH or C_2H_5OH. The yield is 73.1% [3]. A patent describes the preparation from $(C_6H_5)_3SbO$ and acrylic acid by heating to 130 °C for 15 min; yield 60% after recrystallization from CH_3OH [5].

The melting point is 152 to 153 °C. ^{1}H NMR ($CDCl_3$): δ = 5.50, 5.90 (CH_2), 5.90 (CH), 7.25 (m, H-3,4), 7.75 (q, H-2) ppm [3]. ^{13}C NMR ($CDCl_3$): δ = 128.85 (CH_2; J(C, H) = 161.1 Hz), 129.24 (C-3; J(C, H) = 162.4 Hz), 131.04 (C-4; J(C, H) = 159.9 Hz), 131.04 (CH; J(C, H) = 166.0 Hz), 133.77 (C-2; J(C, H) = 168.5 Hz), 137.91 (C-1), 169.83 (CO) ppm [4].

The compound is soluble in CH_3OH, C_2H_5OH, $(CH_3)_2CO$, C_6H_6, $CHCl_3$, and CCl_4, and insoluble in petroleum ether [3]. Bactericidal and fungicidal activities are claimed [5], and a copolymer with $CH_2{=}C(CH_3)CO_2CH_3$ was prepared [6].

$(C_6H_5)_3Sb(O_2CC(CH_3){=}CH_2)_2$

See the previous compound for the preparation from $(C_6H_5)_3SbO$ in CH_3OH. The yield is 75.5% [3].

The melting point is 160 to 161 °C. ^1H NMR ($CDCl_3$): $\delta = 1.80$ (CH_3), 5.30, 5.90 (CH_2), 7.35 (m, H-3,4), 7.90 (q, H-2) ppm [3]. ^{13}C NMR ($CDCl_3$): $\delta = 123.74$ (CH_2; J(C, H) = 128.2 Hz), 129.14 (C-3; J(C, H) = 162.4 Hz), 130.89 (C-4; J(C, H) = 159.9 Hz), 133.68 (C-2; J(C, H) = 164.8 Hz), 138.21 (C-1), 138.60 (CCH_3), 171.05 (CO) ppm [4]. The same solubility as for the previous compound is observed [3].

$(C_6H_5)_3Sb(O_2CR')_2$ (R' = $CH{=}CHCO_2CH_3$, $CCl{=}CClCO_2CH_3$, and $CBr{=}CBrCO_2CH_3$)

Preparation of the compounds from $(C_6H_5)_3SbO$ and the corresponding maleic acid derivative in C_6H_6 is mentioned. The following melting points, ^1H NMR, and IR data are given [7, 8]:

compound	m.p. in °C	^1H NMR ($CDCl_3$) δ in ppm	IR (KBr) $\nu C{=}O$ in cm^{-1}
$(C_6H_5)_3Sb(O_2CCH{=}CHCO_2CH_3)_2$	107 to 109	3.30 (s, 6H), 5.90 (s, 4H), 7.43 (m, 9H), 8.00 (m, 6H)	1720, 1630
$(C_6H_5)_3Sb(O_2CCCl{=}CClCO_2CH_3)_2$	126 to 127	3.27 (s, 2H), 7.24 (m, 3H), 7.92 (m, 2H)	1717, 1678
$(C_6H_5)_3Sb(O_2CCBr{=}CBrCO_2CH_3)_2$	155.5 to 157	3.27 (s, 2H), 7.48 (m, 3H), 7.98 (m, 2H)	1716, 1653

$(C_6H_5)_3Sb(O_2CCH{=}CHR'')_2$ and **$(4\text{-}CH_3C_6H_4)_3Sb(O_2CCH{=}CHR'')_2$** (R'' = CH_3 and C_6H_5)

The compounds are prepared from $(C_6H_5)_3SbBr_2$ or $(4\text{-}CH_3C_6H_4)SbBr_2$ with the appropriate acid in C_6H_6. The reactants are maintained the first 2 h at room temperature and then 2 h at reflux in the presence of $N(C_2H_5)_3$. Concentration of the filtrate gives 70 to 90% yields. The compounds are recrystallized from petroleum ether or petroleum ether/CH_3CN [9].

The melting points and IR spectra are as follows [9]:

compound	m.p. in °C	IR (KBr/Nujol) in cm^{-1} $\nu_{as}O_2C$	ν_sO_2C
$(C_6H_5)_3Sb(O_2CCH{=}CHCH_3)_2$	146	1651s	1330s, 1280w
$(C_6H_5)_3Sb(O_2CCH{=}CHC_6H_5)_2$	192	1640s	1250m
$(4\text{-}CH_3C_6H_4)_3Sb(O_2CCH{=}CHCH_3)_2$	167	1650s	1250w, 1230s
$(4\text{-}CH_3C_6H_4)_3Sb(O_2CCH{=}CHC_6H_5)_2$	112	1680s	1250s

The compounds are monomeric in C_6H_6, nonconducting in CH_3CN, and stable against hydrolysis at room temperature [9].

References on p. 178

References:

[1] K. Bajpai, M. Srivastava, R. C. Srivastava (Indian J. Chem. A **20** [1981] 736/7).
[2] Nitto Chemical Industry Co., Ltd. (Japan. 81-30453 [1981]; C.A. **95** [1981] No. 63222).
[3] J. Havranek, J. Mleziva, A. Lycka (J. Organometal. Chem. **157** [1978] 163/6).
[4] J. Havranek, A. Lycka (Sb. Ved. Pr. Vys. Sk. Chemickotechnol. Pardubice **43** [1980] 123/7).
[5] J. R. Leebrick, M & T Chemicals Inc. (U.S. 3287210 [1962/66]; C.A. **66** [1967] No. 85070).
[6] J. Musher, K. Su (U.S. 3939190 [1972/76]; C.A. **84** [1976] No. 181136).
[7] M. M. Cheng, K. Su, J. I. Musher (AD-757345 [1973] 1/8).
[8] M. M. Cheng, K. Su, J. I. Musher (Israel J. Chem. **12** [1974] 967/70).
[9] K. Bajpai, R. Singhal, R. C. Srivastava (Indian J. Chem. A **18** [1979] 73/5).

2.5.1.1.12.4 $R_3Sb(O_2CR')_2$ Compounds with R' = Aryl

The compounds of this type are described in Table 40. They are prepared by the following methods:

Method I: A suspension or a solution of the triorganoantimony dibromide in C_6H_6 is treated with a slight excess of the appropriate acid and $N(C_2H_5)_3$ with stirring for about 60 min at ambient temperature. After filtration of the $[NH(C_2H_5)_3]Br$, the filtrate is evaporated at 60 °C in a vacuum. The crude products are recrystallized from CH_3CN, or if too soluble in this solvent, from ligroin or petroleum ether. The yields are about 60 to 80% after one recrystallization [1, 2].

Method II: a. A suspension of R_3SbO in CH_3OH is reacted with an excess of the corresponding benzoic acid. The mixture is boiled for 10 min and filtered hot. Upon cooling, the compounds precipitate in good yields [8].

b. $(C_6H_5)_3SbO$ and the corresponding substituted benzoic acid in C_6H_6 are refluxed on a water separator [3, 14, 15]. Upon cooling, crystals deposit [3].

General Remark. The compounds are stable in dry air [1, 15] and in aqueous and organic solvents [15].

Table 40
Triorganoantimony Dibenzoates $R_3Sb(O_2CR')_2$.
Further information on numbers preceded by an asterisk is given at the end of the table.
For explanations, abbreviations, and units, see p. X.

No.	R'	method of preparation (yield in %) properties and remarks [Ref.]
with R = CH_3		
1	C_6H_5	I [1, 2], II b [3] colorless crystals, m.p. 154° (from C_6H_6) [3], 157° [1] 1 H NMR ($CDCl_3$): 2.10 (CH_3), 7.45, 7.79 (C_6H_5) [1] ^{13}C NMR ($CHCl_3$ vs. C_6H_6): 0.1 (C-3,5), 1.8 (C-2,6), 4.1 (C-1,4), 13.1 (CH_3 vs. TMS), 42.3 (CO) [2] IR (mull): 1642s, 1631s (νC=O), 1323s, 1299s (νC-O) [3] see also [1] soluble in most common organic solvents [3] useful as a polymerization catalyst to give colorless polyethylene terephthalate [4]

References on p. 185

Table 40 [continued]

No.	R′	method of preparation (yield in %) properties and remarks [Ref.]
2	C_6H_4Cl-4	I [1, 2] ^1H NMR ($CDCl_3$): 2.07 (CH_3), 7.34, 7.90 (C_6H_4) [1] ^{13}C NMR ($CHCl_3$ vs. C_6H_6): 0.7 (C-3,5), 3.6 (C-1), 3.7 (C-2,6), 10.9 (C-4), 13.3 (CH_3 vs. TMS), the values are compared with calculated ones [2]
3	$C_6H_4NH_2$-2	I m.p. 184° IR (Nujol): 3450 (ν_{as}NH), 3340 (ν_sNH), 303 (νSbO) [1]
4	$C_6H_4NH_2$-3	I m.p. 166° IR (Nujol): 3400 (ν_{as}NH), 3340 (ν_sNH), 303 (νSbO) [1]
5	$C_6H_4NH_2$-4	I m.p. 200° ^1H NMR ($CDCl_3$): 2.04 (CH_3), 6.60, 7.79 (C_6H_4) IR (Nujol): 3430 (ν_{as}NH), 3320 (ν_sNH), 1603 ($\nu_{as}O_2C$), 1343 (ν_sO_2C), 302 (νSbO) [1]
6	$C_6H_4NO_2$-2	I m.p. 148° IR (Nujol): 303 (νSbO), 298 [1]
7	$C_6H_4NO_2$-3	I m.p. 200° IR (Nujol): 302 (νSbO), 297 [1]
8	$C_6H_4NO_2$-4	I m.p. 200° ^1H NMR ($CDCl_3$): 2.07 (CH_3), 8.19 (C_6H_4) IR (Nujol): 1653 ($\nu_{as}O_2C$), 1320 (ν_sO_2C), 303 (νSbO), 297 [1]
9	$C_6H_4CH_3$-2	I m.p. 95° IR (Nujol): 302 (ν_sSbO) [1]
10	$C_6H_4CH_3$-3	I m.p. 74° IR (Nujol): 302 (ν_sSbO) [1]
11	$C_6H_4CH_3$-4	I m.p. 141° ^1H NMR ($CDCl_3$): 2.05 (CH_3Sb), 2.37 (CH_3-4), 7.24, 7.83 (C_6H_4) IR (Nujol): 1630 ($\nu_{as}O_2C$), 1330 (ν_sO_2C), 303 (ν_sSbO) [1]
with R = C_2H_5		
12	C_6H_5	I IR (Nujol): 1643 (ν_sO_2C), 1324 ($\nu_{as}O_2C$) [1]

References on p. 185

Table 40 [continued]

No.	R′	method of preparation (yield in %) properties and remarks [Ref.]
13	C_6H_4Cl-4	I ^{1}H NMR ($CDCl_3$): 1.58, 2.66 (C_2H_5), 7.40, 7.97 (C_6H_4) IR (Nujol): 1643 (ν_sO_2C), 1326 ($\nu_{as}O_2C$) [1]
14	$C_6H_4NH_2$-2	I m.p. 131° IR (Nujol): 3440 ($\nu_{as}NH$), 3330 (ν_sNH), 292 (νSbO) [1]
15	$C_6H_4NH_2$-3	I m.p. 133° IR (Nujol): 3400 ($\nu_{as}NH$), 3320 (ν_sNH), 302 (νSbO) [1]
16	$C_6H_4NH_2$-4	I m.p. 139° IR (Nujol): 3440 ($\nu_{as}NH$), 3340 (ν_sNH), 1605 ($\nu_{as}O_2C$), 1325 (ν_sO_2C), 298 (νSbO) [1]
17	$C_6H_4NO_2$-4	I m.p. 200° ^{1}H NMR ($CDCl_3$): 1.60, 2.71 (C_2H_5), 8.20 (C_6H_4) IR (Nujol): 1644 ($\nu_{as}NH$), 1321 (ν_sNH), 301 (νSbO), 280 [1]
18	$C_6H_4CH_3$-2	I m.p. 58° IR (Nujol): 280 (νSbO) [1]
19	$C_6H_4CH_3$-4	I m.p. 55° ^{1}H NMR ($CDCl_3$): 1.57, 2.64 (C_2H_5), 2.38 (CH_3-4), 7.21, 7.90 (C_6H_4) IR (Nujol): 1643 ($\nu_{as}O_2C$), 1325 (ν_sO_2C), 1313, 280 (νSbO) [1]
with R = C_3H_7		
*20	C_6H_5	see further information IR Nujol: 1636 (νO_2C) [5]
with R = C_4H_9		
21	C_6H_5	mentioned as a catalyst for the polymerization of organoisocyanates [6]
with R = C_6H_5		
*22	C_6H_5	I [1,2], II a (70) [8] m.p. 173° [1], 171.5° (from C_6H_6/petroleum ether) [7], 176 to 177° [8] ^{1}H NMR ($CDCl_3$): 7.35 (m, 15H), 7.85 (q, 4H), 8.05 (q, 6H) [8], see also [1, 2] ^{13}C NMR ($CHCl_3$ vs. C_6H_6): 0.5 (C-3,5 in C_6H_5C), 1.8 (C-3,5 in C_6H_5Sb), 2.3 (C-2,6 in C_6H_5C), 3.6 (C-4 in C_6H_5Sb), 5.1 (C-4 in C_6H_5C), 6.3 (C-2,6 in C_6H_5Sb), 4.5 (C-1 in C_6H_5C), 10.7 (C-1 in C_6H_5Sb), 42.5 (CO) [2] IR (Nujol): 1607 ($\nu_{as}O_2C$), 1341 (ν_sO_2C), 1478, 1431, 456 (νC_6H_5), 298 (νSbO) [1]

References on p. 185

Table 40 [continued]

No.	R′	method of preparation (yield in %) properties and remarks [Ref.]
23	C_6H_4Cl-2	I [1], II a [8] m.p. 162° [1], 170 to 171° [8] ^{1}H NMR ($CDCl_3$): 7.25, 8.05 (C_6H_4), 7.49, 8.21 (C_6H_5) [1] see also [8] IR (Nujol): 1638 ($\nu_{as}O_2C$), 1330 (ν_sO_2C), 1480, 1440, 462 (νC_6H_5) [1] solubility like No. 22 [8]
24	C_6H_4Cl-3	I ^{1}H NMR ($CDCl_3$): 7.27, 7.85, 7.97 (C_6H_4), 7.04, 8.20 (C_6H_5) IR (Nujol): 1638 ($\nu_{as}O_2C$), 1320 (ν_sO_2C), 1477, 1436, 462 (νC_6H_5) [1]
25	C_6H_4Cl-4	I m.p. 200° ^{1}H NMR ($CDCl_3$): 7.29, 7.86 (C_6H_4), 7.56, 8.16 (C_6H_5) IR (Nujol): 1638 ($\nu_{as}O_2C$), 1320 (ν_sO_2C), 1480, 1440, 460 (νC_6H_5) [1]
26	C_6H_4OH-2	I m.p. 176° [1]
27	C_6H_4OH-4	I [1], II b [14, 15] m.p. 160° [1], 198 to 199° [14, 15] ^{1}H NMR ($OS(CD_3)_2$): 6.56 (d, 4H; J=8), 7.30 to 7.78 (m, 19H), 9.76 (s, 2H) IR (Nujol): 3120 (νOH), 1628 (νC=O) [14, 15]
28	$C_6H_4OCH_3$-4	I m.p. 190° IR (Nujol): 1480, 1435, 455 (νC_6H_5) [1]
29	$C_6H_4NH_2$-2	I m.p. 161° IR (Nujol): 3460, 3340 (νNH_2), 1610 ($\nu_{as}O_2C$), 1345 (ν_sO_2C), 1480, 1441, 457 (νC_6H_5), 290 (νSbO) [1]
30	$C_6H_4NH_2$-3	I m.p. 105° IR (Nujol): 3400, 3320 (νNH_2), 1600 ($\nu_{as}O_2C$), 1330 (ν_sO_2C), 1480, 1442, 463 (νC_6H_5), 298 (νSbO) [1]
31	$C_6H_4NH_2$-4	I [1], II [8] m.p. 200° [1], 237 to 239° [8] ^{1}H NMR ($CDCl_3$): 7.10, 7.87 (C_6H_4), 7.53, 8.11 (C_6H_5) [1]; 3.80 (d, NH_2), 6.50, 7.70 (C_6H_4), 7.40, 8.00 (C_6H_5) [8] IR (Nujol): 3450, 3340 (νNH_2), 1600 ($\nu_{as}O_2C$), 1335 (ν_sO_2C), 1480, 1442, 460 (νC_6H_5), 305 (νSbO) [1] solubility like No. 22 [8]
32	$C_6H_4NO_2$-2	I m.p. 179° ^{1}H NMR ($CDCl_3$): 7.54, 8.11 (C_6H_4), 7.61, 8.11 (C_6H_5) IR (Nujol): 1660 ($\nu_{as}O_2C$), 1320 (ν_sO_2C), 1472, 1440, 463 (νC_6H_5), 300 (νSbO) [1]

References on p. 185

Table 40 [continued]

No.	R′	method of preparation (yield in %) properties and remarks [Ref.]
33	$C_6H_4NO_2$-3	I m.p. 198° ^{1}H NMR ($CDCl_3$): 7.28, 8.34, 8.76 (C_6H_4), 7.56, 8.18 (C_6H_5) IR (Nujol): 1645 ($\nu_{as}O_2C$), 1310 (ν_sO_2C), 1471, 1436, 462 (νC_6H_5), 298 (νSbO) [1]
34	$C_6H_4NO_2$-4	I m.p. 200° ^{1}H NMR ($CDCl_3$): 7.48, 8.14 (C_6H_5), 8.14 (C_6H_4) IR (Nujol): 1640 ($\nu_{as}O_2C$), 1315 (ν_sO_2C), 1470, 1435, 466 (νC_6H_5), 302 (νSbO) [1]
35	$C_6H_4CH_3$-2	I m.p. 123° ^{1}H NMR ($CDCl_3$): 2.16 (CH_3), 7.13, 8.18 (C_6H_5), 7.27, 7.79 (C_6H_4) IR (Nujol): 1630 ($\nu_{as}O_2C$), 1318 (ν_sO_2C), 1478, 1432, 460 (νC_6H_5), 299 (νSbO) [1]
36	$C_6H_4CH_3$-3	I m.p. 167° ^{1}H NMR ($CDCl_3$): 2.31 (CH_3), 7.27, 7.77 (C_6H_4), 7.50, 8.14 (C_6H_5) IR (Nujol): 1635 ($\nu_{as}O_2C$), 1328 (ν_sO_2C), 1479, 1432, 463 (νC_6H_5), 297 (νSbO) [1]
37	$C_6H_4CH_3$-4	I m.p. 200° ^{1}H NMR ($CDCl_3$): 2.32 (CH_3), 7.15, 7.86 (C_6H_4), 7.50, 8.14 (C_6H_5) IR (Nujol): 1630 ($\nu_{as}O_2C$), 1325 (ν_sO_2C), 1477, 1432, 463 (νC_6H_5), 306 (νSbO) [1]
with R = 4-ClC_6H_4		
38	C_6H_5	I ^{1}H NMR ($CDCl_3$): 7.39, 7.96 (C_6H_4), 7.52, 8.10 (C_6H_5) IR (Nujol): 1613 ($\nu_{as}O_2C$), 1346 (ν_sO_2C) [1]
39	C_6H_4Cl-4	I ^{1}H NMR ($CDCl_3$): 7.54, 8.07 (C_6H_4Sb), 7.36, 7.91 (C_6H_4Cl) IR (Nujol): 1641 ($\nu_{as}O_2C$), 1323 (ν_sO_2C) [1]
40	$C_6H_4NH_2$-4	I ^{1}H NMR ($CDCl_3$): 3.88 (NH_2), 6.60, 8.74 (C_6H_4Sb), 7.45, 8.06 (C_6H_4N) IR (Nujol): 1593 ($\nu_{as}O_2C$), 1331 (ν_sO_2C) [1]
41	$C_6H_4NO_2$-4	I ^{1}H NMR ($CDCl_3$): 7.60, 8.27 (C_6H_4Sb), 8.11 (C_6H_4N) IR (Nujol): 1649 ($\nu_{as}O_2C$), 1308 (ν_sO_2C) [1]
42	$C_6H_4CH_3$-4	I ^{1}H NMR ($CDCl_3$): 2.39 (CH_3), 7.22, 7.91 (C_6H_4C), 7.50, 8.10 (C_6H_4Sb) IR (Nujol): 1633 ($\nu_{as}O_2C$), 1325 (ν_sO_2C) [1]

References on p. 185

Table 40 [continued]

No.	R′	method of preparation (yield in %) properties and remarks [Ref.]
with R = 2-$CH_3C_6H_4$		
43	C_6H_5	I ^{1}H NMR ($CDCl_3$): 2.64 (CH_3), 7.33, 7.83 (C_6H_5), 7.43, 8.48 (C_6H_4) IR (Nujol): 1653 ($\nu_{as}O_2C$), 1308 (ν_sO_2C) [1]
44	C_6H_4Cl-4	I ^{1}H NMR ($CDCl_3$): 2.57 (CH_3), 7.25, 7.70 (C_6H_4Cl), 7.43, 8.40 (C_6H_4Sb) IR (Nujol): 1659 ($\nu_{as}O_2C$), 1307 (ν_sO_2C) [1]
45	$C_6H_4NH_2$-4	I ^{1}H NMR ($CDCl_3$): 2.56 (CH_3), 6.56, 7.77 (C_6H_4N), 7.45, 8.49 (C_6H_4Sb) IR (Nujol): 1621, 1603 ($\nu_{as}O_2C$), 1312, 1293 (ν_sO_2C) [1]
46	$C_6H_4NO_2$-4	I ^{1}H NMR ($CDCl_3$): 2.62 (CH_3), 7.48, 8.40 (C_6H_4Sb), 7.93, 8.04 (C_6H_4N) IR (Nujol): 1670 ($\nu_{as}O_2C$), 1293 (ν_sO_2C) [1]
47	$C_6H_4CH_3$-4	I ^{1}H NMR ($CDCl_3$): 2.52 (CH_3-4), 2.62 (CH_3-2), 7.12, 7.72 (C_6H_4C), 7.24, 8.49 (C_6H_4Sb) IR (Nujol): 1643 ($\nu_{as}O_2C$), 1309, 1302 (ν_sO_2C) [1]
with R = 3-$CH_3C_6H_4$		
48	C_6H_5	I ^{1}H NMR ($CDCl_3$): 2.42 (CH_3), 7.21, 7.93 (C_6H_5), 7.43, 7.98 (C_6H_4) IR (Nujol): 1629 ($\nu_{as}O_2C$), 1335 (ν_sO_2C) [1]
49	C_6H_4Cl-4	I ^{1}H NMR ($CDCl_3$): 2.37 (CH_3), 7.34, 7.89 (C_6H_4Cl), 7.37, 7.89 (C_6H_4Sb) IR (Nujol): 1633 ($\nu_{as}O_2C$), 1332 (ν_sO_2C) [1]
50	$C_6H_4NH_2$-4	I ^{1}H NMR ($CDCl_3$): 2.35 (CH_3), 6.56, 7.80 (C_6H_4N), 7.40, 7.92 (C_6H_4Sb) IR (Nujol): 1591 ($\nu_{as}O_2C$), 1332 (ν_sO_2C) [1]
51	$C_6H_4NO_2$-4	I ^{1}H NMR ($CDCl_3$): 2.40 (CH_3), 7.40, 7.91 (C_6H_4Sb), 8.13 (C_6H_4N) IR (Nujol): 1644 ($\nu_{as}O_2C$), 1329 (ν_sO_2C) [1]
52	$C_6H_4CH_3$-4	I ^{1}H NMR ($CDCl_3$): 2.36 (CH_3-4), 2.42 (CH_3-3), 7.17, 7.85 (C_6H_4C), 7.32, 7.94 (C_6H_4Sb) IR (Nujol): 1633, 1611 ($\nu_{as}O_2C$), 1334 (ν_sO_2C) [1]
with R = 4-$CH_3C_6H_4$		
53	C_6H_5	I [1, 2] ^{1}H NMR ($CDCl_3$): 2.30 (CH_3), 7.27, 7.96 (C_6H_4), 7.38, 8.05 (C_6H_5) IR (Nujol): 1640 ($\nu_{as}O_2C$), 1367 (ν_sO_2C) [1] ^{13}C NMR ($CHCl_3$, vs. C_6H_6): 2.2 (C-3,5 of $C_6H_4CH_3$), 6.3 (C-2,6 of $C_6H_4CH_3$), 6.8 (C-1 of $C_6H_4CH_3$), 13.5 (C-4 of $C_6H_4CH_3$) [2]

References on p. 185

Table 40 [continued]

No.	R′	method of preparation (yield in %) properties and remarks [Ref.]
54	C_6H_4Cl-4	I ^{1}H NMR ($CDCl_3$): 2.38 (CH_3), 7.34, 7.91 (C_6H_4Cl), 7.34, 8.03 (C_6H_4Sb) IR (Nujol): 1648 ($\nu_{as}O_2C$), 1308 (ν_sO_2C) [1]
55	$C_6H_4NH_2$-4	I ^{1}H NMR ($CDCl_3$): 2.34 (CH_3), 6.51, 7.76 (C_6H_4N), 7.25, 7.98 (C_6H_4Sb) IR (Nujol): 1604 ($\nu_{as}O_2C$), 1331, 1315 (ν_sO_2C) [1]
56	$C_6H_4NO_2$-4	I ^{1}H NMR ($CDCl_3$): 2.42 (CH_3), 7.39, 8.01 (C_6H_4Sb), 8.12 (C_6H_4N) IR (Nujol): 1659 ($\nu_{as}O_2C$), 1303 (ν_sO_2C) [1]
57	$C_6H_4CH_3$-4	I ^{1}H NMR ($CDCl_3$): 2.35 (CH_3), 7.14, 7.85 (C_6H_4C), 7.27, 8.01 (C_6H_4Sb) IR (Nujol): 1641 ($\nu_{as}O_2C$), 1324 (ν_sO_2C) [1]
58	$C_6H_4CO_2CH_3$-2	IIb m.p. 215 to 216° ^{1}H NMR ($CDCl_3$): 3.26 (s, CH_3, 6H), 7.23 to 7.83 (m, 23H) IR (KBr): 1716, 1621 (νC=O) [14, 15]
59	$C_6Cl_4CO_2CH_3$-2	IIb m.p. 219 to 220° ^{1}H NMR ($CDCl_3$): 3.27 (s, 2H), 7.52 (m, 3H), 8.05 (m, 2H) IR (KBr): 1720, 1650 (νC=O) [14, 15]
60	$C_6Br_4CO_2CH_3$-2	IIb m.p. 220 to 221.5° (dec.) ^{1}H NMR ($CDCl_3$): 3.27 (s, 2H), 7.33 (m, 3H), 7.86 (m, 2H) IR (KBr): 1720, 1640 (νC=O) [14, 15]

* Further information:

$(C_3H_7)_3Sb(O_2CC_6H_5)_2$ (Table **40**, No. **20**) is prepared from $(C_3H_7)_3Sb{=}NSO_2C_6H_4CH_3$-4, in CH_2Cl_2, and benzoic acid. After cooling, the precipitated 4-$CH_3C_6H_4SO_2NH_2$ is filtered. Evaporation of the solvent from the filtrate gives 85% yield. The compound dissolves as a monomer in dioxane [5].

$(C_6H_5)_3Sb(O_2CC_6H_5)_2$ (Table **40**, No. **22**) can be also prepared by reaction of $(C_6H_5)_3SbCl_2$ with silver benzoate by shaking 12 h in $CHCl_3$. Evaporation of the solvent from the filtrate gives the compound. Perbenzoic acid and $Sb(C_6H_5)_3$ heated in petroleum ether for 20 min also give the substance in good yield [7].

The crystals are orthorhombic, space group Pccn$-D_{2h}^{10}$ (No. 56) with a=15.834(7), b=19.854(11), and c=16.827(7) Å; d_c=1.5 g/cm³ and Z=8. Refinement gives a final R value of 0.036. The slightly distorted trigonal bipyramidal molecule is shown in **Fig. 17**. The O-Sb-O angle is 175(1)° [9].

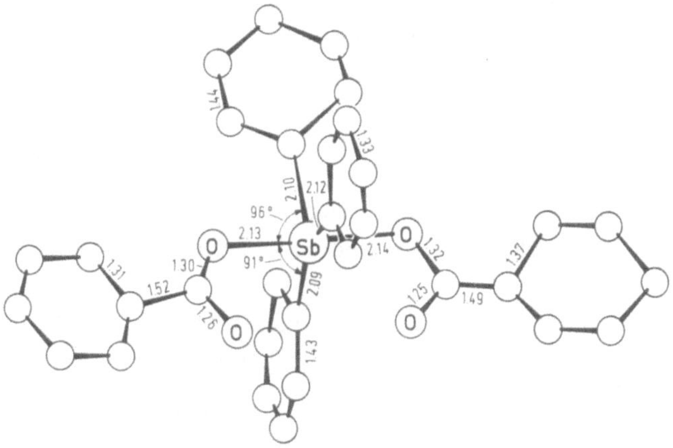

Fig. 17. Molecular structure of $(C_6H_5)_3Sb(O_2CC_6H_5)_2$ [9].

The compound is soluble in CH_3OH, C_2H_5OH, $(CH_3)_2CO$, C_6H_6, $CHCl_3$, and CCl_4, and insoluble in petroleum ether [8]. It reacts with hot aqueous HCl to form $(C_6H_5)_3SbCl_2$ and benzoic acid [7].

The compound was tested as a retarding agent in the burning of epoxy resins [10] and halogen-containing polyester resins [11]. Polymers and copolymers of vinyl chloride are stabilized against discoloration by UV light upon addition of 0.1 to 5% of the compound [12]. It prevents the discoloration of copolymer spinning solutions [13].

References:

[1] A. Ouchi, M. Nakatani, Y. Takahashi, S. Kitazima, T. Sugihara, M. Matsumoto, T. Uehiro, K. Kitano, K. Kawashima, H. Honda (Sci. Papers Coll. Gen. Educ. Univ. Tokyo **25** [1975] 73/99).

[2] A. Ouchi, T. Uehiro, Y. Yoshino (J. Inorg. Nucl. Chem. **37** [1975] 2347/9).

[3] M. Shindo, R. Okawara (J. Organometal. Chem. **5** [1966] 537/44).

[4] H. Terada, T. Imaida, O. Nakagawa, S. Takashima, Mitsubishi Rayon Co., Ltd. (Japan. 69-19270 [1969]; C.A. **71** [1969] No. 125191).

[5] Z. I. Kuplennik, Zh. N. Belaya, A. M. Pinchuk (Zh. Obshch. Khim. **51** [1981] 2711/15; J. Gen. Chem. [USSR] **51** [1981] 2339/43).

[6] S. Herbstman, Stauffer Chemical Co. (U.S. 3278492 [1964/66]; C.A. **66** [1967] No. 66142).

[7] F. Challenger, V. K. Wilson (J. Chem. Soc. **1927** 209/13).

[8] J. Havranek, J. Mleziva, A. Lycka (J. Organometal. Chem. **157** [1978] 163/6).

[9] V. A. Lebedev, R. I. Bochkova, L. F. Kuzubova, E. A. Kuz'min, V. V. Sharutin, N. V. Belov (Dokl. Akad. Nauk SSSR **265** [1982] 332/5).

[10] J. Havranek, J. Mleziva (Angew. Makromol. Chem. **84** [1980] 105/17).

[11] B. O. Schoepfle, B. S. Marks, P. Robitschek, Hooker Chemical Corp. (U.S. 2913428 [1959]; C.A. **1960** 5162/3).

[12] J. K. Fincke, Monsanto Chemical Co. (U.S. 2556420 [1951]; C.A. **1951** 8813).

[13] H. Tanaka, A. Tomioka, A. Odaira, K. Okamura, K. Satokawa, M. Yonetani (Japan. 1479 [1962]; C.A. **60** [1964] 10868).

[14] M. M. Y. Cheng, K. Su, J. I. Musher (AD-757345 [1973] 1/8).

[15] M. M. Y. Cheng, K. Su, J. I. Musher (Israel. J. Chem. **12** [1974] 967/70).

2.5.1.1.12.5 $(\text{-}R_3SbO_2CR'CO_2\text{-})_n$ Compounds

General Remarks. These compounds are described in the original literature either with molecular formulas or as polymers. The polymeric nature of these substances is not doubted.

$(CH_3)_3SbO_2CCO_2$

The compound is prepared by reaction of $(CH_3)_3SbBr_2$ with AgO_2CCO_2Ag in H_2O [1], or by reaction of $(CH_3)_3SbBr_2$ with the free acid in C_6H_6 in the presence of $N(C_2H_5)_3$ in roughly 30% yield [2].

The observed IR vibrations (in KBr, in cm^{-1}) are 3035w (ν_{as}CH), 2950m (ν_sCH), 2570w; 1665vs (νC=O, B_1), 1620s, sh (νC=O, A_1), 1375 m (νC-O, A_1), 1250vs, 1227, 1215 (νC-O, B_1), 855s (ϱCH_3), 827s, sh (νCC, A_1), 755s (δ_{as}O-C=O, B_1), 585w (δ_{as}C-C-O, B_1), 575m (ν_{as}SbC), 525s (δ_sC-C-O, A_1), 420s (δ_sO-C=O, A_1). These values confirm a polymeric structure with bridging oxalate groups. The compound is soluble in H_2O, but not in CH_3OH, $(CH_3)_2CO$, nor CH_3CN [1].

$(CH_3)_3SbO_2CCH_2CO_2$, $(C_2H_5)_3SbO_2CCO_2$, $(C_2H_5)_3SbO_2CCH_2CO_2$, $(C_2H_5)_3SbO_2CC_6H_4CO_2$-2, and $(C_4H_9)_3SbO_2CCO_2$

The compounds are prepared in roughly 30% yield from R_3SbBr_2 with the corresponding free acid in C_6H_6 in the presence of $N(C_2H_5)_3$. They were identified by elemental analysis [2]. $(C_2H_5)_3SbO_2CCH_2CO_2$ is claimed to be a fire retardant for burning polyesters [3].

$(C_4H_9)_3SbO_2CC_6(CH_3)_4CO_2$-4

The compound is obtained in only 2% yield by reaction of $(C_4H_9)_3SbCl_2$ with disodium tetramethylterephthalate in CCl_4/H_2O at 25 °C, stirred for 30 s [5].

$(C_6H_5)_3SbO_2CCO_2$

Triphenylantimony dihalide and AgO_2CCO_2Ag are stirred in H_2O for one day [2], or in CH_3OH [4], to give the compound in a yield of about 30% [2]. A better method for preparation is the addition of $(C_6H_5)_3SbCl_2$ in CCl_4 to NaO_2CCO_2Na in H_2O at 25 °C, and stirring for 30 s [5]. The yield of polymer in this case is 27 to 39% depending on the concentrations. The molecular weight was determined as 8800 [5, 6].

The compound melts with decomposition at 160 to 161 °C. IR absorptions (in mull) of the oxalate group were found at 1740s, 1655s, 1360s, 1200s, 970m, 780m, 760s, and 345m cm^{-1}. These values confirm a nonionic polymeric structure [4].

$(C_6H_5)_3SbO_2C(CH_2)_nCO_2$ (n = 1 to 4), $(C_6H_5)_3SbO_2CC_6H_4CO_2$-2, $(C_6H_5)_3SbO_2CC_6H_4CO_2$-3, and $(C_6H_5)_3SbO_2CC_6H_4CO_2$-4

The compounds are prepared by reaction of the corresponding silver carboxylates with $(C_6H_5)_3SbBr_2$, stirred in H_2O in a dark place for one day at ambient temperature. The yield is roughly 30%. The compounds were identified by elemental analysis, and no further data are given [2]. Triphenylantimony terephthalate is also prepared from $(C_6H_5)_3SbCl_2$ in CCl_4. The reactants are added to a stirred solution of neutralized (with NaOH) terephthalic acid (molar ratio 1:1) in aqueous solution at 25 °C and stirred for 30 s. The yield varies from 6 to 46%, depending on the concentrations employed. The polymer has a molecular weight of 27000 [5, 6]. It is used as a fire retardant in polymers [7]. Substituted terephthalates are listed in Table 41.

References on p. 188

$(C_6H_5)_3SbO_2CCH_2NHCH_2CO_2$

$(C_6H_5)_3SbCl_2$, $HO_2CCH_2NHCH_2CO_2H$, and NaOH are reacted in aqueous THF/C_2H_5OH mixture for 4 h under reflux to give the compound. It is formulated in the original as a chelate $(C_6H_5)_3Sb(\text{-}O_2CCH_2NHCH_2CO_2\text{-})$. It is an insecticide and a potential dentifrice component [8].

$(C_6H_5)_3SbO_2CR'CO_2$ Compounds with Other R' Groups

These compounds are listed in Table 41 together with the yields and molecular weights. They are prepared by addition of 1.00 mmol $(C_6H_5)_3SbCl_2$ in 25 ml CCl_4 to stirred (20500 rpm) solutions of neutralized (with NaOH) diacid (1.00 mmol) in 25 ml of aqueous solution at 25 °C for 30 s stirring time. In Table 41 the yields in parentheses are for analogous reaction conditions except that 15 ml of each phase are employed [5, 6].

The limiting viscosity numbers and refractive index increments in $PO(N(CH_3)_2)_3$ are given in the originals [5, 6]. The bacterial and fungal inhibiting activity of Nos. 7 and 10 was tested [10].

Table 41
Other Compounds of the Type $(\text{-}(C_6H_5)_3SbO_2CR'CO_2\text{-})_n$ [5, 6].
See text for explanations.

No.	$O_2CR'CO_2$	yield in %	molecular weight
1	$O_2CCH(SH)CH_2CO_2$	(31)	–
2	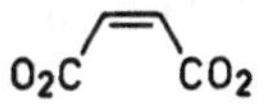	3 (0)	–
3	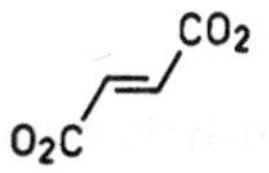	93 (28)	80000
4	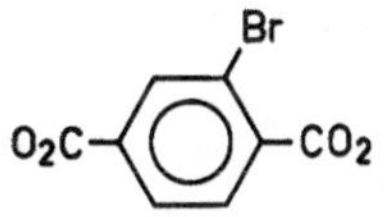	7 (0)	9000
5	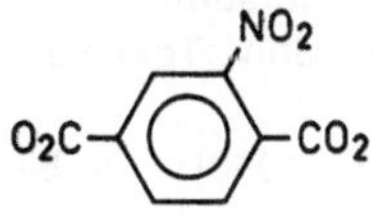	19	71000
6	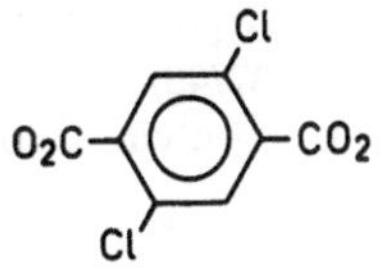	18 (17)	98000
7	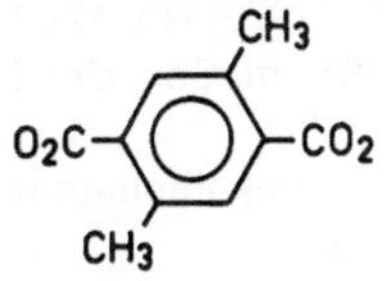	3 (0)	–

References on p. 188

Table 41 [continued]

No.	$O_2CR'CO_2$	yield in %	molecular weight
8	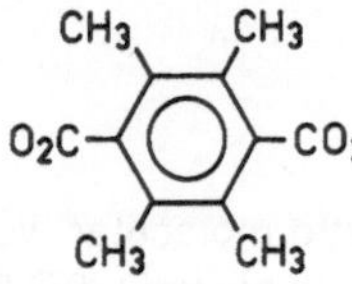	56 (49)	3400
9	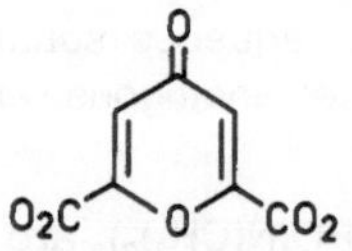	(7)	–
10	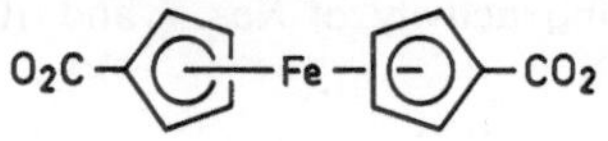	40 (4)	70000

$[(C_6H_5)_3Sb(1\text{-}O_2CC_5H_4CoC_5H_4CO_2\text{-}1')]_n^+ PF_6^-$ (n = 3 to 10)

1,1′-Dicarboxycobalticinium hexafluorophosphate in 0.05 M NaOH is added to a solution of $(C_6H_5)_3SbCl_2$ in CCl_4. The mixture is stirred for 30 s. The organic layer is separated, and the polymer precipitates upon addition of hexane. It is then washed with water and hexane. The yields vary between 50 and 90%, depending on the stirring rate and time. With Br^- and NO_3^-, instead of PF_6^-, much lower yields of 2 to 7% are achieved.

An IR spectrum of the compound in KBr shows absorptions (in cm^{-1}) at 3400 (Sb-OH end groups), 3120, 1640, 1615, 1470, 1435, 1360, 1355, 1340, 1320, 1170, 1080, 1015, 875, 850s (νPF), 790, 745, and 700. The ratio of the areas of the peaks at 1470 and 850 cm^{-1} can be used to tell whether the PF_6^- ion has been replaced by Cl^- or other anions [9].

The polymer is insoluble in organic solvents. A thermogravimetric analysis shows that the compound degrades at 200 °C in air and at 325 to 800 °C in N_2 atmosphere. Differential scanning calorimetry shows exothermic transitions at 310, 375, and 460 °C [9].

References:

[1] H. C. Clar, R. G. Goel (Inorg. Chem. **5** [1966] 998/1003).

[2] A. Ouchi, M. Nakatani, Y. Takahashi, S. Kitazima, T. Sugihara, M. Matsumoto, T. Uehiro, K. Kitano, K. Kawashima, H. Honda (Sci. Papers Coll. Gen. Educ. Univ. Tokyo **25** [1975] 73/99).

[3] E. Eimers, L. Goerden, Farbenfabriken Bayer A.-G. (Ger. 1089967 [1960]; C.A. **55** [1961] 17082/3).

[4] R. G. Goel, P. N. Joshi, D. R. Ridley, R. E. Beaumont (Can. J. Chem. **47** [1969] 1423/7).

[5] C. E. Carraher Jr., H. S. Blaxall (Angew. Makromol. Chem. **83** [1979] 37/45).

[6] C. E. Carraher Jr., H. S. Blaxall (Polym. Prepr. Am. Chem. Soc. Div. Polym. Chem. **16** [1975] 261/3).

[7] J. Musher, K. Su (U.S. 3939190 [1972/76]; C.A. **84** [1976] No. 181136).

[8] H. G. Langer, Dow Chemical Co. (U.S. 3442922 [1964/69]; C.A. **72** [1970] No. 12880).

[9] J. E. Sheats, C. H. Carraher Jr., H. S. Blaxall (Polym. Prepr. Am. Chem. Soc. Div. Polym. Chem. **16** [1975] 655/8).

[10] C. E. Carraher Jr., D. J. Giron, D. R. Cerutis, W. R. Burt, R. S. Venkatachalam, T. J. Gehrke, S. Tsuji, H. S. Blaxall (ACS Symp. Ser. No. 186 [1982] 13/25).

2.5.1.1.13 Triorganoantimony Bis(organylphosphinates) $R_3Sb(OP(O)(H)R')_2$

The compounds are summarized in Table 42. They are prepared by the following method:

R_3SbBr_2 ($R=CH_3$, C_6H_5), 2 molar equivalents of the corresponding organophosphinic acid, and $N(C_2H_5)_3$ are stirred in C_6H_6 for one hour at room temperature. Evaporation of the solvent from the filtrate of the mixture and recrystallization of the residue from C_6H_6/C_6H_{14} give the compounds. Using pyridine as an acceptor and refluxing in C_6H_6 for 20 h led to the reaction of about $^2/_3$ of the starting materials to give a practically equimolar mixture of $R_3Sb(Br)OP(O)(H)R'$ and $R_3Sb(OP(O)(H)R')_2$.

Table 42
Compounds of the Type $R_3Sb(OP(O)(H)R')_2$.
For explanations, abbreviations, and units, see p. X.

No.	R	R′	properties and remarks
1	CH_3	$CH{=}CHC_6H_5$	oil ^1H NMR ($CDCl_3$): 2.17 (s, CH_3), 6.57 (m, CH), 7.31 (d, PH; J(P, H) = 548), 7.41 (m, C_6H_5) ^{31}P NMR ($CDCl_3$): 15.45 (s)
2	CH_3	C_6H_5	75% yield, white solid m.p. 86 to 89° ^1H NMR ($CDCl_3$): 2.08 (s, CH_3), 7.52 (d, PH; J(P, H) = 547), 7.46, 7.74 (m, C_6H_5) ^{31}P NMR ($CDCl_3$): 15.49 (s)
3	C_6H_5	$CH{=}CHC_6H_5$	19% yield, white solid m.p. 158 to 160° ^1H NMR ($CDCl_3$): 5.99 (m, CH), 7.11 (d, PH, J(P, H) = 546), 7.31, 7.67, 8.23 (m, C_6H_5) ^{31}P NMR ($CDCl_3$): 13.73 (s)
4	C_6H_5	C_6H_5	61% yield, white solid m.p. 142 to 144° ^1H NMR ($CDCl_3$): 7.26 (d, PH, J(P, H) = 547), 7.29, 7.55, 8.13 (m, C_6H_5) ^{31}P NMR ($CDCl_3$): 13.56 (s)

Reference:

G. E. Graves, J. R. Van Wazer (J. Organometal. Chem. **131** [1977] 31/4).

2.5.1.1.14 Triorganoantimony Diamides and Imides

2.5.1.1.14.1 $R_3Sb(NR'_2)_2$ and $R_3Sb(N{=}P(C_6H_5)_3)_2$ Compounds

The compounds are summarized in Table 43. NR'_2 is in all known cases a heterocyclic ring. The following methods of preparation are used:

Method I: R_3SbBr_2, 2 equivalents of the appropriate amine, and $N(C_2H_5)_3$ as a HBr acceptor are stirred together in C_6H_6 for 2 h at room temperature. To ensure completion of the reaction, the mixture is refluxed for an additional hour. $[NH(C_2H_5)_3]Br$

is filtered, and the filtrate is concentrated in vacuum to give the products which are recrystallized from petroleum ether (40 to 60 °C) or CH_3CN/petroleum ether. Oily products may be solidified by scratching the wall of the vessel with a glass rod and cooling. The reactions are generally quantitative [1 to 3].

Method II: R_3SbCl_2 and $(CH_3)_3SiN{=}P(C_6H_5)_3$ are stirred together in refluxing toluene for ca. 5 h. The volatiles are removed in vacuum to leave an oily residue which solidifies when triturated with C_6H_{14}. The compounds are recrystallized from C_6H_6/C_6H_{14} [5].

General Remarks. The compounds are stable under aerobic conditions, monomeric in C_6H_6, and nonionic [1, 2, 5]. They are soluble in organic solvents such as $(CH_3)_2CO$ [1], $CHCl_3$, C_6H_6 [1, 5], and CH_2Cl_2 [5], but insoluble in H_2O [1]. The triaryl compounds are stable towards moisture [2].

Table 43
Triorganoantimony Diamides $R_3Sb(NR'_2)_2$ and $R_3Sb(N{=}PR'_3)_2$.
Further information on numbers preceded by an asterisk is given at the end of the table.
For explanations, abbreviations, and units, see p. X.

No.	NR'_2 or $N{=}P(C_6H_5)_3$	method of preparation (yield in %) properties and remarks [Ref.]
with R = CH_3		
1	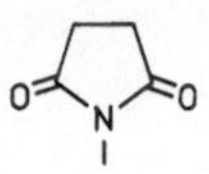	I m.p. 116 to 117° insecticide against cockroaches [1]
2	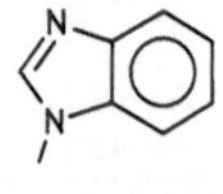	I m.p. 115° strong bactericide against Bacillus subtilis and Sorsena lutea [1]
3	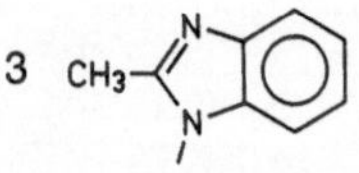	I m.p. 80° insecticide against cockroaches [1]
4	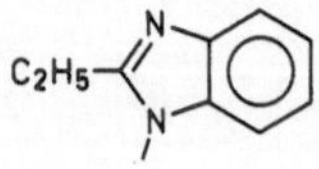	I m.p. 141° [1]
with R = c-C_6H_{11}		
5	$N{=}P(C_6H_5)_3$	II (72) m.p. 65 to 67° ^1H NMR ($CDCl_3$): 1.68 (m, C_6H_{11}), 7.30 (m, C_6H_5) IR (Nujol): 1170s (νP=N) [5]

References on p. 193

Table 43 [continued]

No.	NR'_2 or $N{=}P(C_6H_5)_3$	method of preparation (yield in %) properties and remarks [Ref.]
with R = C_6H_5		
*6	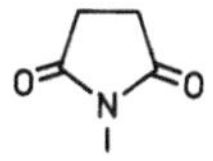	I m. p. 114° (from petroleum ether 40 to 60°) [2], 177 to 180° with dec. [4] ^{1}H NMR ($CDCl_3$): 2.27 (s, CH_2), 7.28 to 7.58 (m, C_6H_5), 7.80 to 8.03 (m, C_6H_5) [4] IR (Nujol): 1288 s (ν_sCO), 1685 s (ν_{as}CO) [2]
7	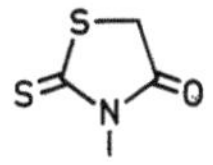	I m. p. 115° [3]
8	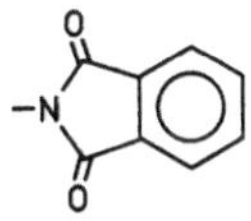	I m. p. 170° [2] IR (Nujol): 1760 s (ν_{as}CO), 1318 s (ν_sCO) stable in refluxing CH_3OH [2]
9	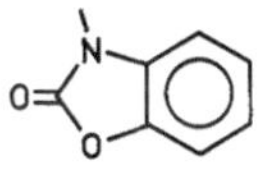	I m. p. 179° IR (Nujol): 1700 (ν_{as}CO), 1290 m, 1244 s (ν_sCO) [2]
10	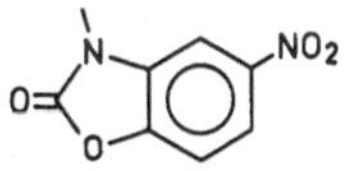	I m. p. 126° IR (Nujol): 1620 m (ν_{as}CO), 1328 s, 1275 s (ν_sCO) [2]
11	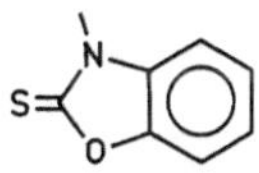	I m. p. 118 to 120° [2]
12	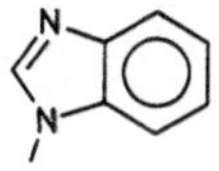	I m. p. 112 to 114° [2]
13	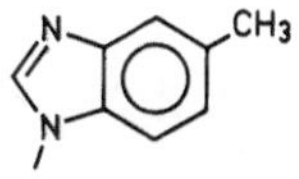	I m. p. 140° [2]
14	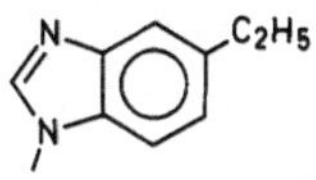	I m. p. 158° ^{1}H NMR ($CDCl_3$): 1.85 (t, CH_3), 3.35 (q, CH_3), 6.4 (s, CH), 7.5 (m, C_6H_5) stable towards H_2O at room temperature [2]
15	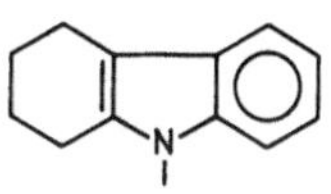	I m. p. 110° no reaction is observed in refluxing CH_3OH within 2 h [3]
16	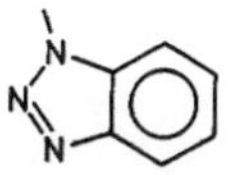	I m. p. 76° [2]
*17	$N{=}P(C_6H_5)_3$	II (80), see further information m. p. 110 to 112° IR (Nujol): 1190 (νP=N) [5]

References on p. 193

Table 43 [continued]

No.	NR'_2 or $N{=}P(C_6H_5)_3$	method of preparation (yield in %) properties and remarks [Ref.]
with R = 4-$CH_3C_6H_4$		
18	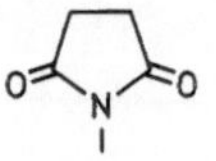	I m.p. 118° ^{1}H NMR ($CDCl_3$): 2.2 (m, CH_2), 2.55 (s, CH_3), 7.25 (m, C_6H_4) IR (Nujol): 1680s (ν_{as}CO), 1295s (ν_sCO) [2]
19	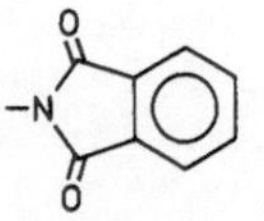	I m.p. 214° IR (Nujol): 1730s (ν_sCO), 1308m (ν_{as}CO) [2]
20	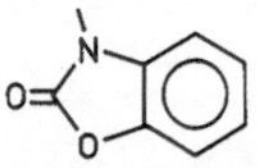	I m.p. 120° IR (Nujol): 1710m (ν_sCO), 1292w (ν_{as}CO) [2]
21	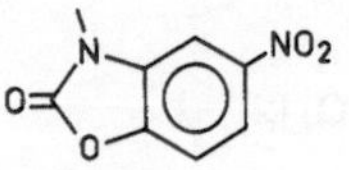	I m.p. 102° IR (Nujol): 1618m (ν_sCO), 1300s, 1250m (ν_{as}CO) [2]
22	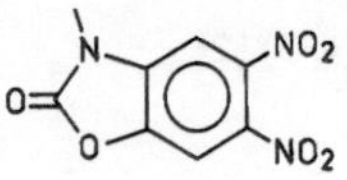	I m.p. 198° (dec.) IR (Nujol): 1720s (ν_sCO), 1305m (ν_{as}CO) [2]
23	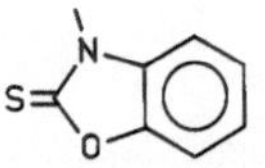	I m.p. 130° [2]
24	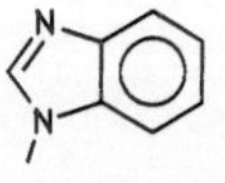	I m.p. 122° stable in refluxing CH_3OH [2]
25	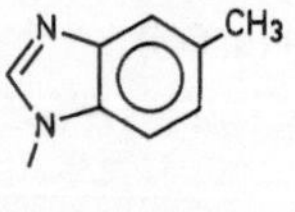	I m.p. 132° [2]
26	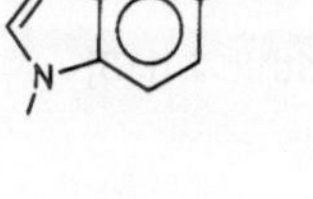	I m.p. 146° reaction with $TeCl_4$ in refluxing C_6H_6 gives (4-$CH_3C_6H_4)_3SbCl_2$ [2]
*27	$N{=}P(C_6H_5)_3$	see further information m.p. 144 to 145° ^{1}H NMR ($CDCl_3$): 1.40 (s, CH_3), 6.90 (m, C_6H_4) IR (Nujol): 1185 (νP=N) [5]

* Further information:

$(C_6H_5)_3Sb(NC_4H_4O_2)_2$ (Table **43**, No. **6**) is also prepared by reaction of $(C_6H_5)_3SbBr_2$ and silver succinimide (molar ratio 1:2) in $CHCl_3$. The suspension is shaken at room temperature for 25 min. The filtrate of the mixture is concentrated. Crystals precipitate in a yield of 68.5% upon storing for 12 h at −5 °C [4].

Addition of $(C_6H_5)_3SbX_2$ (X = Cl, Br) gives an equilibrium with $(C_6H_5)_3SbX(NC_4H_4O_2)$ [4]. Reaction with Br_2 in $CHCl_3$ gives $(C_6H_5)_3SbBr_2$ and 1-bromosuccinimide [2].

$(C_6H_5)_3Sb(N{=}P(C_6H_5)_3)_2$ (Table **43**, No. **17**) is also prepared by adding $(CH_3)_3SiN{=}P(C_6H_5)_3$ (4 mmol) to a solution of $(C_6H_5)_3Sb(OCH_3)_2$ (2 mmol) in CH_3OH. The reaction mixture is heated for ca. 2 h at reflux. Removal of the volatiles under reduced pressure and crystallization of the residue from C_6H_6/C_6H_{14} afford the compound in 75% yield [5].

The compound is stable towards alcoholysis and does not react with $C_6H_5C{\equiv}CH$; however, it reacts with $C_6H_5TeCl_3$ in refluxing C_6H_6 to give $C_6H_5Te(N{=}P(C_6H_5)_3)Cl_2$ [5].

$(4\text{-}CH_3C_6H_4)_3Sb(N{=}P(C_6H_5)_3)_2$ (Table **43**, No. **27**). $(4\text{-}CH_3C_6H_4)_3SbBr_2$ (1 mmol) and $(CH_3)_3SiN{=}P(C_6H_5)_3$ (2 mmol) are dissolved in C_6H_6 (20 mL) containing CH_3OH (ca. 5 mL) and $N(C_2H_5)_3$ (ca. 1 mL). The reaction mixture is stirred at room temperature for about 5 h during which time a white precipitate of $[NH(C_2H_5)_3]Br$ separates. The volatiles are removed in vacuum, and the residue is extracted with C_6H_6. Concentration of the C_6H_6 extract to ca. 2 mL, addition of C_6H_{14} (ca. 10 mL), and cooling gave colorless needles of the compound in 75% yield [5].

The compound reacts with $TeCl_4$ when refluxed for 4 h in C_6H_6 to give a 75% yield of $(4\text{-}CH_3C_6H_4)_3SbCl_2$ and a black residue [5].

References:

[1] K. Bajpai, M. Srivastava, R. C. Srivastava (Indian J. Chem. A **20** [1981] 736/7).
[2] K. Bajpai, R. C. Srivastava (Syn. Reactiv. Inorg. Metal.-Org. Chem. **9** [1979] 557/64).
[3] P. Raj, A. Ranjan, A. K. Saxena (Indian J. Chem. A **22** [1983] 120/3).
[4] J. Dahlmann, K. Winsel (J. Prakt. Chem. **321** [1979] 370/8).
[5] K. Bajpai, R. C. Srivastava (Syn. Reactiv. Inorg. Metal.-Org. Chem. **12** [1982] 47/54).

2.5.1.1.14.2 $(\text{-}R_3SbNR'\text{-}R''\text{-}NR'''\text{-})_n$ Compounds

These polymeric compounds are summarized in Table 44. They are prepared by an interfacial technique. Triorganoantimony dihalide, $(CH_3)_3SbCl_2$ for Nos. 1 and 2, $(C_6H_5)_3SbCl_2$ for Nos. 3 to 18, and $(C_6H_5)_3SbBr_2$ for Nos. 5 and 19, dissolved in CCl_4, is added to a stirred (18000 rpm, no load) solution of the appropriate diamine and NaOH in H_2O (molar ratios 1:1:2) for 30 s at 25 °C.

Structural characterization was accomplished by elemental analysis, light-scattering photometry, IR spectroscopy, and mass spectroscopy. Control reactions were performed by excluding one of the starting materials. The polymers exhibit mild inhibition to a wide range of bacteria and to HeLa, BHK-21, and L929 cancer-related cell lines.

Table 44
Compounds of the Type $(-R_3SbNR'-R''-NR'''-)_n$.
The values in parentheses are for the products obtained from $(C_6H_5)_3SbBr_2$.

No.	NR'H-R''-NR'''H reactant	yield in %	molecular weight	IR (νSbN) in cm^{-1}
$R=CH_3$				
1	2,6-diamino-8-purinol	93	4.4×10^5	
2	4,4'-diaminodiphenylsulfone	40.6	3.7×10^6	
$R=C_6H_5$				
3	1,6-diaminohexane	36		1175
4	1,12-diaminododecane	(elemental analysis given)		
5	2,6-diamino-8-purinol	39 (62)	3.7×10^3 (1.0×10^3)	1100
6	2,6-diaminoanthraquinone	40		
7	2,4-diamino-5-nitropyrimidine	14		1180
8	2,4-diamino-5-(3,4-dimethoxybenzyl)pyrimidine	21	4.8×10^3	1150
9	4,4'-diaminodiphenylsulfone	18	4.5×10^3	
10	adenine	9	2.6×10^3	1125
11	1,4-diaminobenzene	12		
12	1,4-diamino-2-methoxybenzene	76		
13	1,4-diamino-2,3,5,6-tetramethylbenzene	8		
14	1,4-diamino-2,5-dichlorobenzene	39 or 40	4.3×10^3 or 9.3×10^3	1175
15	1,4-diamino-2-nitrobenzene	23		
16	4,4'-methylenedianiline	54		
17	4,4'-diaminobenzanilide	43		
18	$[SC(S)NH(CH_2)_2NHC(S)S]^{2-}Zn^{2+}$	36		
19	4,4'-diaminodiphenyl	(51)	(1.2×10^3)	
20	4,6-diamino-5-nitrosopyrimidine	(no bacteriological activity)		
21	4,6-diamino-2-methyl-5-nitrosopyrimidine	(tested for bacteriological activity and on cancer cell lines)		
22	2,6-diaminopyridine			

Reference:

C. E. Carraher Jr., M. D. Naas, D. J. Giron, D. R. Cerutis (J. Macromol. Sci. Chem. A **19** [1983] 1101/20).

2.5.1.1.14.3 $R_3Sb{=}NR'$ Compounds

$(C_6H_5)_3Sb{=}NH$

The preparation of the compound is mentioned as a reference [1] in a publication, concerning its reaction with NOCl in ether at 20 °C to give $(C_6H_5)_3SbO$, N_2, and $[(C_6H_5)_3SbNH_2]Cl$ [2].

$R_3Sb{=}NCOR''$ (R = aryl)

A solution of SbR_3 (R = aryl) in dry $(CH_3)_2CO$ is treated with the sodium salt of the appropriate unsubstituted or substituted N-bromoacetamide $Na[XCH_2C(O)NBr]$. The reaction is initiated by several drops of concentrated HCl. The mixture is then heated for half an hour to 50 °C and filtered; the filtrate is concentrated in vacuum at room temperature. Cooling to −5 to −10 °C gives the compounds shown in Table 45. They are recrystallized from ether [3, 5, 6].

These compounds, dissolved in $CHCl_3$ [3, 5] or $(CH_3)_2CO$ [6], react with dilute ethanolic (5 to 6%) solutions of $HgCl_2$ or $CuCl_2$ upon standing for 12 h to give the corresponding 1:1 adducts. The yields and melting points of the adducts are also given in Table 45 [3, 5, 6].

Table 45
$R_3Sb{=}NCOR''$ Compounds and Their 1:1 Adducts with $HgCl_2$ (A) and $CuCl_2$ (B).

No.	compound yield in %	m.p. in °C	adduct A yield in % m.p. in °C	adduct B yield in % m.p. in °C	Ref.
1	$(C_6H_5)_3Sb{=}NCOCH_3$ 75	157 to 159	no yield 126 to 128	no yield 138 to 140	[3]
2	$(C_6H_5)_3Sb{=}NCOCH_2Cl$ 70	174 to 176	43.1 130 to 132	41.4 150 to 152	[5]
3	$(C_6H_5)_3Sb{=}NCOCH_2Br$ 68	193 to 195	67.5 138 to 140	61.8 153 to 155	[5]
4	$(C_6H_5)_3Sb{=}NCOCH_2I$ 67	208 to 210	64.1 140 to 142	50.9 156 to 158	[5]
5	$(2\text{-}CH_3C_6H_4)_3Sb{=}NCOCH_3$ 72	110 to 112	74.8 162 to 164	71.3 154 to 156	[6]
6	$(3\text{-}CH_3C_6H_4)_3Sb{=}NCOCH_3$ 69	83 to 85	69.2 130 to 132	68.6 138 to 139	[6]
7	$(4\text{-}CH_3C_6H_4)_3Sb{=}NCOCH_3$ 72	132 to 134	81 208 to 210	82.2 190 to 192	[6]
8	$(2\text{-}CH_3C_6H_4)_3Sb{=}NCOCH_2Cl$ 87	148 to 150	86.6 166 to 168	78.6 157 to 159	[6]
9	$(3\text{-}CH_3C_6H_4)_3Sb{=}NCOCH_2Cl$ 73	123 to 125	72.1 133 to 135	70.3 141 to 143	[6]
10	$(4\text{-}CH_3C_6H_4)_3Sb{=}NCOCH_2Cl$ 90	173 to 175	85.1 210 to 212	83.2 193 to 195	[6]

References on p. 199

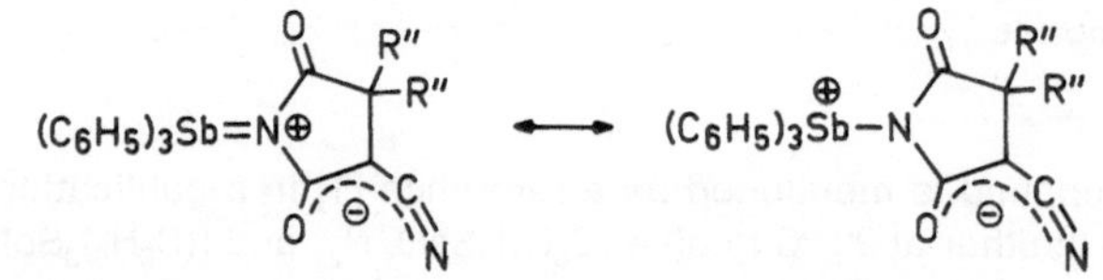

I

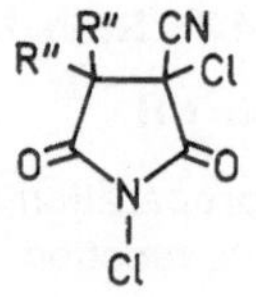

II

$(C_6H_5)_3SbC_5N_2O_2R''_2$ (Formula I, $R'' = CH_2C_6H_5$ and C_6H_5)

Both compounds are prepared by treating the appropriate pyrrolidinedione II with $Sb(C_6H_5)_3$ (ratio 1:2) in boiling C_6H_6. The yields are about 50% [7].

The compounds melt at 142 and 120 °C, respectively. IR vibrations are observed at 2180 (νCN), 1730 (νCO), and 1600 (νC=C) cm^{-1} for the compound with $R'' = C_6H_5$, and at 2170 (νCN), 1728 (νCO), and 1605 (νC=C) cm^{-1} for the compound with $R'' = CH_2C_6H_5$. A mechanism of formation of the compounds is discussed [7].

$R_3Sb{=}NSO_2R''$ (R = alkyl)

The trialkylantimony arylsulfonylimides are summarized in Table 46. They are obtained by slow addition of an ether solution of $R''SO_2N_3$ to a solution of the appropriate trialkylstibine SbR_3 in the same solvent. Nitrogen is vigorously evolved. The precipitating compounds are filtered and recrystallized from C_6H_6. The yields are about 60 to 70% of the pure compounds [8].

Table 46
Compounds of the Type $R_3Sb{=}NSO_2R''$ [8].

No.	compound	m.p. in °C
1	$(C_2H_5)_3Sb{=}NSO_2C_6H_5$	106 to 108
2	$(C_2H_5)_3Sb{=}NSO_2C_6H_4CH_3$-4	136 to 138
3	$(C_3H_7)_3Sb{=}NSO_2C_6H_5$	163
4	$(C_3H_7)_3Sb{=}NSO_2C_6H_4CH_3$-4	172 to 173

The compounds dissolve as monomers in C_6H_6. Dissolved in $(CH_2Cl)_2$, the compounds No. 1 to 4 react with H_2O to give the corresponding R_3SbO and $R''SO_2NH_2$. With gaseous HCl in the same solvent, R_3SbCl_2 and $R''SO_2NH_2$ are quantitatively formed. Nos. 3 and 4 were reacted with glacial acetic acid, and No. 4 with benzoic acid, to give the corresponding tripropylantimony dicarboxylates and the sulfonamide. With $CHF_2(CF_2)_3CH_2OH$, Nos. 3 and 4 form about 80% of the tripropylantimony diolate. Nos. 1 and 3 react with $CH_3OC_6H_4CHO$ in $(CH_2Cl)_2$ to give the corresponding R_3SbO. With SnX_4 (X = Cl, Br), the compounds No. 1 and 3 form R_3SbX_2 and $C_6H_5SO_2NSnX_2$ [8].

$(C_6H_5)_3Sb{=}NSO_2CF_3$

The compound is obtained quantitatively from $Sb(C_6H_5)_3$ and $CF_3SO_2N_3$ in petroleum ether at 20 °C as a precipitate. It melts at 173 to 175 °C when recrystallized from C_6H_6. Characteristic IR absorptions are given at 1585s, 1485s, 1445m, 1345s, 1285s, 1195w, 1155m, 1090m, 1040s, 1015s, 970s, and 700s cm^{-1} [9].

References on p. 199

$(C_6H_5)_3Sb{=}NSO_2C_6H_5$

Triphenylantimony phenylsulfonylimide is obtained in a yield of 81% by reacting $Sb(C_6H_5)_3$ with $Na[C_6H_5SO_2NCl]$ in $CHCl_3$ under reflux. The filtrate of the mixture is concentrated in vacuum, and yellow crystals precipitate upon cooling to 0 °C. They melt at 168 to 170 °C, recrystallized from ether [10]. A compound which melts at 198 °C (from CH_3OH or C_2H_5OH) is obtained from $Sb(C_6H_5)_3$ and $C_6H_5SO_2NHCl$ when heated for 15 min in C_2H_5OH and then to about 70 °C [11]. $Sb(C_6H_5)_3$ in CCl_4 and $C_6H_5SO_2NCl_2$ (ratio 2:1) give the compound in a yield of 88% after evaporation of the solvent in vacuum, extraction of the residue with ether, and evaporation of the latter. $Sb(C_6H_5)_3$ and $C_6H_5SO_2N_3$ dissolved in C_6H_6 form the title compound quantitatively with evolution of N_2. It is isolated by evaporating the solvent in a vacuum and is recrystallized from C_6H_6/c-C_6H_{12} mixtures [12]. A further preparation of the compound by reaction of $Sb(C_6H_5)_5$ and $C_6H_5SO_2NCl_2$, at temperatures of 130 to 140 °C with formation of C_6H_5Cl as a byproduct, is mentioned [13].

An IR spectrum of the compound shows characteristic absorptions at 1190, 1042 ($\nu_{as}SO_2$), 1138, 1135 (ν_sSO_2), and 920 (νSbN) cm^{-1}. UV maxima (in CH_3OH) are observed at 256, 261, and 266.5 nm [11].

The compound reacts with gaseous HCl in CH_2Cl_2 to form $(C_6H_5)_3SbCl_2$ and $C_6H_5SO_2NH_2$ [12]. With BX_3 (X=Cl, Br), under similar conditions, it forms triphenylantimony dihalide and $(C_6H_5SO_2NBX)_n$ quantitatively [14]. $GeCl_4$ or SnX_4 (X=Cl, Br) and the compound, reacted in $(CH_2Cl)_2$ for 30 min at 40 to 50 °C, form $(C_6H_5)_3SbX_2$ and $C_6H_5SO_2N{=}GeCl_2$ or $C_6H_5SO_2N{=}SnX_2$ nearly quantitatively. Similarly, with $(CH_3)_3SiCl$ (1:3) or $(CH_3)_2SiCl_2$ (1:1), the compounds $C_6H_5SO_2N(Si(CH_3)_3)_2$, $(C_6H_5SO_2NSi(CH_3)_2)_2$, and $(C_6H_5)_3SbCl_2$ are formed. With RCOCl (R= C_6H_5, 4-ClC_6H_4, CCl_3) in a ratio of 1:1 heated for 30 min in $(CH_2Cl)_2$ it forms $[(C_6H_5)_3Sb$-$N(COR)SO_2C_6H_5]Cl$ in yields of 90, 95, and 45%, respectively. Refluxing with C_6H_5COCl in a ratio of 1:2 for 8 h in $(CH_2Cl)_2$ gives 55% $C_6H_5SO_2N(COC_6H_5)_2$ and $(C_6H_5)_3SbCl_2$ [15]. Alcohols ROH (R=$CHF_2(CF_2)_5CH_2$, C_6H_5, 4-ClC_6H_4, 4-BrC_6H_4, 4-$NO_2C_6H_4$, 2,4-$Br_2C_6H_3$, and C_6Cl_5) and the compound give the corresponding triphenylantimony diolates in yields of 80 to 95% upon standing at room temperature in CH_2Cl_2. C_6H_5CHO reacts under similar conditions to give $(C_6H_5)_3SbO$ quantitatively [12]. $(C_6H_5)_3Sb{=}NSO_2C_6H_5$ forms 1:1 complexes with $HgCl_2$ or $CuCl_2$ [10].

For toxicity tests against mice and tests of biological activity against Sarcina lutea and Staphylococcus epidermis, and its antimalarial activity in birds, see [11].

$(C_6H_5)_3Sb{=}NSO_2C_6H_4X$-4 (X=Cl, Br)

The compounds are formed by reacting $Sb(C_6H_5)_3$ and the corresponding 4-$XC_6H_4SO_2NCl_2$ (X=Cl, Br, ratio 2:1) in CCl_4. After evaporation of the solvent in a vacuum the compounds are extracted from the byproduct, $(C_6H_5)_3SbCl_2$, with ether; after evaporation of the latter, the product is recrystallized from C_6H_6/c-C_6H_{12}. The yield is 90 to 95% [12].

The compounds react with gaseous HCl in CH_2Cl_2 to form $(C_6H_5)_3SbCl_2$ and the corresponding arylsulfonylamide [12]. $(C_6H_5)_3SbNSO_2C_6H_4Cl$-4 reacts with BX_3 (X=Cl, Br) in CH_2Cl_2 to give $(C_6H_5)_3SbX_2$ and $(4\text{-}ClC_6H_4SO_2NBX)_n$ quantitatively [14].

$(C_6H_5)_3Sb{=}NSO_2C_6H_4NO_2$-4

The compound is obtained quantitatively from $Sb(C_6H_5)_3$ and 4-$NO_2C_6H_4SO_2N_3$ in C_6H_6, upon standing 15 to 20 h at room temperature. After evaporation of the solvent, the residue is recrystallized from C_6H_6/c-C_6H_{12} [12].

References on p. 199

It reacts with gaseous HCl in CH_2Cl_2 to form $(C_6H_5)_3SbCl_2$ and the corresponding arylsulfonylamide. It gives with Cl_2 in the same solvent $(C_6H_5)_3SbCl_2$ and $4\text{-}NO_2C_6H_4SO_2NCl_2$ in an exothermic reaction. With alcohols ROH (R = $CHF_2(CF_2)_5CH_2$, C_6H_5, $4\text{-}ClC_6H_4$, $4\text{-}BrC_6H_4$, $4\text{-}NO_2C_6H_4$, $2,4\text{-}Br_2C_6H_3$, and C_6Cl_5) in CH_2Cl_2, 12 h at 20 °C, the corresponding triphenylantimony diolates are formed [12].

$(C_6H_5)_3Sb{=}NSO_2C_6H_4CH_3\text{-}4$

$Sb(C_6H_5)_3$ and $Na[4\text{-}CH_3C_6H_4SO_2NCl]$ are refluxed in $CHCl_3$. The filtrate of the mixture is concentrated in vacuum and cooled to 0 °C to give crystals in a yield of 81% which melt at 182 to 184 °C [10]. The same reaction, performed in refluxing CH_3CN, gives after workup an impure product in a yield of 72% which melts at 150 to 154 °C [16]. A compound with a melting point of 180 °C (from C_2H_5OH or CH_3OH) is obtained from $Sb(C_6H_5)_3$ and $4\text{-}CH_3C_6H_4SO_2NHCl$ in C_2H_5OH after heating for 15 min at 60 to 70 °C and then standing for 12 h at room temperature [11]. The compound is formed nearly quantitatively from $Sb(C_6H_5)_3$ and $4\text{-}CH_3C_6H_4SO_2NCl_2$ (ratio 2:1 in CCl_4) or with $4\text{-}CH_3C_6H_4SO_2N_3$ (ratio 1:1 in C_6H_6) in analogy to the corresponding phenyl compound (see above) [12].

IR bands were found at 1200, 1045 ($\nu_{as}SO_2$), 1160, 1138 (ν_sSO_2), and 925 (νSbN) cm^{-1}. UV maxima (in CH_3OH) were found at 256, 261, and 266 nm [11].

The title compound is hydrolyzed in CH_3CN to give $(C_6H_5)_3SbO$ [16]. With HCl in CH_2Cl_2, $4\text{-}CH_3C_6H_4SO_2NH_2$ and $(C_6H_5)_3SbCl_2$ are formed [12]. BX_3 (X = Cl, Br) and the compound react in CH_2Cl_2 to give $(C_6H_5)_3SbCl_2$ and $(4\text{-}CH_3C_6H_4SO_2NBX)_n$ [14]. SnX_4 (X = Cl, Br) or $GeCl_4$ give $(C_6H_5)_3SbX_2$ and $4\text{-}CH_3C_6H_4SO_2N{=}SnX_2$ or $4\text{-}CH_3C_6H_4SO_2N{=}GeCl_2$ in good yields. With $(CH_3)_2SiCl_2$ in $(CH_2Cl)_2$, the compound forms $(C_6H_5)_3SbCl_2$ and $(4\text{-}CH_3C_6H_4SO_2\text{-}NSi(CH_3)_2)_2$ [15]. $HgCl_2$ or $CuCl_2$ give 1:1 adducts with the title compound [10]. Reacted with C_6H_5COCl in a 1:1 ratio, the compound forms $[4\text{-}CH_3C_6H_4SO_2N(COC_6H_5)Sb(C_6H_5)_3]Cl$ in a yield of 75%. With $(CH_3)_2NCCl_3$ in $(CH_2Cl)_2$ at about 40 °C, $4\text{-}CH_3C_6H_4SO_2N{=}CClN(CH_3)_2$ and $(C_6H_5)_3SbCl_2$ are obtained quantitatively [15]. ROH (R = $CHF_2(CF_2)_5CH_2$, C_6H_5, $4\text{-}ClC_6H_4$, $4\text{-}BrC_6H_4$, $4\text{-}NO_2C_6H_4$, $2,4\text{-}Br_2C_6H_3$, and C_6Cl_5) and the compound, reacted in CH_2Cl_2 for 12 h at room temperature, give $(C_6H_5)_3Sb(OR)_2$ in yields of 80 to 95%. With benzaldehyde under similar conditions, $(C_6H_5)_3SbO$ is formed [12]. The compound reacts with LiC_6H_5 in a molar ratio of 2:7 for two hours in ether to give $Sb(C_6H_5)_5$ in a yield of 85% [16, 17].

For biological activity against Sarcina lutea and Staphylococcus epidermis, and antimalarial activities in birds as well as toxicity tests, see [11].

$(C_6H_5)_3Sb{=}NSO_2C_6H_4C_2H_5\text{-}4$

The compound is prepared from $Sb(C_6H_5)_3$ and $4\text{-}C_2H_5C_6H_4SO_2NCl_2$ in C_2H_5OH. The reaction is first maintained for 15 min at 60 to 70 °C, and then for 12 h standing at room temperature. The resulting precipitate melts at 190 °C, recrystallized from C_2H_5OH or CH_3OH [11].

It shows IR absorptions at 1195, 1037 ($\nu_{as}SO_2$), 1132 (ν_sSO_2), and 921 (νSbN) cm^{-1}. The UV spectrum (in CH_3OH) shows maxima at 251 and 267 nm. Toxicity and biological activity tests of the compound were made [11].

$(C_6H_5)_3Sb{=}NP(O)(C_6H_5)_2$

The compound is obtained from $Sb(C_6H_5)_3$ and $(C_6H_5)_2P(O)N_3$ in boiling triglyme. It is useful as an insecticide, as a stabilizer against light in polymers, and as a fire retardant [4].

References:

[1] W. Heinzelmann (Diss. Heidelberg 1962).
[2] J. MacCordick, R. Appel (Z. Naturforsch. **24b** [1969] 938).
[3] L. P. Petrenko (Zh. Obshch. Khim. **24** [1954] 520/1; J. Gen. Chem. [USSR] **24** [1954] 531/2).
[4] R. M. Washburn, R. A. Baldwin, American Potash & Chemical Corp. (U.S. 3189564 [1961/65]; C.A. **63** [1965] 9991).
[5] L. P. Petrenko (Tr. Voronezhsh. Gos. Univ. **49** [1958] 19/23).
[6] L. P. Petrenko (Tr. Voronezhsh. Gos. Univ. **49** [1958] 25/9).
[7] D. Leguern, G. Morel, A. Foucaud (Tetrahedron Letters **1974** 955/8).
[8] Z. I. Kuplennik, Z. N. Belaya, A. M. Pinchuk (Zh. Obshch. Khim. **51** [1981] 2711/5; J. Gen. Chem. [USSR] **51** [1981] 2339/43).
[9] O. A. Radchenko, V. P. Nazaretyan, L. M. Yagupol'skii (Zh. Obshch. Khim. **46** [1976] 565/8; J. Gen. Chem. [USSR] **46** [1976] 561/4).
[10] L. P. Petrenko (Tr. Voronezhsh. Gos. Univ. **57** [1959] 145/7; C.A. **55** [1961] 6425).

[11] J. J. Shah (J. Tennessee Acad. Sci. **51** [1976] 130/4).
[12] A. M. Pinchuk, Z. I. Kuplennik, Zh. N. Belaya (Zh. Obshch. Khim. **46** [1976] 2242/6; J. Gen. Chem. [USSR] **46** [1976] 2155/8).
[13] A. M. Pinchuk, A. M. Khmaruk, L. A. Feshchenko, T. V. Kovalevskaya, Z. I. Kuplennik, G. K. Bespal'ko (Zh. Obshch. Khim. **46** [1976] 2744/5; J. Gen. Chem. [USSR] **46** [1976] 2618).
[14] A. M. Pinchuk, G. K. Bespal'ko, T. V. Khimchenko, Z. I. Kuplennik (Zh. Obshch. Khim. **47** [1977] 2153/4; J. Gen. Chem. [USSR] **47** [1977] 1964).
[15] Z. I. Kuplennik, A. M. Pinchuk (Zh. Obshch. Khim. **49** [1979] 155/60; J. Gen. Chem. [USSR] **49** [1979] 135/9).
[16] G. Wittig, D. Hellwinkel (Chem. Ber. **97** [1964] 789/93).
[17] G. Wittig, D. Hellwinkel (Angew. Chem. **74** [1962] 76).

2.5.1.1.15 Triorganoantimony Sulfides R_3SbS

$(CH_3)_3SbS$

Trimethylantimony sulfide is obtained by reacting $Sb(CH_3)_3$ with sulfur in ether under reflux. Glittering flakes are obtained by evaporating the solvent from $(CH_3)_3SbO$ and H_2S in aqueous solution [1]. The compound is also obtained from the same reactants in methanolic solution [2].

It melts at 176 °C when recrystallized from CH_3OH. IR absorptions (in Nujol) are at 855s, 785w, 555s, 531m, 431s, 180m, 149s cm^{-1} and in $CHCl_3$ at 838s, 774w [2], 551s (ν_{as}SbC), 531m (ν_sSbC), and 433s (νSbS) cm^{-1} [2, 33]. UV spectra were measured in C_6H_{14} with $\lambda_{max}=279$ nm, and in CH_3CN with $\lambda_{max}=267$ nm and log $\varepsilon=3.6$; a figure is shown in [3]. A ^{13}C NMR resonance (in $CHCl_3$) at 136.5 Hz is discussed in connection with a formal positive charge of 0.67 on the Sb atom [5].

Molecular weight determinations show that the substance dissolves as a monomer in C_6H_6 or $CHCl_3$ [2]. Trimethylantimony sulfide can be desulfurized with Sn_2R_6 ($R=C_6H_5$, $C_6H_5CH_2$) in refluxing $CHCl_3$ to give $Sb(CH_3)_3$ and $(R_3Sn)_2S$ [6]. With acylhalides RCOX ($R=CH_3$, C_6H_5; X=Cl, Br) reacted in $CHCl_3$ in a ratio of 1:1, it forms $(CH_3)_3Sb(X)SCOR$ quantitatively [7]. Benzylbromide and the compound, reacted for 15 min in $CHCl_3$ in a nitrogen atmosphere, give 89% $Sb(CH_3)_3$ and 81% $(C_6H_5CH_2S)_2$ after distillation. With RI ($R=CH_3$, C_2H_5, $C_6H_5CH_2$) under similar conditions, but in the presence of air for 7 d, $(RS)_2$ and

$[(CH_3)_3SbI]_2O$ are formed [3]. CCl_3CO_2H and the compound, dissolved in CH_2Cl_2 for one hour at 20 °C, react to form $(CH_3)_3Sb(O_2CCCl_3)_2$ and H_2S quantitatively [8].

$(CH_3)_3SbS$ forms 2:1 complexes with ZnX_2, CdX_2, HgX_2, and CoX_2 (X=halogen). With $CdBr_2$ or HgI_2, 1:1 complexes are formed [9]. $M(NO_3)_2$ (M=Co, Zn, Cd) and $(CH_3)_3SbS$ give complexes like $M(NO_3)_2(SSb(CH_3)_3)_4$ and $M(NO_3)_2(SSb(CH_3)_3)_2$ [10]. $(CH_3)_3SbS$ and R_2InX (R=CH_3, C_2H_5, X=Cl, Br, I) react in CH_3OH at 0 °C to give the 1:1 adducts which subsequently decompose upon heating in the same solvent to yield RInS and $R(CH_3)_3SbX$. The analogous reaction occurs with CH_3InCl_2. With $InCl_3$ the products are $(CH_3)_3SbCl_2$, InSCl, and In_2S_3 [11]. Complex formation with $SnCl_4$ or organotin chlorides and sulfur-halogen exchange reactions are described in [2]. The behavior of mixtures of $(CH_3)_3SbS$ with R_3SnCl (R=CH_3 and C_6H_5) [4] or R_2SnCl_2 (R=CH_3 [33], C_6H_5 [4]) in solution was followed by IR and ^1H NMR spectroscopy. There is an equilibrium with $(CH_3)_3SbCl_2$ and $(R_3Sn)_2S$. The equilibrium constants at 20 °C were determined [4].

$(C_2H_5)_3SbS$

Triethylantimony sulfide is prepared by reacting $Sb(C_2H_5)_3$ and sulfur in an inert atmosphere without a solvent [12, 13] or in ether [12] or C_2H_5OH [15] solution. Upon concentrating the solutions, the compound precipitates quantitatively. Recrystallized from C_2H_5OH it melts at 118 °C [14]. The compound is also obtained from $(C_2H_5)_3SbBr_2$ and $Na_2S \cdot nH_2O$ in CH_3OH solution. Recrystallized from petroleum ether or CH_3OH, it melts at 119 to 120 °C [2].

An IR spectrum [2, 13] shows characteristic vibrations (mull) at 706s, 529m, and 506s cm^{-1}, and at 421s (νSbS) cm^{-1} [2]. The bands are compared with those of corresponding P and As compounds in [16].

Like all the other trialkylantimony sulfides studied the compound decomposes upon standing for 2 to 3 weeks with formation of antimony sulfides [14]. It reacts with aqueous $CuSO_4$ to give $(C_2H_5)_3SbSO_4$ and with aqueous KCN to give $Sb(C_2H_5)_3$ and KSCN [12, 15, 17].

$(C_3H_7)_3SbS$

$Sb(C_3H_7)_3$ and sulfur are reacted under N_2 for one week. The filtrate of the mixture is extracted with ether. Cooling this extract gives 70 to 90% yield of the compound [13]. From the same reactants in refluxing C_6H_6, the compound is obtained quantitatively after evaporation of the solvent. The very hygroscopic substance melts at 35 °C [14]. A given melting point of 88 °C of a product prepared from $(C_3H_7)_3SbO$ and H_2S in C_2H_5OH [18] is probably that of a hydrate of the compound [14]. $Sb(C_3H_7)_3$ and sulfur, heated together under H_2O, form the $(C_3H_7)_3SbS \cdot Sb_2S_3$ adduct [18]. Traces of dissolved sulfur are removed by repeated cooling and filtration [14].

The IR absorptions for νSbS were found at 439 cm^{-1} in CCl_4 and at 434 cm^{-1} for the pure liquid [13]. They are compared with those of the corresponding P and As compounds in [16].

$(C_4H_9)_3SbS$

$Sb(C_4H_9)_3$ and sulfur are reacted in a nitrogen atmosphere either without a solvent [14] for one week or in refluxing C_6H_6 for 4.5 h [13]. Workup of the mixtures by filtration, dilution with ether, and evaporation of the latter gives nearly quantitative yields of the liquid compound. Last traces of dissolved sulfur are removed by repeated cooling and filtration [14].

References on p. 203

The refractive index is $n_D^{20}=1.5528$ [14]. In the IR spectrum, the νSbS vibrations are observed at 440 (in CCl_4) or 434 (liquid) cm^{-1} [13, 16].

The compound inserts CS_2 to give $(C_4H_9)_3Sb(CS_3)$ [19]. It is useful as a catalyst for the polymerization of organoisocyanates [20].

$(C_5H_{11})_3SbS$

$Sb(C_5H_{11})_3$ and sulfur are mixed and stored for one week under N_2. The filtrate of the mixture is treated with ether, and the extract is cooled to −78 °C. The compound precipitates in a yield of 70 to 90%. It is a liquid with a refractive index of $n_D^{20}=1.5325$. The IR spectrum shows absorptions at 438 (CCl_4) or 434 (liquid) cm^{-1} [13, 16].

$(c\text{-}C_6H_{11})_3SbS$

$Sb(C_6H_{11}\text{-}c)_3$ and an excess (10%) of sulfur are refluxed for 3 h in C_6H_6. Concentration of the filtrate of the mixture gives the compound quantitatively. It melts at 144 °C, recrystallized from C_2H_5OH [14]. From the same reactants without a solvent the compound is obtained upon standing for a week under N_2 atmosphere [13]. $(c\text{-}C_6H_{11})_3SbBr_2$ and $Na_2S \cdot nH_2O$, reacted in CH_3OH, give the compound with the same melting point when recrystallized from CH_3OH or petroleum ether [2].

The IR spectrum shows νSbS vibrations at 435 (CCl_4) or 440 (KBr) cm^{-1} [13, 16]. UV spectra were measured (see original figure) and show absorption maxima at 282 nm (log $\varepsilon=3.7$, in C_6H_{14}) or at 274 nm (log $\varepsilon=3.8$, in CH_3CN) [3]. The compound dissolves as a monomer in C_6H_6 [2]. For the reactions of the substance with R_nSnX_{4-n} compounds see [2].

$(C_6H_5)_3SbS$

Triphenylantimony sulfide is obtained from $(C_6H_5)_3SbBr_2$ dissolved in C_2H_5OH and saturated with NH_3 and H_2S is passed through this solution until a weak yellow color is observed (note: an excess of H_2S must be avoided). The precipitating colorless crystals are filtered and recrystallized from C_2H_5OH. The yield is 80 to 100% [21 to 26]. Similarly, the compound is formed from $(C_6H_5)_3Sb(O_2CCH_3)_2$ in C_2H_5OH containing Na_2S by passing H_2S through the solution. Concentration of the filtrate to a small volume and addition of ether give the compound in 62% yield [27]. $(R_3Sn)_2S$ (R = C_4H_9 or C_6H_5) reacts with equivalent amounts of $(C_6H_5)_3SbX_2$ (X = Cl, Br) in $CHCl_3$ at −5 °C to give the compound in yields of about 80 to 90% along with R_3SnCl. The two substances are separated by evaporating the solvent and extracting R_3SnX with CH_3OH from the residue [28]. 5-Ethoxy-1,2,3,4-thiatriazole and $Sb(C_6H_5)_3$, reacted for 12 h at room temperature in ether, form the compound quantitatively [29]. From $(C_6H_5)_2ICl$, Na_2S, and antimony powder shaken in H_2O/ether, the compound is obtained from the ether phase [27].

Melting points of the substance are given as 119 to 120 °C [21, 22, 26, 30], 119 °C (from C_2H_5OH) [31], 112 °C (from C_2H_5OH) [23], and 110 to 112 °C [28].

The IR spectrum (in KBr) shows vibrations for the C_6H_5 substituents at 1574w, 1479m, 1435s, 1330w, 1312w, 1178vw, 1155vw, 1065m, 1019w, 996m, 729vs, 692vs, 454s, and 446s cm^{-1} [34]. High-energy photoelectron spectroscopy with Al Kα as exciting radiation leads to binding energies of Sb 3d(3/2) = 538.4, Sb 3d(5/2) = 529.3, and S 2p = 165.0 eV [35]. A ^{121}Sb Mössbauer spectrum at 4.2 K gives an isomer shift of $\delta=-5.85(5)$ vs. $Ba^{121}SnO_3$ with $e^2qQ=-1.8(4)$, and $\Gamma=2.8$ mm/s. The asymmetry parameter is $\eta=0.2$ [36].

The compound crystallizes as a monoclinic solid in the space group $P2_1/n-C_{2h}^5$ (No. 14) with a = 11.0, b = 9.74, c = 14.8 Å, and $\beta=92°$; $d_c=1.59$ and $d_m=1.6$ g/cm³; Z = 4 [38].

References on p. 203

A reinvestigation of the X-ray structure gave the following data: monoclinic, space group $P2_1/n-C^5_{2h}$ (No. 14) with a = 1109.8(2), b = 980.4(5), c = 1501.4(4) pm, and β = 94.31(2)°; Z = 4. The structure was determined from 2260 independent reflections, of which 1998 were used for calculations. A final R value of 2.5% was reached [36].

The molecular structure is shown in **Fig. 18**. The crystal is built up by discrete, tetrahedral molecules. The Sb-S distance is very short which indicates significant Sb=S π bonding. Dihedral angles of 4.8° (plane I), 40.5° (plane II), and 60.4° (plane III) for the C_6H_5 groups towards the S-Sb-C planes are observed. A reason for this is the intermolecular S···H interaction with the H-3 atom of the neighboring phenyl ring III. This distance is with 280 pm smaller than the van der Waals distance of 305 pm [36].

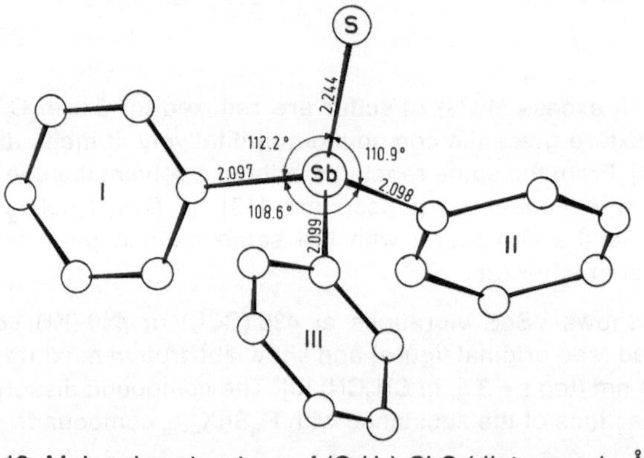

Fig. 18. Molecular structure of $(C_6H_5)_3SbS$ (distances in Å) [36].

The compound is very soluble in C_6H_6, $CHCl_3$, and glacial acetic acid [21, 26], soluble in C_2H_5OH, and slightly soluble in ether and petroleum ether [21]. It dissolves as a monomer in C_6H_6 [23]. Its dipole moment in this solvent at 25 °C is 5.40 D [31] or 5.66 D. From this value a molar Kerr constant of -523×10^{-12} was determined, and a twisting at the C_6H_5 substituents of 41° was suggested [32].

A mass spectrum (70 eV) shows the fragments $[M]^+$, $[M-H]^+$, $[M-S]^+$, $[M-C_6H_5]^+$, $[M-C_6H_5-2H]^+$, $[M-C_6H_5-S]^+$, $[M-C_6H_5-S-2H]^+$, $[C_6H_5SbH]^+$, $[C_6H_5Sb]^+$ (base peak), $[C_6H_4Sb]^+$, $[Sb]^+$, $[(C_6H_5)_2-nH]^+$ (n = 0, 1, 2), $[C_6H_6]^+$, and $[C_6H_5]^+$ [27]. An enthalpy of combustion of ΔH = −965 kcal/mol was determined and compared with those of other organometallic compounds [37].

The compound reacts in the presence of acids such as CH_3CO_2H or H_2S (see preparation!) in C_2H_5OH to give $Sb(C_6H_5)_3$ and sulfur [21, 22, 26]. It is also desulfurized by $Sn_2(C_6H_5)_6$ [6] or Pb_2R_6 (R = C_6H_5, 4-$CH_3C_6H_5$) [39] in refluxing $CHCl_3$ to $Sb(C_6H_5)_3$ and $(R_3M)_2S$ (M = Sn, Pb). With tetrachloro orthoquinone, refluxed for 10 h in C_6H_6, the compound forms sulfur and compound I [40]. CuI_2 and the compound give a 1:2 adduct; HgI_2 and the compound give a 1:1 adduct [41]. $(Ru(CO)_2I_2)_n$ and $(C_6H_5)_3SbS$ react in C_6H_6 to form $Ru(CO)_2(Sb(C_6H_5)_3)_2I_2$ and some $Ru(CO)(Sb(C_6H_5)_3)_3I_2$ [42].

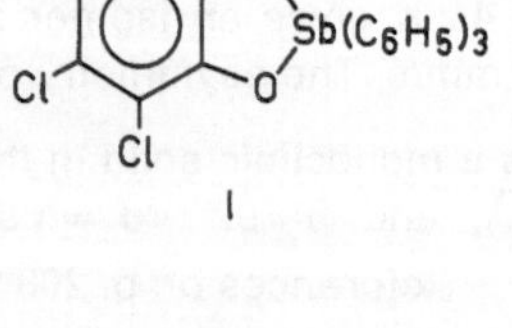

I

Patents claim that the compound is a catalyst together with $(C_2H_5)_3Al_2Cl_3$ and $TiCl_3$ for the polymerization of 4-methyl-1-pentene [43]. The compound is useful as a catalyst for the polycondensation of bis(2-hydroxyethyl)terephthalate [44] and as an antioxidant in polypropylene, nylon-6, or polystyrene [45]. It is not an antiknock material in motor fuels [37].

$(3\text{-}CH_3C_6H_4)_3SbS$

The compound is obtained as a powder from $(3\text{-}CH_3C_6H_4)_3SbO$ and H_2S in C_2H_5OH, saturated with NH_3. The product is recrystallized from C_2H_5OH/C_6H_6 and melts at 162 to 163 °C [46].

$(4\text{-}CH_3C_6H_4)_3SbS$

The compound is obtained from $(4\text{-}CH_3C_6H_4)_3SbBr_2$ in C_2H_5OH, saturated with NH_3, by passing H_2S through the mixture. The yield is 82%; the compound melts at 111.5 °C, recrystallized from C_2H_5OH [47]. $(R_3Sn)_2S$ ($R=C_6H_5$ or C_4H_9) and $(4\text{-}CH_3C_6H_4)_3SbX_2$ (X=Cl, Br) react in $CHCl_3$ at −5 °C to give, after evaporation of the solvent and extraction of the residue with CH_3OH, 80 to 90% yield of the compound, with a melting point of 118 to 120 °C [28].

The compound crystallizes as a monoclinic solid with a=9.5, b=19.5, c=10.5 Å, and β=106°; space group $P2_1/n-C_{2h}^5$ (No. 14). Z=4 gives $d_c=1.50$ and $d_m=1.52$ g/cm³. Isomorphy and morphotropy with R_3PX compounds ($R=C_6H_5$, X=S; $R=4\text{-}CH_3C_6H_4$, X=S, Se) and with $(C_6H_5)_3SbS$ were proved [38].

The compound reacts with Pb_2R_6 ($R=C_6H_5$, $4\text{-}CH_3C_6H_4$) in ratios of 1:1 in refluxing $CHCl_3$ to form $(R_3Pb)_2S$ and $Sb(C_6H_5)_3$ in yields of about 75% [39].

$(4\text{-}C_6H_5C_6H_4)_3SbS$

A compound of this composition with a melting point of 173 °C is mentioned without details. It is possibly prepared from $(4\text{-}C_6H_5C_6H_4)_3Sb(OH)_2$ and H_2S in C_2H_5OH [48].

References:

[1] H. Landolt (J. Prakt. Chem. **84** [1861] 328/339).
[2] M. Shindo, Y. Matsumura, R. Okawara (J. Organometal. Chem. **11** [1968] 299/305).
[3] J. Otera, R. Okawara (J. Organometal. Chem. **16** [1969] 335/8).
[4] M. Shindo, Y. Matsumura, R. Okawara (Bull. Chem. Soc. Japan **42** [1969] 265/6).
[5] J. Otera, R. Okawara (Inorg. Nucl. Chem. Letters **6** [1970] 855/7).
[6] J. Otera, T. Kadowaki, R. Okawara (J. Organometal. Chem. **19** [1969] 213/4).
[7] J. Otera, R. Okawara (J. Organometal. Chem. **17** [1969] 353/7).
[8] T. Okada, R. Okawara (J. Organometal. Chem. **42** [1972] 117/21).
[9] T. Saito, J. Otera, R. Okawara (Bull. Chem. Soc. Japan. **43** [1970] 1733/6).
[10] R. Okawara, J. Otera, T. Osaki (Inorg. Chem. **10** [1971] 402/4).

[11] T. Maeda, G. Yoshida, R. Okawara (J. Organometal. Chem. **44** [1972] 237/41).
[12] C. Löwig, E. Schweizer (Liebigs Ann. Chem. **75** [1850] 315/55).
[13] G. N. Chremos, R. A. Zingaro (J. Organometal. Chem. **22** [1970] 637/46).
[14] R. A. Zingaro, A. Merijanian (J. Organometal. Chem. **1** [1964] 369/73).
[15] G. B. Buckton (J. Chem. Soc. **13** [1861] 115/21).
[16] G. N. Chremos, R. A. Zingaro (J. Organometal. Chem. **22** [1970] 647/51).
[17] G. B. Buckton (Jahresber. Fortschr. Chem. **1860** 371/4).
[18] W. J. C. Dyke, W. J. Jones (J. Chem. Soc. **1930** 1921/7).
[19] A. W. Breindel, S. Herbstman (U.S. 3317575 [1963/67]; C.A. **67** [1967] No. 32780).

[20] S. Herbstman, Stauffer Chemical Co. (U.S. 3278492 [1964/66]; C.A. **66** [1967] No. 66142).

[21] L. Kaufmann (Ber. Deut. Chem. Ges. **41** [1980] 2762/6).
[22] L. Kaufmann (C. **1908** II, 1260/1).
[23] W. J. Lile, R. J. Menzies (J. Chem. Soc. **1950** 617/21).
[24] L. Kaufmann (Brit. 18896 [1909]; C.A. **1911** 2905).
[25] L. Kaufmann (Ger. 223694 [1908]; C.A. **1910** 3122).
[26] L. Kaufmann (U.S. 1060765 [1913]; C.A. **1913** 2094).
[27] C. Glidewell (J. Organometal. Chem. **116** [1976] 199/209).
[28] S. N. Bhattacharya, P. Raj, A. K. Saxena (Indian J. Chem. A **16** [1978] 1071/4).
[29] K. A. Jensen, A. Holm, E. Huge-Jensen (Acta Chem. Scand. **23** [1969] 2919/20).
[30] R. B. Sandin, F. T. McClure, F. Irwin (J. Am. Chem. Soc. **61** [1939] 2944/6).

[31] K. A. Jensen (Z. Anorg. Allgem. Chem. **250** [1943] 268/76).
[32] M. J. Aroney, R. J. W. Le Fevre, J. D. Saxby (J. Chem. Soc. **1964** 6180/5).
[33] M. Shindo, R. Okawara (Inorg. Nucl. Chem. Letters **3** [1967] 75/7).
[34] K. A. Jensen, P. H. Nielsen (Acta Chem. Scand. **17** [1963] 1875/85).
[35] T. Birchall, J. A. Connor, I. H. Hillier (J. Chem. Soc. Dalton Trans. **1975** 2003/6).
[36] J. Pebler, F. Weller, K. Dehnicke (Z. Anorg. Allgem. Chem. **492** [1982] 139/47).
[37] H. W. Charch, E. Mark, E. Board (Ind. Eng. Chem. **18** [1926] 339).
[38] G. S. Zhdanov, V. A. Pospelov, M. M. Umanskii, V. P. Glushkova (Dokl. Akad. Nauk SSSR **92** [1953] 983/5; C.A. **49** [1955] 12075).
[39] S. N. Bhattacharya, A. K. Saxena (Indian J. Chem. A **17** [1979] 307/9).
[40] M. M. Sidky, M. R. Mahran, W. M. Abdou (Phosphorus Sulfur **15** [1983] 129/35).

[41] M. G. King, G. P. McQuillan (J. Chem. Soc. A **1967** 898/901).
[42] W. Hieber, P. John (Chem. Ber. **103** [1970] 2161/77).
[43] H. W. Coover Jr., F. B. Joyner, Eastman Kodak Co. (Brit. 1000348 [1965]; C.A. **64** [1966] 2191).
[44] O. K. Carlson, J. A. Price, FMC Corp. (U.S. 3415787 [1966/8]; C.A. **70** [1969] No. 29547).
[45] T. Ohseki, M. Watanabe, Mitsubishi Rayon Co., Ltd (Japan. 71-22104 [1967/71]; C.A. **76** [1972] No. 154856).
[46] A. Michaelis, U. Genzken (Liebigs Ann. Chem. **242** [1887] 164/188).
[47] V. P. Glushkova, T. V. Talalaeva, Z. P. Razmanova, G. S. Zhdanov, K. A. Kocheshkov (Sb. Statei Obshch. Khim. Akad. Nauk SSSR **2** [1953] 992/6).
[48] D. E. Worrall (J. Am. Chem. Soc. **52** [1930] 2046/50).

2.5.1.1.16 Triorganoantimony Dithiolates, Bis(carbothioates), Bis(carbonodithioates), and Bis(carbamodithioates)

2.5.1.1.16.1 $R_3Sb(SR')_2$ Compounds with R′ = Alkyl and Aryl

The compounds of this type are shown in Table 47. The trimethyl derivatives are prepared by the following method:

Equivalent amounts of the appropriate mercaptan and $(CH_3)_3SbCl_2$ are dissolved in $(CH_3)_2CO$, cooled to −60 °C, and mixed. A slight excess of $N(C_2H_5)_3$ is added, and the mixture is stirred for 6 h at −25 to −30 °C. The precipitated $[NH(C_2H_5)_3]Cl$ is filtered with cooling, and the filtrate is concentrated in a vacuum. The cooled residues are dissolved in −10 to 0 °C cold solvent mixtures and cooled to −60 °C where the compounds crystallize [1].

The triarylantimony dithiolates are published in the patent literature without details of their preparation [2].

Table 47
Compounds of the Type $R_3Sb(SR')_2$.
For explanations, abbreviations, and units see p. X.

No.	R′	yield in % properties [Ref.]
with R = CH_3		
1	CH_3	74 colorless shiny needles (from ether/C_5H_{12}) m.p. 39 to 49° (dec.) ^{1}H NMR (C_6H_6): 1.23 (s, CH_3Sb), 1.5 (s, CH_3S) [1]
2	C_2H_5	78 colorless needles (from $CH_3C_6H_5$/ether/C_5H_{12}) m.p. 46 to 48° (dec.) ^{1}H NMR (C_6H_6): 0.80 (t, CH_3C), 1.23 (s, CH_3Sb), 1.925 (q, CH_2; $^3J(H,H)=7$) [1]
3	$CH_2C_6H_5$	79 m.p. 61 to 63° (dec., from $CH_3C_6H_5$/ether) ^{1}H NMR (C_6D_6): 0.95 (s, CH_3Sb), 2.97 (s, CH_2), 6.44 (m, C_6H_5) [1]
4	C_6H_5	79 m.p. 85 to 87° (dec.) (from $CH_3C_6H_5$/ether) ^{1}H NMR (C_6D_6): 0.99 (s, CH_3Sb), 6.4 (m, C_6H_5) [1]
with R = C_6H_5		
5	$C_{12}H_{25}$	useful as synergists for phenolic antioxidants in polyethylene or polypropylene [2]; No. 9 is also useful as a blowing catalyst for producing polyurethane foams [3] (Nos. 5 to 9)
6	$CH_2CO_2C_8H_{17}$	
7	$CH_2CH_2CO_2C_{12}H_{25}$	
with R = 4-$CH_3OC_6H_4$		
8	$C_{12}H_{25}$	
with R = 4-$CH_3C_6H_4$		
9	$CH_2CH_2CO_2C_{12}H_{25}$	

General Remarks. The $(CH_3)_3Sb(SR')_2$ compounds decompose upon melting and in C_6H_6 solution at 30 °C to form $Sb(CH_3)_3$ and RSSR. A ^{1}H NMR study of this decomposition of compound No. 1 was made and shows that the reaction is probably of first order [1].

References:

[1] H. Schmidbaur, K. H. Mitschke (Chem. Ber. **104** [1971] 1842/6).
[2] T. Ohseki, M. Watanabe, Mitsubishi Rayon Co., Ltd. (Japan. 71-05209 [1968/71]; C.A. **76** [1972] No. 4551).
[3] R. N. Haszeldine, B. O. West (J. Chem. Soc. **1957** 3880/4).

2.5.1.1.16.2 $R_3Sb(SR')_2$ Compounds with $R'=C(O)R''$, $C(S)OR''$, $C(S)NR''_2$, and $C(S)N(-R''-)$

$(CH_3)_3Sb(SC(O)CH_3)_2$

$(CH_3)_3Sb(OC_2H_5)_2$ and $CH_3C(O)SH$ are stirred together in C_6H_6 below 5 °C for 15 min. The compound deposits in the form of colorless needles in a yield of 71%. They melt at 51 to 52 °C and dissolve as monomers in $CHCl_3$ at 25 °C. In an IR spectrum, νCO vibrations are observed at 1639 and 1634 cm^{-1} for the solid and at 1645 cm^{-1} in CCl_4 solution [1].

The compound reacts with $(CH_3)_3SbX_2$ (X=Cl, Br) in a ratio of 1:1 in $CHCl_3$ at room temperature to give $(CH_3)_3Sb(SC(O)CH_3)X$ quantitatively [2].

$(CH_3)_3Sb(SC(O)C_6H_5)_2$

$(CH_3)_3Sb(OC_2H_5)_2$ and $C_6H_5C(O)SH$ react in C_6H_6 at room temperature for 15 min under N_2 to produce a 92% yield of colorless crystals. The compound melts at 108 °C and dissolves as a monomer in $CHCl_3$ at 25 °C. A νCO absorption in the IR spectrum is observed at 1621 cm^{-1} in the solid and in CCl_4 solution [1]. An X-ray powder diagram is shown in [2].

Refluxing the compound for 7 h in C_6H_6 under N_2 leads to an equilibrium with about 60% $Sb(CH_3)_3$ and $(C_6H_5CO)_2S_2$. This equilibrium is reached from both sides [1]. The compound reacts with $(CH_3)_3SbX_2$ (X=Cl, Br) in $CHCl_3$ for 30 min at room temperature to give $(CH_3)_3Sb(SC(O)C_6H_5)X$ quantitatively [2].

$(CH_3)_3Sb(SC(S)OR'')_2$ ($R''=C_3H_7$, i-C_3H_7, i-C_4H_9, and C_6H_5)

The compounds are prepared from $(CH_3)_3SbBr_2$ and equivalent amounts of the appropriate $NaSC(S)OR''$ in an i-C_3H_7OH/H_2O (1:1) mixture. Stirring for two hours and removing the solvent gives a residue which is washed with cold water to remove the NaBr. The remaining residue is recrystallized from ether. The yield is about 60%. Melting points and spectroscopic data are given in Table 48 [3].

Table 48
Physical Data of $(CH_3)_3Sb(SC(S)OR'')_2$ Compounds [3].
For explanations, abbreviations, and units, see p. X.

No.	R″	melting points, ¹H NMR, and IR data
1	C_3H_7	m.p. 79° (dec.) ¹H NMR ($CDCl_3$): 1.03, 1.87, 4.48 (C_3H_7), 2.40 (CH_3) IR: 1188, 1128 (νCS), 1052, 1036 (νCOC), 848 (νC_3Sb), 365 (νSbS)
2	i-C_3H_7	m.p. 98° (dec.) ¹H NMR ($CDCl_3$): 1.45, 5.70 (C_3H_7), 2.37 (CH_3) IR: 1200, 1139 (νCS), 1080 to 1030 (νCOC), 846 (νC_3Sb), 369 (νSbS)
3	i-C_4H_9	m.p. 71° (dec.) ¹H NMR ($CDCl_3$): 1.03, 2.10, 4.25 (C_4H_9), 2.36 (CH_3) IR: 1195, 1170 (νCS), 1055 (νCOC), 848 (νC_3Sb), 357 (νSbS)
4	$CH_2C_6H_5$	m.p. 69° (dec.) ¹H NMR ($CDCl_3$): 2.22 (CH_3), 5.49, 7.35 ($CH_2C_6H_5$) IR: 1211, 1158 (νCS), 1047 (νCOC), 835 (νC_3Sb), 369 (νSbS)

References on p. 208

$(CH_3)_3Sb(SC(S)N(CH_3)_2)_2$

$(CH_3)_3SbBr_2$ in H_2O is added to an aqueous solution of $NaSC(S)N(CH_3)_2$. The resulting precipitate is filtered, dried, and recrystallized from C_6H_6. The compound melts at 146 to 147.5 °C. The same compound is obtained if 0.9 mol $(CH_3)_2NC(S)SSC(S)N(CH_3)_2$ is added to a distilled ethereal solution of $Sb(CH_3)_3$ (1 mol), which results from a Grignard reaction of $SbCl_3$ and CH_3MgI. A spontanous reaction occurs, and the compound precipitates. It was recrystallized twice from C_6H_6 [4].

The substance shows in the IR spectrum (KBr) a νC-N vibration at 1500 cm^{-1}. The ^1H NMR spectrum (in $CDCl_3$) has resonances at $\delta=2.48$ (s, $SbCH_3$) and 3.45 (s, NCH_3) ppm [4]. ^{121}Sb Mössbauer data are given in [4, 5]: $\delta=2.5\pm0.1$ (at 4.2 K vs. InSb), $e^2qQ=-17.4\pm1.0$, $\Gamma=1.11\pm0.1$ mm/s. The asymmetry parameter is $\eta=0.4\pm0.2$ [5].

The compound crystallizes in the orthorhombic space group $Cmc2_1-C_{2v}^{12}$ (No. 36) with a = 7.448(4), b = 18.071(7), and c = 12.475(9) Å; Z = 4. From 1359 independent reflections, 1213 were used for structure determination, and refinements lead to a final R value of 0.06. Bond distances and angles are given in **Fig. 19**.

The structure can be described as a slightly deformed trigonal bipyramid with three methyl groups in the equatorial plane and one sulfur of each dithiocarbamato ligand, S(1) and S(4), at apical positions. The remaining sulfur atoms, S(2) and S(3), are situated at 3.274 and 3.15 Å from the Sb atom, indicating that there is a weak Sb-S(2) and Sb-S(3) interaction, at least in the solid state. The sum of the van der Waals radii is 4.05 Å. The distance of 3.84 Å for S(2)-S(3) is a normal van der Waals contact and is almost twice as large as an S-S bonding distance in thiuram disulfide (2.00 Å). Sb, S(1), S(2), S(3),

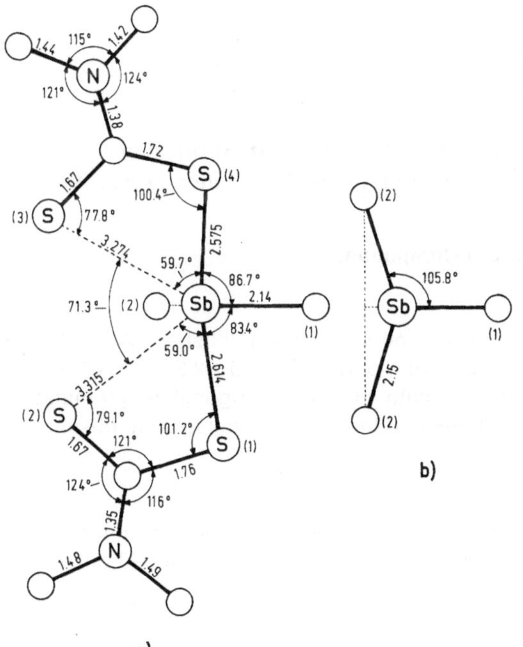

Fig. 19. Molecular structure of $(CH_3)_3Sb(SC(S)N(CH_3)_2)_2$, a) parallel and b) perpendicular to the SbC_3 plane [4].

References on p. 208

S(4), and C(1) are positioned in the mirror plane 0yz. The distances of these atoms to the projection point of the two equivalent C(2) atoms on the mirror plane are equal within 0.07 Å. Therefore, the structure can also be described as an IF_7-like pentagonal bipyramid with the Sb atom 0.52 Å out of the center [4].

$(CH_3)_3Sb(SC(S)NC_4H_8)_2$ (NC_4H_8 = 1-pyrrolidinyl)

The compound is obtained from $(CH_3)_3SbBr_2$ and equivalent amounts of $NaSC(S)NC_4H_8$ in a 5:1 solvent mixture of C_6H_6/CH_3CN. The mixture is stirred for one hour at 20 °C, and the NaBr precipitate is filtered. The filtrate is concentrated in a vacuum below 60 °C, and the residue is dissolved in $CHCl_3$. Upon addition of petroleum ether and cooling to 0 °C, the compound slowly precipitates. The yield is about 50% [3]. (Note: an excess of thiocarbamate must be avoided to prevent reduction of Sb^V to Sb^{III}.)

The following physical properties are given: m.p. 110 °C (dec.); IR: 1450 (νCN), 1243, 1160, 998 (νCS), 947 (νCNC), 850 (νC_3Sb), 407 (νSbS) cm^{-1}; ^1H NMR ($CDCl_3$): δ=2.01, 3.66 (CH_2), and 2.49 (CH_3) ppm [3].

$(CH_3)_3Sb(SC(S)NC_5H_{10})_2$ (NC_5H_{10} = 1-piperidinyl)

The compound is prepared like the previous one, but with $NaSC(S)NC_5H_{10}$ [3].

Physical properties: m.p. 109 °C (dec.); IR: 1468 (νCN), 1223, 1113, 987 (νCS), 946 (νCNC), 848 (νC_3Sb), 405 (νSbS) cm^{-1}; ^1H NMR ($CDCl_3$): δ=1.68, 4.12 (CH_2), and 2.03 (CH_3) ppm [3].

References:

[1] Y. Matsumura, M. Shindo, R. Okawara (Inorg. Nucl. Chem. Letters **3** [1967] 219/22).
[2] J. Otera, R. Okawara (J. Organometal. Chem. **17** [1969] 353/7).
[3] A. Ouchi, M. Shimoi, F. Ebina, T. Uehiro, Y. Yoshino (Bull. Chem. Soc. Japan **51** [1978] 3511/3).
[4] J. A. Cras, J. Willemse (Rec. Trav. Chim. **97** [1978] 28/9).
[5] J. G. Stevens, J. M. Trooster (J. Chem. Soc. Dalton Trans. **1979** 740/4).

2.5.1.1.16.3 $R_3Sb(-SR'S-)$ Compounds

$(CH_3)_3SbS_2C{=}C(CN)_2$

The preparation of the compound is not published. A ^{121}Sb Mössbauer spectrum (at 4.2 K vs. InSb) gives the following parameters: $\delta = 2.3 \pm 0.1$, $e^2qQ = -18.7 \pm 1.0$, $\Gamma = 1.35 \pm 0.1$ mm/s; η is ca. 0.0. The data indicate a trigonal bipyramidal structure with the CH_3 groups in equatorial positions and the sulfur atoms in axial positions [1].

$(C_6H_5)_3SbS_2C{=}NC{\equiv}N$

A patent describes the preparation of the compound from $(C_6H_5)_3SbCl_2$ and $K_2[S_2CNCN]$ in a 10:1 mixture of $HCON(CH_3)_2/H_2O$, agitated for 10 min [2].

$(C_6H_5)_3SbS_2C{=}C(CN)_2$

The preparation of this compound is achieved from $K_2[S_2C{=}C(CN)_2]$ and $(C_6H_5)_3SbCl_2$ at room temperature in $HCON(CH_3)_2$. The compound melts at 129.5 to 130.5 °C and shows antibacterial and fungicidal activities [3].

References:

[1] J. G. Stevens, J. M. Trooster (J. Chem. Soc. Dalton Trans. **1979** 740/4).
[2] W. L. Mosby, American Cyanamid Co. (U.S. 3365478 [1965/68]; C.A. **68** [1968] No. 95978).
[3] W. L. Mosby, American Cyanamid Co. (U.S. 3429905 [1965/69]; C.A. **70** [1969] No. 115333).

2.5.1.1.17 Triorganoantimony Bis(diorganylphosphinodithioates) $R_3Sb(SP(S)R'_2)_2$

$(CH_3)_3Sb(SP(S)(C_2H_5)_2)_2$

Only one compound is mentioned in the literature. Its preparation is not published. However, a ^{121}Sb Mössbauer spectrum is given, and a discussion of the structure on the basis of the measured values was made: $\delta = 2.6 \pm 0.1$ (vs. InSb at 4.2 K), $e^2qQ = -18.7 \pm 1.0$, $\Gamma = 1.20 \pm 0.1$ mm/s. The asymmetry parameter is $\eta = 0.2 \pm 0.2$.

Reference:

J. G. Stevens, J. M. Trooster (J. Chem. Soc. Dalton Trans. **1979** 740/4).

2.5.1.1.18 Triorganoantimony Selenides R_3SbSe_n (n = 1, 2)

$(CH_3)_3SbSe_2$

$Sb(CH_3)_3$ and an excess of selenium are refluxed in C_6H_6 in an N_2 atmosphere. Upon concentration of the filtrate, pale yellow, very air sensitive crystals are obtained [1].

The compound appeared to be stable if stored under C_6H_6. The elemental analysis corresponds reasonably well with the formula $(CH_3)_3SbSe_2$, and the cyclic structure I is proposed [1].

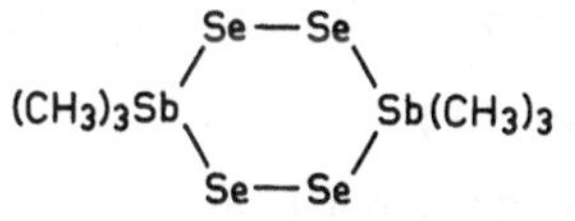

I

$(C_2H_5)_3SbSe$

Triethylantimony selenide is obtained by reacting $Sb(C_2H_5)_3$ with a 10% excess of Se in refluxing C_2H_5OH for two hours. The filtrate of the mixture is concentrated, whereby the compound deposits quantitatively. It melts at 124 °C [1]. A preparation from the same starting materials in ether is mentioned without details [2]. Without a solvent, the reaction proceeds upon standing for one week under N_2. Workup of the mixture with ether gives 70 to 90% yield [3].

In the IR spectrum a νSb=Se vibration is observed at 272 cm^{-1} [3]. This value is compared with those of the corresponding P and As compounds in [4].

Other R_3SbSe Compounds (R = C_3H_7, C_4H_9, C_5H_{11}, and c-C_6H_{11})

The compounds are obtained by reacting the corresponding stibine with a slight excess (10%) of Se upon standing for one week under a nitrogen atmosphere. The filtrate of the mixture is treated with ether. On cooling to −78 °C, the compounds crystallize in yields of about 70 to 90% [3]. The compounds $(C_5H_{11})_3SbSe$ and (c-$C_6H_{11})_3SbSe$ were also prepared from the same reactants in refluxing C_6H_6 for 3 to 9 h. After evaporation of the solvent from the filtrate in a vacuum, the liquid products remain in quantitative yield [1].

The physical data are given below. The IR vibrations were compared with those of the corresponding P and As compounds [4].

Table 49
Other R_3SbSe Compounds.

No.	compound	n_D^{25} [Ref.]	IR (in cm^{-1}) $\nu Sb{=}Se$ [Ref.]
1	$(C_3H_7)_3SbSe$	1.5846 [1]	300 [3]
2	$(C_4H_9)_3SbSe$	1.5534 [1]	294 [3]
3	$(C_5H_{11})_3SbSe$	1.5455 [3]	296 [3]
4	$(c\text{-}C_6H_{11})_3SbSe$	1.5455 [3]	290 [3]

$(C_6H_5)_3SbSe$

The unisolated compound is mentioned as a byproduct (35% yield) during the oxidation of $Sb(C_6H_5)_3$ with SeO_2 in C_6H_6 [5].

References:

[1] R. A. Zingaro, A. Merijanian (J. Organometal. Chem. **1** [1964] 369/73).
[2] C. Löwig, E. Schweizer (Liebigs Ann. Chem. **75** [1850] 315/55).
[3] G. N. Chremos, R. A. Zingaro (J. Organometal. Chem. **22** [1970] 637/46).
[4] G. N. Chremos, R. A. Zingaro (J. Organometal. Chem. **22** [1970] 647/51).
[5] N. N. Mel'nikov, M. S. Rokitskaya (J. Gen. Chem. [USSR] **8** [1938] 834/8).

2.5.1.1.19 Triorganoantimonio-bis(organylsilanes) $R_3Sb(SiR'_3)_2$

$(CH_3)_3Sb(Si(C_6H_5)_3)_2$, $(C_6H_5)_3Sb(Si(CH_3)_3)_2$, and $(C_6H_5)_3Sb(Si(C_6H_5)_3)_2$

These compounds are mentioned in a patent as useful catalysts for the polycondensation of terephthalic acid or dimethyl terephthalate with ethylene glycol [1].

Reference:

H. Terada, T. Imaida, S. Takashima, Mitsubishi Rayon Co., Ltd. (Japan. 70-00511 [1965/70]; C.A. **72** [1970] No. 112668).

Empirical Formula Index

In the following index the compounds are listed by their empirical formula in the order of increasing carbon content (first column). The second column contains the substance formulas wherein cyclic ligands are partly also written as empirical formula. Formulas of ionic compounds are given in brackets; ions as well as components of solvates and adducts are separated by a period.

In the third column, page references are printed in ordinary type, table numbers in bold face, and compound numbers within the table in italics.

$\mathbf{C}_3Br_2F_9Sb$	$(CF_3)_3SbBr_2$	73
$C_3Cl_2F_9Sb$	$(CF_3)_3SbCl_2$	23
C_3D_9OSb	$(CD_3)_3SbO$	115
$C_3H_6Br_2Cl_3Sb$	$(CH_2Cl)_3SbBr_2$	73
$C_3H_9Br_2Sb$	$(CH_3)_3SbBr_2$	60/9
$C_3H_9Cl_2O_8Sb$	$(CH_3)_3Sb(ClO_4)_2$	106
$C_3H_9Cl_2Sb$	$(CH_3)_3SbCl_2$	11
$C_3H_9CrO_4Sb$	$(CH_3)_3SbCrO_4$	111
$C_3H_9F_2Sb$	$(CH_3)_3SbF_2$	1/2
$C_3H_9N_2O_6Sb$	$(CH_3)_3Sb(NO_3)_2$	107
$C_3H_9N_3Sb$	$(CH_3)_3Sb(N_3)_2$	102
C_3H_9OSb	$(CH_3)_3SbO$	113/4
$C_3H_9O_2Sb$	$(CH_3)_3SbO_2$	130
$C_3H_9O_4SSb$	$(CH_3)_3SbSO_4$	110
$C_3H_9O_4SbSe$	$(CH_3)_3SbSeO_4$	111
C_3H_9SSb	$(CH_3)_3SbS$	199/200
$C_3H_9SbSe_2$	$(CH_3)_3SbSe_2$	209
$C_3H_{11}O_2Sb$	$(CH_3)_3Sb(OH)_2$	114/5
$C_3H_{11}O_4Sb$	$(CH_3)_3Sb(OOH)_2$	131
$\mathbf{C}_4H_9O_3Sb$	$(CH_3)_3SbCO_3$	111
$C_5H_9N_2O_2Sb$	$(CH_3)_3Sb(NCO)_2$	102
$C_5H_9N_2S_2Sb$	$(CH_3)_3Sb(NCS)_2$	104/5
$C_5H_9O_4Sb$	$(CH_3)_3Sb(O_2CCO_2)$	186
$C_5H_{11}O_3Sb$	$(CH_3)_3Sb(OCH_2CO_2)$	143, **32**, *1*
$C_5H_{11}O_4Sb$	$(CH_3)_3Sb(O_2CH)_2$	163/4
$C_5H_{15}O_2Sb$	$(CH_3)_3Sb(OCH_3)_2$	131
$C_5H_{15}S_2Sb$	$(CH_3)_3Sb(SCH_3)_2$	205, **47**, *1*
$\mathbf{C}_6H_6Br_2Cl_3Sb$	$(cis\text{-}ClCH{=}CH)_3SbBr_2$	76
$C_6H_6Cl_5Sb$	$(ClCH{=}CH)_3SbCl_2$	30
	$(cis\text{-}ClCH{=}CH)_3SbCl_2$	29/30
	$(trans\text{-}ClCH{=}CH)_3SbCl_2$	28/9
$C_6H_6F_{11}Sb$	$(CF_3CH_2)_3SbF_2$	3
$C_6H_9Br_2Sb$	$(CH_2{=}CH)_3SbBr_2$	75
$C_6H_9Cl_2Sb$	$(CH_2{=}CH)_3SbCl_2$	26/7
$C_6H_9F_2Sb$	$(CH_2{=}CH)_3SbF_2$	4
$C_6H_9I_2Sb$	$(CH_2{=}CH)_3SbI_2$	96
$C_6H_{11}O_4Sb$	$(CH_3)_3Sb(O_2CCH_2CO_2)$	186
$C_6H_{15}Br_2Sb$	$(C_2H_5)_3SbBr_2$	70

$C_6H_{15}Cl_2Sb$	$(C_2H_5)_3SbCl_2$	20/1
$C_6H_{15}F_2Sb$	$(C_2H_5)_3SbF_2$	3
$C_6H_{15}I_2Sb$	$(C_2H_5)_3SbI_2$	94/5
$C_6H_{15}N_2O_6Sb$	$(C_2H_5)_3Sb(NO_3)_2$	109
$C_6H_{15}OSb$	$(C_2H_5)_3SbO$	115/6
$C_6H_{15}O_4SSb$	$(C_2H_5)_3SbSO_4$	110
$C_6H_{15}SSb$	$(C_2H_5)_3SbS$	200
$C_6H_{15}SbSe$	$(C_2H_5)_3SbSe$	209
$C_6H_{17}O_2Sb$	$(C_2H_5)_3Sb(OH)_2$	116
$\mathbf{C}_7H_9Cl_6O_4Sb$	$(CH_3)_3Sb(O_2CCCl_3)_2$	171, **37**, *4*
$C_7H_9D_6O_4Sb$	$(CH_3)_3Sb(O_2CCD_3)_2$	164
$[C_7H_9D_{12}O_2S_2Sb]^{2+}$	$[(CH_3)_3Sb((CD_3)_2SO)_2]^{2+} \cdot 2[ClO_4]^-$	107, **24**, *2*
$C_7H_9F_6O_4Sb$	$(CH_3)_3Sb(O_2CCF_3)_2$	171, **37**, *1*
$C_7H_9F_{12}N_2O_2Sb$	$(CH_3)_3Sb(ON(CF_3)_2)_2$	161
$C_7H_9N_2S_2Sb$	$(CH_3)_3Sb(S_2C{=}C(CN)_2)$	208
$C_7H_{11}Br_4O_4Sb$	$(CH_3)_3Sb(O_2CCHBr_2)_2$	171, **37**, *7*
$C_7H_{11}Cl_4O_4Sb$	$(CH_3)_3Sb(O_2CCHCl_2)_2$	171, **37**, *5*
$C_7H_{11}F_4O_4Sb$	$(CH_3)_3Sb(O_2CCHF_2)_2$	171, **37**, *2*
$C_7H_{13}Br_2O_4Sb$	$(CH_3)_3Sb(O_2CCH_2Br)_2$	171, **37**, *8*
$C_7H_{13}Cl_2O_4Sb$	$(CH_3)_3Sb(O_2CCH_2Cl)_2$	171, **37**, *6*
$C_7H_{13}F_2O_4Sb$	$(CH_3)_3Sb(O_2CCH_2F)_2$	171, **37**, *3*
$C_7H_{15}O_2S_2Sb$	$(CH_3)_3Sb(SC(O)CH_3)_2$	206
$C_7H_{15}O_3Sb$	$(C_2H_5)_3SbCO_3$	111
$C_7H_{15}O_4Sb$	$(CH_3)_3Sb(O_2CCH_3)_2$	164
$C_7H_{19}O_2Sb$	$(CH_3)_3Sb(OC_2H_5)_2$	132
$C_7H_{19}S_2Sb$	$(CH_3)_3Sb(SC_2H_5)_2$	205, **47**, *2*
$[C_7H_{21}O_2S_2Sb]^{2+}$	$[(CH_3)_3Sb((CH_3)_2SO)_2]^{2+} \cdot 2[ClO_4]^-$	107, **24**, *1*
$\mathbf{C}_8H_8Cl_6Sb$	(cis-ClCH=CH)(trans-ClCH=CH))$_2$SbCl$_2$	30/1
$C_8H_{15}N_2O_2Sb$	$(C_2H_5)_3Sb(NCO)_2$	103, **23**, *1*
$C_8H_{15}O_4Sb$	$(C_2H_5)_3Sb(O_2CCO_2)$	186
$\mathbf{C}_9H_{13}N_2O_4Sb$	$(CH_3)_3Sb(O_2CCH_2CN)_2$	171, **37**, *9*
$C_9H_{13}O_2Sb$	$(CH_3)_3Sb(2\text{-}OC_6H_4O)$	143, **32**, 2
$(C_9H_{14}N_5OSb)_n$	$((CH_3)_3Sb(NH\text{-}C_6H_2N_3(OH)NH))_n$	194, **44**, *1*
$C_9H_{15}Br_2Sb$	$(CH_2{=}C(CH_3))_3SbBr_2$	76
	$(CH_2{=}CHCH_2)_3SbBr_2$	76
	(cis-CH$_3$CH=CH)$_3$SbBr$_2$	75/6
	(trans-CH$_3$CH=CH)$_3$SbBr$_2$	75/6
$C_9H_{15}Cl_2Sb$	$(CH_2{=}C(CH_3))_3SbCl_2$	28
	$(CH_2{=}CHCH_2)_3SbCl_2$	28
	(cis-CH$_3$CH=CH)$_3$SbCl$_2$	27
	(trans-CH$_3$CH=CH)$_3$SbCl$_2$	28
$C_9H_{15}F_2Sb$	$(CH_2{=}CCH_3)_3SbF_2$	5
	(cis-CH$_3$CH=CH)$_3$SbF$_2$	5
	(trans-CH$_3$CH=CH)$_3$SbF$_2$	5
$C_9H_{15}I_2Sb$	$(CH_2{=}C(CH_3))_3SbI_2$	96
	(cis-CH$_3$CH=CH)$_3$SbI$_2$	96
	(trans-CH$_3$CH=CH)$_3$SbI$_2$	97
$C_9H_{17}O_4Sb$	$(C_2H_5)_3Sb(O_2CCH_2CO_2)$	186

$C_9H_{19}O_4Sb$	$(CH_3)_3Sb(O_2CC_2H_5)_2$	164
$C_9H_{21}Br_2Sb$	$(C_3H_7)_3SbBr_2$	70
	$(i\text{-}C_3H_7)_3SbBr_2$	71
$C_9H_{21}Cl_2Sb$	$(C_3H_7)_3SbCl_2$	21
	$(i\text{-}C_3H_7)_3SbCl_2$	22
$C_9H_{21}F_2Sb$	$(i\text{-}C_3H_7)_3SbF_2$	3
$C_9H_{21}I_2Sb$	$(C_3H_7)_3SbI_2$	95
	$(i\text{-}C_3H_7)_3SbI_2$	95
$C_9H_{21}N_2O_2Sb$	$(CH_3)_3Sb(ON{=}C(CH_3)_2)_2$	158, **35**, *1*
$C_9H_{21}N_2S_4Sb$	$(CH_3)_3Sb(SC(S)N(CH_3)_2)_2$	207/8
$C_9H_{21}OSb$	$(C_3H_7)_3SbO$	116/7
$C_9H_{21}SSb$	$(C_3H_7)_3SbS$	200
$C_9H_{21}SbSe$	$(C_3H_7)_3SbSe$	209
$C_9H_{24}Cl_5SbSi_3$	$(Cl(CH_3)_2SiCH_2)_3SbCl_2$	24
$C_9H_{27}O_2SbSi_2$	$(CH_3)_3Sb((OSi(CH_3)_3))_2$	162
$C_{10}H_{21}O_4Sb$	$(C_2H_5)_3Sb(O_2CCH_3)_2$	165
$C_{10}H_{25}O_2Sb$	$(C_2H_5)_3Sb(OC_2H_5)_2$	132
$C_{11}H_{17}N_2O_4Sb$	$(CH_3)_3Sb(C_4H_4NO_2)_2$	190, **43**, *1*
$C_{11}H_{19}O_4Sb$	$(CH_3)_3Sb(O_2CCH{=}CHCH_3)_2$	176
$C_{11}H_{23}O_2S_4Sb$	$(CH_3)_3Sb(SC(S)OC_3H_7)_2$	206, **48**, *1*
	$(CH_3)_3Sb(SC(S)OCH(CH_3)_2)_2$	206, **48**, *2*
$C_{11}H_{23}O_4Sb$	$(CH_3)_3Sb(O_2CC_3H_7)_2$	164
$C_{11}H_{25}N_2O_2Sb$	$(CH_3)_3Sb(ON{=}C(CH_3)C_2H_5)_2$	158, **35**, *2*
$C_{11}H_{27}O_2Sb$	$(CH_3)_3Sb(OC_4H_9\text{-}t)_2$	132
$C_{11}H_{27}O_4Sb$	$(CH_3)_3Sb(O_2C_4H_9\text{-}t)_2$	155, **34**, *1*
$C_{11}H_{29}P_2S_4Sb$	$(CH_3)_3Sb(SP(S)(C_2H_5)_2)_2$	209
$C_{12}H_9Br_2S_3Sb$	$(C_4H_3S)_3SbBr_2$	91
$C_{12}H_9Cl_2O_3Sb$	$(C_4H_3O)_3SbCl_2$	58
$C_{12}H_9Cl_2S_3Sb$	$(C_4H_3S)_3SbCl_2$	58
$C_{12}H_9OS_3Sb$	$(2\text{-}C_4H_3S)_3SbO$	129
$C_{12}H_{20}NO_2SSb$	$(C_2H_5)_3Sb({=}NSO_2C_6H_5)$	196, **46**, *1*
$C_{12}H_{27}Br_2Sb$	$(C_4H_9)_3SbBr_2$	71
	$(i\text{-}C_4H_9)_3SbBr_2$	72
$C_{12}H_{27}Cl_2Sb$	$(C_4H_9)_3SbCl_2$	22
	$(i\text{-}C_4H_9)_3SbCl_2$	22
$C_{12}H_{27}F_2Sb$	$(C_4H_9)_3SbF_2$	3
$C_{12}H_{27}I_2Sb$	$(C_4H_9)_3SbI_2$	95
	$(i\text{-}C_4H_9)_3SbI_2$	96
$C_{12}H_{27}N_2O_2Sb$	$(C_2H_5)_3Sb(ON{=}C(CH_3)_2)_2$	159, **35**, *9*
$C_{12}H_{27}OSb$	$(C_4H_9)_3SbO$	117
	$(i\text{-}C_4H_9)_3SbO$	117
$C_{12}H_{27}SSb$	$(C_4H_9)_3SbS$	200/1
$C_{12}H_{27}SbSe$	$(C_4H_9)_3SbSe$	209
$C_{12}H_{29}O_2Sb$	$(C_4H_9)_3Sb(OH)_2$	117
$C_{12}H_{33}Br_2SbSi_3$	$((CH_3)_3SiCH_2)_3SbBr_2$	73
$C_{12}H_{33}Cl_2SbSi_3$	$((CH_3)_3SiCH_2)_3SbCl_2$	24

C$_{13}$H$_{22}$NO$_2$SSb	$(C_2H_5)_3Sb(=NSO_2C_6H_4CH_3\text{-}4)$	196, **46**, *2*
C$_{13}$H$_{25}$N$_2$S$_4$Sb	$(CH_3)_3Sb(SC(S)NC_4H_8)_2$	208
C$_{13}$H$_{27}$OS$_2$Sb	$(C_4H_9)_3SbCOS_2$	112, **26**, *3*
	$(i\text{-}C_4H_9)_3SbCOS_2$	112, **26**, *4*
C$_{13}$H$_{27}$O$_2$S$_4$Sb	$(CH_3)_3Sb(SC(S)OCH_2CH(CH_3)_2)_2$	206, **48**, *3*
C$_{13}$H$_{27}$O$_3$Sb	$(C_4H_9)_3SbCO_3$	112, **26**, *1*
	$(i\text{-}C_4H_9)_3SbCO_3$	112, **26**, *2*
C$_{13}$H$_{27}$O$_4$Sb	$(C_3H_7)_3Sb(O_2CCH_3)_2$	165
	$(CH_3)_3Sb(O_2CCH_2CH(CH_3)_2)_2$	164
C$_{13}$H$_{27}$S$_3$Sb	$(C_4H_9)_3SbCS_3$	112, **26**, *5*
C$_{13}$H$_{29}$N$_2$O$_2$Sb	$(CH_3)_3Sb(ON=C(C_2H_5)_2)_2$	159, **35**, *4*
	$(CH_3)_3Sb(ON=C(CH_3)C_3H_7)_2$	159, **35**, *3*
C$_{14}$H$_{19}$O$_4$Sb	$(C_2H_5)_3Sb(O_2CC_6H_4CO_2\text{-}2)$	186
C$_{14}$H$_{27}$N$_2$O$_2$Sb	$(C_4H_9)_3Sb(NCO)_2$	103, **23**, *2*
	$(i\text{-}C_4H_9)_3Sb(NCO)_2$	103, **23**, *3*
C$_{14}$H$_{27}$O$_4$Sb	$(C_4H_9)_3Sb(O_2CCO_2)$	186
C$_{14}$H$_{31}$N$_2$O$_2$Sb	$(C_2H_5)_3Sb(ON=C(CH_3)C_2H_5)_2$	159, **35**, *10*
C$_{14}$H$_{33}$O$_2$Sb	$(C_2H_5)_3Sb(OC_4H_9)_2$	132
C$_{14}$H$_{33}$O$_4$Sb	$(C_2H_5)_3Sb(O_2C_4H_9\text{-}t)_2$	155, **34**, *4*
	$(C_2H_5)_3Sb(OCH_2CH_2OC_2H_5)_2$	132
C$_{15}$H$_{17}$Br$_2$O$_2$Sb	$(CH_3)_3Sb(OC_6H_4Br\text{-}4)_2$	137, **31**, *5*
C$_{15}$H$_{17}$Cl$_2$O$_2$Sb	$(CH_3)_3Sb(OC_6H_4Cl\text{-}2)_2$	136, **31**, *2*
	$(CH_3)_3Sb(OC_6H_4Cl\text{-}3)_2$	136, **31**, *3*
	$(CH_3)_3Sb(OC_6H_4Cl\text{-}4)_2$	137, **31**, *4*
C$_{15}$H$_{17}$N$_2$O$_6$Sb	$(CH_3)_3Sb(OC_6H_4NO_2\text{-}2)_2$	137, **31**, *6*
	$(CH_3)_3Sb(OC_6H_4NO_2\text{-}3)_2$	137, **31**, *7*
	$(CH_3)_3Sb(OC_6H_4NO_2\text{-}4)_2$	137, **31**, *8*
(C$_{15}$H$_{19}$N$_2$O$_2$SSb)$_n$	$((CH_3)_3Sb(4\text{-}NHC_6H_4SO_2C_6H_4NH\text{-}4'))_n$	194, **44**, *2*
C$_{15}$H$_{19}$O$_2$Sb	$(CH_3)_3Sb(OC_6H_5)_2$	136, **31**, *1*
C$_{15}$H$_{19}$S$_2$Sb	$(CH_3)_3Sb(SC_6H_5)_2$	205, **47**, *4*
C$_{15}$H$_{21}$O$_4$P$_2$Sb	$(CH_3)_3Sb(OP(O)(H)C_6H_5)_2$	189, **42**, *2*
C$_{15}$H$_{26}$NO$_2$SSb	$(C_3H_7)_3Sb(=NSO_2C_6H_5)$	196, **46**, *3*
C$_{15}$H$_{27}$Br$_2$Sb	$(c\text{-}C_5H_9)_3SbBr_2$	72
C$_{15}$H$_{27}$O$_7$Sb	$(i\text{-}C_3H_7O_2CCH_2)_3SbO$	118
C$_{15}$H$_{29}$N$_2$O$_2$Sb	$(CH_3)_3Sb(ON=C(CH_2)_5)_2$	159, **35**, *5*
C$_{15}$H$_{29}$N$_2$S$_4$Sb	$(CH_3)_3Sb(SC(S)NC_5H_{10})_2$	208
C$_{15}$H$_{32}$NO$_2$Sb	$(C_4H_9)_3Sb(CO_2N(C_2H_5))$	112, **26**, *6*
C$_{15}$H$_{33}$Br$_2$Sb	$(C_5H_{11})_3SbBr_2$	72
C$_{15}$H$_{33}$Cl$_2$Sb	$(C_5H_{11})_3SbCl_2$	23
C$_{15}$H$_{33}$F$_2$Sb	$(C_5H_{11})_3SbF_2$	3
C$_{15}$H$_{33}$I$_2$Sb	$(C_5H_{11})_3SbI_2$	96
C$_{15}$H$_{33}$N$_2$O$_6$Sb	$(C_5H_{11})_3Sb(NO_3)_2$	109
C$_{15}$H$_{33}$OSb	$(C_5H_{11})_3SbO$	117/8
C$_{15}$H$_{33}$O$_4$SSb	$(C_5H_{11})_3SbSO_4$	110
C$_{15}$H$_{33}$SSb	$(C_5H_{11})_3SbS$	201
C$_{15}$H$_{33}$SbSe	$(C_5H_{11})_3SbSe$	209
C$_{16}$H$_{18}$NO$_2$Sb	$(CH_3)_3Sb(2\text{-}OC_6H_4CH=NC_6H_4O\text{-}2)$	143, **32**, *3*
C$_{16}$H$_{28}$NO$_2$SSb	$(C_3H_7)_3Sb(=NSO_2C_6H_4CH_3\text{-}4)$	196, **46**, *4*
C$_{16}$H$_{33}$O$_4$S$_2$Sb	$(C_4H_9)_3Sb(O_2CCH_2SH)_2$	172

$C_{16}H_{33}O_4Sb$	$(C_4H_9)_3Sb(O_2CCH_3)_2$	165
	$(i\text{-}C_4H_9)_3Sb(O_2CCH_3)_2$	165
$C_{16}H_{37}O_2Sb$	$(C_4H_9)_3Sb(OC_2H_5)_2$	132
C$_{17}H_{17}Cl_2O_4Sb$	$(CH_3)_3Sb(O_2CC_6H_4Cl\text{-}4)_2$	179, **40**, *2*
$C_{17}H_{17}N_2O_8Sb$	$(CH_3)_3Sb(O_2CC_6H_4NO_2\text{-}2)_2$	179, **40**, *6*
	$(CH_3)_3Sb(O_2CC_6H_4NO_2\text{-}3)_2$	179, **40**, *7*
	$(CH_3)_3Sb(O_2CC_6H_4NO_2\text{-}4)_2$	179, **40**, *8*
$C_{17}H_{19}N_4Sb$	$(CH_3)_3Sb(C_7H_5N_2)_2$	190, **43**, *2*
$C_{17}H_{19}O_2S_2Sb$	$(CH_3)_3Sb(SC(O)C_6H_5)_2$	206
$C_{17}H_{19}O_2S_4Sb$	$(CH_3)_3Sb(SC(S)OC_6H_5)_2$	206, **48**, *4*
$C_{17}H_{19}O_4Sb$	$(CH_3)_3Sb(O_2CC_6H_5)_2$	178, **40**, *1*
$C_{17}H_{21}N_2O_4Sb$	$(CH_3)_3Sb(O_2CC_6H_4NH_2\text{-}2)_2$	179, **40**, *3*
	$(CH_3)_3Sb(O_2CC_6H_4NH_2\text{-}3)_2$	179, **40**, *4*
	$(CH_3)_3Sb(O_2CC_6H_4NH_2\text{-}4)_2$	179, **40**, *5*
$C_{17}H_{23}N_4O_2Sb$	$(CH_3)_3Sb(ON{=}C(NH_2)C_6H_5)_2$	159, **35**, *8*
$C_{17}H_{23}O_2Sb$	$(CH_3)_3Sb(OC_6H_4CH_3\text{-}2)_2$	137, **31**, *11*
	$(CH_3)_3Sb(OC_6H_4CH_3\text{-}3)_2$	137, **31**, *12*
	$(CH_3)_3Sb(OC_6H_4CH_3\text{-}4)_2$	137, **31**, *13*
$C_{17}H_{23}O_4Sb$	$(CH_3)_3Sb(OC_6H_4OCH_3\text{-}2)_2$	137, **31**, *9*
	$(CH_3)_3Sb(OC_6H_4OCH_3\text{-}4)_2$	137, **31**, *10*
$C_{17}H_{23}S_2Sb$	$(CH_3)_3Sb(SCH_2C_6H_5)_2$	205, **47**, *3*
C$_{18}Br_2Cl_{15}Sb$	$(C_6Cl_5)_3SbBr_2$	87
$C_{18}Br_2F_{15}Sb$	$(C_6F_5)_3SbBr_2$	87
$C_{18}Cl_2F_{15}O_8Sb$	$(C_6F_5)_3Sb(ClO_4)_2$	107
$C_{18}Cl_2F_{15}Sb$	$(C_6F_5)_3SbCl_2$	51
$C_{18}Cl_{17}Sb$	$(C_6Cl_5)_3SbCl_2$	52
$C_{18}F_{15}N_2O_6Sb$	$(C_6F_5)_3Sb(NO_3)_2$	109
$C_{18}F_{17}Sb$	$(C_6F_5)_3SbF_2$	8
$C_{18}H_9Cl_8Sb$	$(2,4\text{-}Cl_2C_6H_3)_3SbCl_2$	52
$C_{18}H_9F_8Sb$	$(3,4\text{-}F_2C_6H_3)_3SbF_2$	8
$C_{18}H_{12}Br_2Cl_3Sb$	$(4\text{-}ClC_6H_4)_3SbBr_2$	87
$C_{18}H_{12}Br_3Cl_2Sb$	$(4\text{-}BrC_6H_4)_3SbCl_2$	52/3
$C_{18}H_{12}Br_3F_2Sb$	$(4\text{-}BrC_6H_4)_3SbF_2$	9
$C_{18}H_{12}Br_3I_2Sb$	$(4\text{-}BrC_6H_4)_3SbI_2$	100, **22**, *2*
$C_{18}H_{12}Br_5Sb$	$(4\text{-}BrC_6H_4)_3SbBr_2$	87
$C_{18}H_{12}Cl_2I_3Sb$	$(4\text{-}IC_6H_4)_3SbCl_2$	53
$C_{18}H_{12}Cl_2N_3O_6Sb$	$(NO_2C_6H_4)_3SbCl_2$	54
$C_{18}H_{12}Cl_3F_2Sb$	$(4\text{-}ClC_6H_4)_3SbF_2$	8
$C_{18}H_{12}Cl_3I_2Sb$	$(4\text{-}ClC_6H_4)_3SbI_2$	100, **22**, *1*
$C_{18}H_{12}Cl_5Sb$	$(2\text{-}ClC_6H_4)_3SbCl_2$	51
	$(4\text{-}ClC_6H_4)_3SbCl_2$	51/2
$C_{18}H_{12}F_5Sb$	$(4\text{-}FC_6H_4)_3SbF_2$	8
$C_{18}H_{12}N_9OSb$	$(3\text{-}N_3C_6H_4)_3SbO$	128
$C_{18}H_{14}N_3O_8Sb$	$(3\text{-}NO_2C_6H_4)_3Sb(OH)_2$	127
$C_{18}H_{15}Br_2Sb$	$(C_6H_5)_3SbBr_2$	78
$C_{18}H_{15}Cl_2Sb$	$(C_6H_5)_3SbCl_2$	31/50
$C_{18}H_{15}CrO_4Sb$	$(C_6H_5)_3SbCrO_4$	111
$C_{18}H_{15}F_2Sb$	$(C_6H_5)_3SbF_2$	6
$C_{18}H_{15}I_2Sb$	$(C_6H_5)_3SbI_2$	98

$C_{18}H_{15}N_2O_6Sb$	$(C_6H_5)_3Sb(NO_3)_2$	109
$C_{18}H_{15}N_3Sb$	$(C_6H_5)_3Sb(N_3)_2$	102
$C_{18}H_{15}OSb$	$(C_6H_5)_3SbO$	120/3
$C_{18}H_{15}O_2Sb$	$(C_6H_5)_3SbO_2$	130
$C_{18}H_{15}O_4SSb$	$(C_6H_5)_3SbSO_4$	111
$C_{18}H_{15}O_4S_2Sb$	$(C_6H_5)_3SbS_2O_4$	110
$C_{18}H_{15}O_4SbSe$	$(C_6H_5)_3SbSeO_4$	111
$C_{18}H_{15}SSb$	$(C_6H_5)_3SbS$	201/3
$C_{18}H_{15}SbSe$	$(C_6H_5)_3SbSe$	210
$C_{18}H_{16}NSb$	$(C_6H_5)_3Sb(=NH)$	195
$C_{18}H_{17}O_2Sb$	$(C_6H_5)_3Sb(OH)_2$	123/5
$C_{18}H_{17}O_4Sb$	$(C_6H_5)_3Sb(OOH)_2$	131
$C_{18}H_{25}O_2Sb$	$(C_2H_5)_3Sb(OC_6H_5)_2$	138, **31**, *18*
$C_{18}H_{33}Br_2Sb$	$(c\text{-}C_6H_{11})_3SbBr_2$	73
$C_{18}H_{33}Cl_2Sb$	$(c\text{-}C_6H_{11})_3SbCl_2$	23
$C_{18}H_{33}OSb$	$(c\text{-}C_6H_{11})_3SbO$	118
$C_{18}H_{33}SSb$	$(c\text{-}C_6H_{11})_3SbS$	201
$C_{18}H_{33}SbSe$	$(c\text{-}C_6H_{11})_3SbSe$	209
$\mathbf{C}_{19}H_{15}F_3NO_2SSb$	$(C_6H_5)_3Sb(=NSO_2CF_3)$	196
$C_{19}H_{23}N_4O_6Sb$	$(CH_3)_3Sb(ON=C(CH_3)C_6H_4NO_2\text{-}4)_2$	159, **35**, *7*
$C_{19}H_{23}N_4Sb$	$(CH_3)_3Sb(C_7H_4N_2\text{-}CH_3)_2$	190, **43**, *3*
$C_{19}H_{23}O_4Sb$	$(CH_3)_3Sb(O_2CC_6H_4CH_3\text{-}2)_2$	179, **40**, *9*
	$(CH_3)_3Sb(O_2CC_6H_4CH_3\text{-}3)_2$	179, **40**, *10*
	$(CH_3)_3Sb(O_2CC_6H_4CH_3\text{-}4)_2$	179, **40**, *11*
	$(CH_3)_3Sb(O_2CCH_2C_6H_5)_2$	172
$C_{19}H_{25}N_2O_2Sb$	$(CH_3)_3Sb(ON=C(CH_3)C_6H_5)_2$	159, **35**, *6*
$C_{19}H_{25}O_4P_2Sb$	$(CH_3)_3Sb(OP(O)(H)CH=CHC_6H_5)_2$	189, **42**, *1*
$C_{19}H_{27}F_{16}O_2Sb$	$(C_3H_7)_3Sb(OCH_2(CF_2)_3CHF_2)_2$	132
$C_{19}H_{27}O_2Sb$	$(CH_3)_3Sb(OC_6H_4C_2H_5\text{-}4)_2$	137, **31**, *14*
$C_{19}H_{31}ClNO_2Sb$	$(C_4H_9)_3Sb(CO_2NC_6H_4Cl\text{-}3)$	112, **26**, *9*
$C_{19}H_{32}NO_2Sb$	$(C_4H_9)_3Sb(CO_2NC_6H_5)$	112, **26**, *8*
	$(i\text{-}C_4H_9)_3Sb(CO_2NC_6H_5)$	112, **26**, *10*
$\mathbf{C}_{20}F_{15}H_6O_2Sb$	$(C_6F_5)_3Sb(OCH_3)_2$	135
$C_{20}H_{13}Cl_6O_2Sb$	$(4\text{-}ClC_6H_4)_3Sb(OCH(CCl_3)O)$	147, **32**, *29*
$C_{20}H_{15}N_2O_2Sb$	$(C_6H_5)_3Sb(NCO)_2$	103/4
$C_{20}H_{15}N_2S_2Sb$	$(C_6H_5)_3Sb(NCS)_2$	105
$C_{20}H_{15}O_4Sb$	$(C_6H_5)_3Sb(O_2CCO_2)$	186
$C_{20}H_{16}Cl_3O_2Sb$	$(C_6H_5)_3Sb(OCH(CCl_3)O)$	144, **32**, *6*
$C_{20}H_{16}F_3O_2Sb$	$(C_6H_5)_3Sb(OCH(CF_3)O)$	144, **32**, *5*
$C_{20}H_{17}BrNOSb$	$(C_6H_5)_3Sb(=NCOCH_2Br)$	195, **45**, *3*
$C_{20}H_{17}ClNOSb$	$(C_6H_5)_3Sb(=NCOCH_2Cl)$	195, **45**, *2*
$C_{20}H_{17}INOSb$	$(C_6H_5)_3Sb(=NCOCH_2I)$	195, **45**, *4*
$C_{20}H_{17}O_3Sb$	$(C_6H_5)_3Sb(OCH_2CO_2)$	145, **32**, *12*
$C_{20}H_{17}O_4Sb$	$(C_6H_5)_3Sb(O_2CH)_2$	165
$C_{20}H_{18}NOSb$	$(C_6H_5)_3Sb(=NCOCH_3)$	195, **45**, *1*
$C_{20}H_{21}O_2Sb$	$(C_6H_5)_3Sb(OCH_3)_2$	132
$C_{20}H_{23}Cl_2O_4Sb$	$(C_2H_5)_3Sb(O_2CC_6H_4Cl\text{-}4)_2$	180, **40**, *13*
$C_{20}H_{23}N_2O_8Sb$	$(C_2H_5)_3Sb(O_2CC_6H_4NO_2\text{-}4)_2$	180, **40**, *17*
$C_{20}H_{25}O_4Sb$	$(C_2H_5)_3Sb(O_2CC_6H_5)_2$	179, **40**, *12*

$C_{20}H_{27}N_2O_4Sb$	$(C_2H_5)_3Sb(O_2CC_6H_4NH_2\text{-}2)_2$	180, **40**, *14*
	$(C_2H_5)_3Sb(O_2CC_6H_4NH_2\text{-}3)_2$	180, **40**, *15*
	$(C_2H_5)_3Sb(O_2CC_6H_4NH_2\text{-}4)_2$	180, **40**, *16*
$C_{20}H_{35}O_3Sb$	$(c\text{-}C_6H_{11})_3Sb(OCH_2CO_2)$	144, **32**, *4*
$C_{21}H_{12}Cl_3F_6O_2Sb$	$(4\text{-}ClC_6H_4)_3Sb(OC(CF_3)_2O)$	147, **32**, *30*
$C_{21}H_{17}O_4Sb$	$(C_6H_5)_3Sb(O_2CCH_2CO_2)$	186
$C_{21}H_{18}Cl_5Sb$	$(5\text{-}Cl\text{-}2\text{-}CH_3C_6H_3)_3SbCl_2$	57
$C_{21}H_{18}F_5Sb$	$(3\text{-}F\text{-}4\text{-}CH_3C_6H_3)_3SbF_2$	10
$C_{21}H_{18}N_3O_7Sb$	$(3\text{-}NO_2\text{-}4\text{-}CH_3C_6H_3)_3SbO$	129
$C_{21}H_{18}N_5O_{12}Sb$	$(3\text{-}NO_2\text{-}4\text{-}CH_3C_6H_3)_3Sb(NO_3)_2$	110
$C_{21}H_{21}Br_2O_3Sb$	$(2\text{-}CH_3OC_6H_4)_3SbBr_2$	87
	$(3\text{-}CH_3OC_6H_4)_3SbBr_2$	88
	$(4\text{-}CH_3OC_6H_4)_3SbBr_2$	88
$C_{21}H_{21}Br_2Sb$	$(2\text{-}CH_3C_6H_4)_3SbBr_2$	88
	$(3\text{-}CH_3C_6H_4)_3SbBr_2$	89
	$(4\text{-}CH_3C_6H_4)_3SbBr_2$	89
	$(C_6H_5CH_2)_3SbBr_2$	73
$C_{21}H_{21}Cl_2O_3Sb$	$(2\text{-}CH_3OC_6H_4)_3SbCl_2$	53
	$(3\text{-}CH_3OC_6H_4)_3SbCl_2$	53
	$(4\text{-}CH_3OC_6H_4)_3SbCl_2$	53
$C_{21}H_{21}Cl_2Sb$	$(C_6H_5CH_2)_3SbCl_2$	24
	$(2\text{-}CH_3C_6H_4)_3SbCl_2$	54/5
	$(3\text{-}CH_3C_6H_4)_3SbCl_2$	55
	$(4\text{-}CH_3C_6H_4)_3SbCl_2$	55/7
$C_{21}H_{21}F_2Sb$	$(C_6H_5CH_2)_3SbF_2$	3
	$(2\text{-}CH_3C_6H_4)_3SbF_2$	9
	$(3\text{-}CH_3C_6H_4)_3SbF_2$	9
	$(4\text{-}CH_3C_6H_4)_3SbF_2$	9
$C_{21}H_{21}I_2O_3Sb$	$(2\text{-}CH_3OC_6H_4)_3SbI_2$	100, **22**, *3*
	$(3\text{-}CH_3OC_6H_4)_3SbI_2$	100, **22**, *4*
	$(4\text{-}CH_3OC_6H_4)_3SbI_2$	100, **22**, *7*
$C_{21}H_{21}I_2Sb$	$(C_6H_5CH_2)_3SbI_2$	96
	$(2\text{-}CH_3C_6H_4)_3SbI_2$	100, **22**, *10*
	$(3\text{-}CH_3C_6H_4)_3SbI_2$	100, **22**, *11*
	$(4\text{-}CH_3C_6H_4)_3SbI_2$	99/100
$C_{21}H_{21}N_2O_9Sb$	$(4\text{-}CH_3OC_6H_4)_3Sb(NO_3)_2$	109
$C_{21}H_{21}OSb$	$(C_6H_5CH_2)_3SbO$	118
	$(2\text{-}CH_3C_6H_4)_3SbO$	128
	$(3\text{-}CH_3C_6H_4)_3SbO$	128
	$(4\text{-}CH_3C_6H_4)_3SbO$	128
$C_{21}H_{21}O_4Sb$	$(4\text{-}CH_3OC_6H_4)_3SbO$	127
$C_{21}H_{21}SSb$	$(3\text{-}CH_3\text{-}C_6H_4)_3SbS$	203
	$(4\text{-}CH_3\text{-}C_6H_4)_3SbS$	203
$C_{21}H_{23}O_2Sb$	$(C_6H_5CH_2)_3Sb(OH)_2$	118
	$(2\text{-}CH_3C_6H_4)_3Sb(OH)_2$	128
	$(4\text{-}CH_3C_6H_4)_3Sb(OH)_2$	128
$C_{21}H_{23}O_4Sb$	$(CH_3)_3Sb(O_2CCH{=}CHC_6H_5)_2$	176
$C_{21}H_{27}N_4Sb$	$(CH_3)_3Sb(C_7H_4N_2\text{-}C_2H_5)_2$	190, **43**, *4*
$C_{21}H_{27}O_6Sb$	$(CH_3)_3Sb(O_2C_9H_9O)_2$	155, **34**, *3*
$C_{21}H_{31}O_4Sb$	$(CH_3)_3Sb(O_2C_6H_5(CH_3)_2C)_2$	155, **34**, *2*

$C_{21}H_{44}NO_2Sb$	$(C_4H_9)_3Sb(CO_2N(C_8H_{17}))$	112, **26**, *7*
$C_{21}H_{45}I_2Sb$	$(C_7H_{15})_3SbI_2$	96
$C_{21}H_{45}OSb$	$(C_7H_{15})_3SbO$	118
$C_{22}F_{27}N_2O_2Sb$	$(C_6F_5)_3Sb(ON(CF_3)_2)_2$	162
$C_{22}H_{15}Cl_6O_4Sb$	$(C_6H_5)_3Sb(O_2CCCl_3)_2$	173, **38**, *4*
$C_{22}H_{15}D_6O_4Sb$	$(C_6H_5)_3Sb(O_2CCD_3)_2$	167/8
$[C_{22}H_{15}D_{12}O_2S_2Sb]^{2+}$	$[(C_6H_5)_3Sb((CD_3)_2SO)_2]^{2+} \cdot 2[ClO_4]^-$	107, **24**, *6*
$C_{22}H_{15}F_6O_4Sb$	$(C_6H_5)_3Sb(O_2CCF_3)_2$	173, **38**, *1*
$C_{22}H_{15}N_2S_2Sb$	$(C_6H_5)_3Sb(S_2C{=}C(CN)_2)$	208
$C_{22}H_{17}Br_4O_4Sb$	$(C_6H_5)_3Sb(O_2CCHBr_2)_2$	174, **38**, *7*
$C_{22}H_{17}Cl_4O_4Sb$	$(C_6H_5)_3Sb(O_2CCHCl_2)_2$	173, **38**, *5*
$C_{22}H_{17}F_4O_4Sb$	$(C_6H_5)_3Sb(O_2CCHF_2)_2$	173, **38**, *2*
$C_{22}H_{17}O_4Sb$	$(C_6H_5)_3Sb(O_2CC_2H_2CO_2)$	187, **41**, *2*
	$(C_6H_5)_3Sb(O_2CC_2H_2CO_2)$	187, **41**, *3*
$(C_{22}H_{18}N_5OSb)_n$	$((C_6H_5)_3Sb(HN{-}C_4HN_2(NO){-}NH))_n$	194, **44**, *20*
$(C_{22}H_{18}N_5O_2Sb)_n$	$((C_6H_5)_3Sb(HN{-}C_4HN_2(NO_2){-}NH))_n$	194, **44**, *7*
$C_{22}H_{19}Br_2O_4Sb$	$(C_6H_5)_3Sb(O_2CCH_2Br)_2$	174, **38**, *8*
$C_{22}H_{19}Cl_2O_4Sb$	$(C_6H_5)_3Sb(O_2CCH_2Cl)_2$	173, **38**, *6*
$C_{22}H_{19}F_2O_4Sb$	$(C_6H_5)_3Sb(O_2CCH_2F)_2$	173, **38**, *3*
$(C_{22}H_{19}N_2S_4SbZn)_n$	$((C_6H_5)_3Sb(SC(S)N(CH_2)_2NC(S)SZn))_n$	194, **44**, *18*
$(C_{22}H_{19}O_4SSb)_n$	$((C_6H_5)_3Sb(O_2CCH(SH)CH_2CO_2))_n$	187, **41**, *1*
$C_{22}H_{19}O_4Sb$	$(C_6H_5)_3Sb(O_2C(CH_2)_2CO_2)$	186
$C_{22}H_{20}NO_4Sb$	$(C_6H_5)_3Sb(O_2CCH_2NHCH_2CO_2)$	187
$C_{22}H_{21}O_4Sb$	$(C_6H_5)_3Sb(O_2CCH_3)_2$	165
$C_{22}H_{23}N_2O_2Sb$	$(C_6H_5)_3Sb(ON{=}CHCH_3)_2$	160, **35**, *12*
$C_{22}H_{25}O_2Sb$	$(C_6H_5)_3Sb(OC_2H_5)_2$	134
$C_{22}H_{25}O_4Sb$	$(C_6H_5)_3Sb(OCH_2CH_2OH)_2$	134
$[C_{22}H_{27}O_2S_2Sb]^{2+}$	$[(C_6H_5)_3Sb((CH_3)_2SO)_2]^{2+} \cdot 2[ClO_4]^-$	107, **24**, *5*
$C_{22}H_{29}O_4Sb$	$(C_2H_5)_3Sb(O_2CC_6H_4CH_3{-}2)_2$	180, **40**, *18*
	$(C_2H_5)_3Sb(O_2CC_6H_4CH_3{-}4)_2$	180, **40**, *19*
$C_{22}H_{31}N_2O_2Sb$	$(C_2H_5)_3Sb(ON{=}C(CH_3)C_6H_5)_2$	160, **35**, *11*
$C_{23}H_{17}Cl_3F_3O_3Sb$	$(4{-}ClC_6H_4)_3Sb(OC(CF_3)CH_2C(CH_3)O_2)$	147, **32**, *31*
$(C_{23}H_{17}N_5Sb)_n$	$((C_6H_5)_3Sb(HN{-}C_5HN_4))_n$	194, **44**, *10*
$C_{23}H_{20}F_3O_3Sb$	$(C_6H_5)_3Sb(OC(CF_3)CH_2C(CH_3)OO)$	144, **32**, *8*
$(C_{23}H_{20}N_3Sb)_n$	$((C_6H_5)_3Sb(HN{-}C_5H_3N{-}NH))_n$	194, **44**, *22*
$(C_{23}H_{20}N_5OSb)_n$	$((C_6H_5)_3Sb(HN{-}C_4(CH_3)N_2(NO){-}NH))_n$	194, **44**, *21*
$C_{23}H_{21}N_2S_2Sb$	$(4{-}CH_3C_6H_4)_3Sb(NCS)_2$	105
$C_{23}H_{21}O_4Sb$	$(C_6H_5)_3Sb(O_2C(CH_2)_3CO_2)$	186
$C_{23}H_{22}Cl_3O_2Sb$	$(4{-}CH_3C_6H_4)_3Sb(OCH(CCl_3)O)$	147, **32**, *33*
$C_{23}H_{22}F_3O_2Sb$	$(4{-}CH_3C_6H_4)_3Sb(OCH(CF_3)O)$	147, **32**, *32*
$C_{23}H_{23}ClNOSb$	$(2{-}CH_3C_6H_4)_3Sb({=}NCOCH_2Cl)$	195, **45**, *8*
	$(3{-}CH_3C_6H_4)_3Sb({=}NCOCH_2Cl)$	195, **45**, *9*
	$(4{-}CH_3C_6H_4)_3Sb({=}NCOCH_2Cl)$	195, **45**, *10*
$C_{23}H_{23}O_4Sb$	$(4{-}CH_3C_6H_4)_3Sb(O_2CH)_2$	169
$C_{23}H_{24}NOSb$	$(2{-}CH_3C_6H_4)_3Sb({=}NCOCH_3)$	195, **45**, *5*
	$(3{-}CH_3C_6H_4)_3Sb({=}NCOCH_3)$	195, **45**, *6*
	$(4{-}CH_3C_6H_4)_3Sb({=}NCOCH_3)$	195, **45**, *7*
$C_{23}H_{31}O_4Sb$	$(C_3H_7)_3Sb(O_2CC_6H_5)_2$	180, **40**, *20*
$C_{23}H_{35}O_2Sb$	$(CH_3)_3Sb(OC_6H_4C_4H_9{-}t{-}4)_2$	137, **31**, *15*

$C_{24}H_{12}Br_2Mn_3O_9Sb$	$((CO)_3MnC_5H_4)_3SbBr_2$	77
$C_{24}H_{15}Cl_4O_2Sb$	$(C_6H_5)_3Sb(2\text{-}OC_6Cl_4O)$	145, **32**, *15*
$C_{24}H_{15}F_{12}O_4Sb$	$(C_6H_5)_3Sb(OC(CF_3)_2O)_2$	144, **32**, *7*
$C_{24}H_{19}BrNO_2SSb$	$(C_6H_5)_3Sb(=NSO_2C_6H_4\text{-}Br\text{-}4)$	197
$C_{24}H_{19}ClNO_2SSb$	$(C_6H_5)_3Sb(=NSO_2C_6H_4\text{-}Cl\text{-}4)$	197
$(C_{24}H_{19}Cl_2N_2Sb)_n$	$((C_6H_5)_3Sb(HN\text{-}C_6H_2Cl_2\text{-}NH))_n$	194, **44**, *14*
$C_{24}H_{19}N_2O_2S_4Sb$	$(C_6H_5)_3Sb(NC_3H_2OS_2)_2$	191, **43**, *7*
$C_{24}H_{19}N_2O_4SSb$	$(C_6H_5)_3Sb(=NSO_2C_6H_4\text{-}NO_2\text{-}4)$	197/8
$C_{24}H_{19}O_2Sb$	$((C_6H_5)_3Sb(2\text{-}OC_6H_4O)\cdot 0.5H_2O$	145, **32**, *14*
	$(C_6H_5)_3Sb(2\text{-}OC_6H_4O)$	145, **32**, *13*
$C_{24}H_{20}NO_2SSb$	$(C_6H_5)_3Sb(=NSO_2C_6H_5)$	197
$(C_{24}H_{20}N_3O_2Sb)_n$	$((C_6H_5)_3Sb(HN\text{-}C_6H_3(NO_2)\text{-}NH))_n$	194, **44**, *15*
$(C_{24}H_{20}N_5OSb)_n$	$((C_6H_5)_3Sb(HN\text{-}C_6H_2N_3(OH)NH))_n$	194, **44**, *5*
$C_{24}H_{21}Br_3N_3O_7Sb$	$(2,4\text{-}(CH_3)_2\text{-}5\text{-}NO_2\text{-}6\text{-}BrC_6H)_3SbO$	129
$C_{24}H_{21}Br_5N_3O_6Sb$	$(2,4\text{-}(CH_3)_2\text{-}5\text{-}NO_2\text{-}6\text{-}BrC_6H)_3SbBr_2$	89
$C_{24}H_{21}Cl_2Sb$	$(4\text{-}CH_2{=}CHC_6H_4)_3SbCl_2$	57
$C_{24}H_{21}F_6O_2Sb$	$(4\text{-}CH_3C_6H_4)_3Sb(OC(CF_3)_2O)$	147, **32**, *34*
$(C_{24}H_{21}N_2Sb)_n$	$((C_6H_5)_3Sb(HN\text{-}C_6H_4\text{-}NH))_n$	194, **44**, *11*
$C_{24}H_{21}O_4Sb$	$(C_6H_5)_3Sb(O_2CCH{=}CH_2)_2$	176
$C_{24}H_{23}O_4Sb$	$(C_6H_5)_3Sb(O_2C(CH_2)_4CO_2)$	186
$C_{24}H_{24}Cl_2N_3O_3Sb$	$(4\text{-}CH_3CONHC_6H_4)_3SbCl_2$	54
$C_{24}H_{24}N_3O_4Sb$	$((4\text{-}CH_3CONHC_6H_4)_3SbO\cdot 4H_2O$	127/8
$C_{24}H_{24}N_3O_7Sb$	$(2,4\text{-}(CH_3)_2\text{-}5\text{-}NO_2C_6H_2)_3SbO$	129
$C_{24}H_{24}N_5O_{12}Sb$	$(2,4\text{-}(CH_3)_2\text{-}5\text{-}NO_2C_6H_2)_3Sb(NO_3)_2$	110
$C_{24}H_{25}O_4S_2Sb$	$(C_6H_5)_3Sb(O_2CCH_2SCH_3)_2$	175, **39**, *2*
$C_{24}H_{25}O_4Sb$	$(C_6H_5)_3Sb(O_2CC_2H_5)_2$	168
$C_{24}H_{27}Br_2O_3Sb$	$(2\text{-}C_2H_5OC_6H_4)_3SbBr_2$	88
$C_{24}H_{27}Br_2O_3Sb$	$(4\text{-}C_2H_5OC_6H_4)_3SbBr_2$	88
$C_{24}H_{27}Br_2Sb$	$(2,4\text{-}(CH_3)_2C_6H_3)_3SbBr_2$	89
$C_{24}H_{27}Cl_2O_3Sb$	$(2\text{-}C_2H_5OC_6H_4)_3SbCl_2$	53
	$(4\text{-}C_2H_5OC_6H_4)_3SbCl_2$	54
$C_{24}H_{27}Cl_2Sb$	$(2,4\text{-}(CH_3)_2C_6H_3)_3SbCl_2$	57
	$(2,5\text{-}(CH_3)_2C_6H_3)_3SbCl_2$	57
	$(3,5\text{-}(CH_3)_2C_6H_3)_3SbCl_2$	57
$C_{24}H_{27}I_2O_3Sb$	$(2\text{-}C_2H_5OC_6H_4)_3SbI_2$	100, **22**, *5*
	$(4\text{-}C_2H_5OC_6H_4)_3SbI_2$	100, **22**, *8*
$C_{24}H_{27}N_2O_2Sb$	$(C_6H_5)_3Sb(ON{=}C(CH_3)_2)_2$	160, **35**, *13*
$C_{24}H_{27}N_2O_9Sb$	$(4\text{-}C_2H_5OC_6H_4)_3Sb(NO_3)_2$	109/10
$C_{24}H_{27}O_2Sb$	$(C_6H_5)_3Sb(OC(CH_3)_2C(CH_3)_2O)$	145, **32**, *10*
$(C_{24}H_{29}N_2Sb)_n$	$((C_6H_5)_3Sb(HN\text{-}(CH_2)_6\text{-}NH))_n$	194, **44**, *3*
$C_{24}H_{30}Br_2N_3Sb$	$(4\text{-}(CH_3)_2NC_6H_4)_3SbBr_2$	88
$C_{24}H_{30}Cl_2N_3Sb$	$(4\text{-}(CH_3)_2NC_6H_4)_3SbCl_2$	54
$C_{24}H_{30}F_2N_3Sb$	$(4\text{-}(CH_3)_2NC_6H_4)_3SbF_2$	9
$C_{24}H_{30}I_2N_3Sb$	$(4\text{-}(CH_3)_2NC_6H_4)_3SbI_2$	100, **22**, *9*
$C_{24}H_{33}O_2SbSi_2$	$(C_6H_5)_3Sb((OSi(CH_3)_3))_2$	162
$C_{24}H_{33}SbSi_2$	$(C_6H_5)_3Sb(Si(CH_3)_3)_2$	210
$C_{24}H_{39}O_4Sb$	$(C_4H_9)_3Sb(O_2CC_6(CH_3)_4CO_2\text{-}4)$	186
$C_{24}H_{51}Cl_2Sb$	$(C_8H_{17})_3SbCl_2$	23
$C_{24}H_{51}OSb$	$(C_8H_{17})_3SbO$	118

$(C_{25}H_{17}O_6Sb)_n$	$((C_6H_5)_3Sb(O_2CC_5H_2O_2CO_2))_n$	188, **41**, *9*
$C_{25}H_{21}Cl_6O_4Sb$	$(4\text{-}CH_3C_6H_4)_3Sb(O_2CCCl_3)_2$	175
$C_{25}H_{21}F_6O_4Sb$	$(4\text{-}CH_3C_6H_4)_3Sb(O_2CCF_3)_2$	175
$C_{25}H_{22}NO_2SSb$	$(C_6H_5)_3Sb(=NSO_2C_6H_4\text{-}CH_3\text{-}4)$	198
$(C_{25}H_{23}N_2OSb)_n$	$((C_6H_5)_3Sb(HN\text{-}C_6H_3(OCH_3)\text{-}NH))_n$	194, **44**, *12*
$C_{25}H_{25}Br_2O_4Sb$	$(4\text{-}CH_3C_6H_4)_3Sb(O_2CCH_2Br)_2$	175
$C_{25}H_{25}Cl_2O_4Sb$	$(4\text{-}CH_3C_6H_4)_3Sb(O_2CCH_2Cl)_2$	175
$C_{25}H_{27}O_4Sb$	$(2\text{-}CH_3C_6H_4)_3Sb(O_2CCH_3)_2$	169
	$(4\text{-}CH_3C_6H_4)_3Sb(O_2CCH_3)_2$	169
$(C_{26}H_{17}Cl_2O_4Sb)_n$	$((C_6H_5)_3Sb(O_2CC_6H_2Cl_2CO_2))_n$	187, **41**, *6*
$(C_{26}H_{18}BrO_4Sb)_n$	$((C_6H_5)_3Sb(O_2CC_6H_3BrCO_2))_n$	187, **41**, *4*
$(C_{26}H_{18}NO_6Sb)_n$	$((C_6H_5)_3Sb(O_2CC_6H_3NO_2CO_2))_n$	187, **41**, *5*
$C_{26}H_{19}O_4Sb$	$(C_6H_5)_3Sb(O_2CC_6H_4CO_2\text{-}2)$	186
	$(C_6H_5)_3Sb(O_2CC_6H_4CO_2\text{-}3)$	186
	$(C_6H_5)_3Sb(O_2CC_6H_4CO_2\text{-}4)$	186
$C_{26}H_{23}N_2O_4Sb$	$(C_6H_5)_3Sb(C_4H_4NO_2)_2$	191, **43**, *6*
$C_{26}H_{24}NO_2SSb$	$(C_6H_5)_3Sb(=NSO_2C_6H_4\text{-}C_2H_5\text{-}4)$	198
$C_{26}H_{25}O_4Sb$	$(C_6H_5)_3Sb(O_2CC(CH_3)=CH_2)_2$	177
	$(C_6H_5)_3Sb(O_2CCH=CHCH_3)_2$	177
$C_{26}H_{29}O_4S_2Sb$	$(C_6H_5)_3Sb(O_2CCH_2SC_2H_5)_2$	175, **39**, *3*
$C_{26}H_{29}O_4Sb$	$(C_6H_5)_3Sb(O_2CC_3H_7\text{-}i)_2$	168
$C_{26}H_{31}N_2O_2Sb$	$(C_6H_5)_3Sb(ON=C(CH_3)C_2H_5)_2$	160, **35**, *14*
$C_{26}H_{33}O_2Sb$	$(C_6H_5)_3Sb(OC_4H_9)_2$	134
$C_{26}H_{33}O_4Sb$	$(C_6H_5)_3Sb(O_2C_4H_9\text{-}t)_2$	155, **34**, *9*
	$(C_6H_5)_3Sb(OCH_2CH_2OC_2H_5)_2$	134
$C_{26}H_{35}O_4SbTl_2$	$(C_6H_5)_3Sb(O_2Tl(C_2H_5)_2)_2$	156, **34**, *16*
$C_{26}H_{37}O_4Sb$	$(C_4H_9)_3Sb(O_2CC_6H_5)_2$	180, **40**, *21*
$C_{26}H_{51}N_2O_2Sb$	$(C_8H_{17})_3Sb(NCO)_2$	103, **23**, *4*
$C_{26}H_{51}O_4Sb$	$(C_6H_{11})_3Sb(O_2C_4H_9\text{-}t)_2$	155, **34**, *5*
$C_{27}H_{24}NO_2Sb$	$(C_6H_5)_3Sb(2\text{-}OC_6H_4CH=NCH_2CH_2O)$	145, **32**, *16*
$C_{27}H_{27}Br_2Sb$	$(4\text{-}CH_2=C(CH_3)C_6H_4)_3SbBr_2$	91
	$(4\text{-}CH_3CH=CHC_6H_4)_3SbBr_2$	91
$C_{27}H_{27}Br_8Sb$	$(4\text{-}CH_2BrCHBrCH_2C_6H_4)_3SbBr_2$	89
$C_{27}H_{27}Cl_2O_6Sb$	$(4\text{-}C_2H_5O_2CC_6H_4)_3SbCl_2$	57
$C_{27}H_{27}O_2Sb$	$(CH_3)_3Sb(OC_6H_4C_6H_5\text{-}2)_2$	137, **31**, *16*
	$(CH_3)_3Sb(OC_6H_4C_6H_5\text{-}4)_2$	138, **31**, *17*
$C_{27}H_{33}F_2Sb$	$(2,4,6\text{-}(CH_3)_3C_6H_2)_3SbF_2$	10
$C_{27}H_{33}N_2O_2Sb$	$(4\text{-}CH_3C_6H_4)_3Sb(ON=C(CH_3)_2)_2$	161, **35**, *25*
$C_{27}H_{35}O_8Sb$	$(4\text{-}HOCH_2CH(OH)CH_2C_6H_4)_3Sb(OH)_2$	128
$C_{28}H_{21}Br_4O_8Sb$	$(C_6H_5)_3Sb(O_2CCBr=CBrCO_2CH_3)_2$	177
$C_{28}H_{21}Cl_4O_8Sb$	$(C_6H_5)_3Sb(O_2CCCl=CClCO_2CH_3)_2$	177
$C_{28}H_{23}N_2O_4Sb$	$(C_6H_5)_3Sb(ON=CHC_4H_3O)_2$	160, **35**, *19*
$(C_{28}H_{23}O_4Sb)_n$	$((C_6H_5)_3Sb(O_2CC_6H_2(CH_3)_2CO_2))_n$	187, **41**, *7*
$C_{28}H_{25}O_8Sb$	$(C_6H_5)_3Sb(O_2CCH=CHCO_2CH_3)_2$	177
$C_{28}H_{26}NO_2Sb$	$(C_6H_5)_3Sb(2\text{-}OC_6H_4C(CH_3)=NCH_2CH_2O)$	145, **32**, *17*
	$(C_6H_5)_3Sb(2\text{-}OC_6H_4CH=NCH_2CH(CH_3)O)$	145, **32**, *18*
	$(C_6H_5)_3Sb(2\text{-}OC_6H_4CH=NCH_2CH_2CH_2O)$	146, **32**, *25*
$C_{28}H_{28}NO_2Sb$	$(C_6H_5)_3Sb(2\text{-}OC_6H_4(CH_3)=NCH_2CH_2CH_2O)$	146, **32**, *26*
$(C_{28}H_{29}N_2Sb)_n$	$((C_6H_5)_3Sb(HN\text{-}C_6(CH_3)_4\text{-}NH))_n$	194, **44**, *13*

$C_{28}H_{29}O_6Sb$	$(C_6H_5)_3Sb(O_2CCH_2CH_2COCH_3)_2$	174
$C_{28}H_{30}N_3O_{10}Sb$	$(2,4\text{-}(CH_3)_2\text{-}5\text{-}NO_2C_6H_2)_3Sb(O_2CCH_3)_2$	169
$C_{28}H_{33}O_4S_2Sb$	$(C_6H_5)_3Sb(O_2CCH_2SC_3H_7\text{-}i)_2$	175, **39**, *5*
$C_{28}H_{33}O_4S_2Sb$	$(C_6H_5)_3Sb(O_2CCH_2SC_3H_7)_2$	175, **39**, *4*
$C_{28}H_{35}N_2O_2Sb$	$(C_6H_5)_3Sb(ON{=}C(C_2H_5)_2)_2$	160, **35**, *16*
	$(C_6H_5)_3Sb(ON{=}C(CH_3)C_3H_7)_2$	160, **35**, *15*
C$_{29}H_{26}NO_2Sb$	$(C_6H_5)_3Sb(OC(CH_3){=}CHC(CH_3){=}NC_6H_4O\text{-}2)$	145, **32**, *21*
$C_{29}H_{28}NO_2Sb$	$(C_6H_5)_3Sb(2\text{-}OC_6H_4C(CH_3){=}NCH_2CH(CH_3)O)$	145, **32**, *19*
$C_{29}H_{29}N_2O_4Sb$	$(4\text{-}CH_3\text{-}C_6H_4)_3Sb(C_4H_4NO_2)_2$	192, **43**, *18*
$C_{29}H_{31}O_4Sb$	$(4\text{-}CH_3C_6H_4)_3Sb(O_2CCH{=}CHCH_3)_2$	177
$C_{29}H_{35}O_4Sb$	$(4\text{-}CH_3C_6H_4)_3Sb(O_2CC_3H_7\text{-}i)_2$	169
$C_{29}H_{37}N_2O_2Sb$	$(4\text{-}CH_3C_6H_4)_3Sb(ON{=}C(CH_3)C_2H_5)_2$	161, **35**, *26*
$C_{29}H_{39}O_4Sb$	$(C_6H_5CH_2)_3Sb(O_2C_4H_9\text{-}t)_2$	155, **34**, *6*
C$_{30}H_{15}Cl_{10}O_2Sb$	$(C_6H_5)_3Sb(OC_6Cl_5)_2$	138, **31**, *23*
$C_{30}H_{20}Br_2Cl_3O_2Sb$	$(4\text{-}ClC_6H_4)_3Sb(OC_6H_4Br\text{-}4)_2$	139, **31**, *36*
$C_{30}H_{20}Cl_3N_2O_6Sb$	$(4\text{-}ClC_6H_4)_3Sb(OC_6H_4NO_2\text{-}4)_2$	140, **31**, *37*
$C_{30}H_{20}Cl_5O_2Sb$	$(4\text{-}ClC_6H_4)_3Sb(OC_6H_4Cl\text{-}4)_2$	140, **31**, *35*
$C_{30}H_{21}Br_2Sb$	$(1\text{-}C_{10}H_7)_3SbBr_2$	91
$C_{30}H_{21}Br_4O_2Sb$	$(C_6H_5)_3Sb(OC_6H_3Br_2\text{-}2,4)_2$	138, **31**, *25*
$C_{30}H_{21}Cl_2Sb$	$(1\text{-}C_{10}H_7)_3SbCl_2$	58
	$(2\text{-}C_{10}H_7)_3SbCl_2$	58
$C_{30}H_{21}F_2Sb$	$(1\text{-}C_{10}H_7)_3SbF_2$	10
$C_{30}H_{21}I_2Sb$	$(1\text{-}C_{10}H_7)_3SbI_2$	100, **22**, *13*
$C_{30}H_{21}OSb$	$(1\text{-}C_{10}H_7)_3SbO$	129
$C_{30}H_{22}Cl_3O_2Sb$	$(4\text{-}ClC_6H_4)_3Sb(OC_6H_5)_2$	139, **31**, *34*
$C_{30}H_{23}Br_2O_2Sb$	$(C_6H_5)_3Sb(OC_6H_4Br\text{-}4)_2$	138, **31**, *24*
$C_{30}H_{23}Cl_2O_2Sb$	$(C_6H_5)_3Sb(OC_6H_4Cl\text{-}2)_2$	138, **31**, *20*
	$(C_6H_5)_3Sb(OC_6H_4Cl\text{-}3)_2$	138, **31**, *21*
	$(C_6H_5)_3Sb(OC_6H_4Cl\text{-}4)_2$	138, **31**, *22*
$(C_{30}H_{23}FeO_4Sb)_n$	$((C_6H_5)_3Sb(O_2CC_{10}H_8FeCO_2))_n$	188, **41**, *10*
$C_{30}H_{23}N_2O_6Sb$	$(C_6H_5)_3Sb(OC_6H_4NO_2\text{-}4)_2$	138, **31**, *26*
$C_{30}H_{23}N_6Sb$	$(C_6H_5)_3Sb(C_6H_4N_3)_2$	191, **43**, *16*
$C_{30}H_{25}NOPSb$	$(C_6H_5)_3Sb({=}NP(O)(C_6H_5)_2)$	198
$(C_{30}H_{25}N_2O_2SSb)_n$	$((C_6H_5)_3Sb(4\text{-}NHC_6H_4SO_2C_6H_4NH\text{-}4'))_n$	194, **44**, *9*
$(C_{30}H_{25}N_2Sb)_n$	$((C_6H_5)_3Sb(HN\text{-}C_6H_4C_6H_4\text{-}NH))_n$	194, **44**, *19*
$C_{30}H_{25}O_2Sb$	$(C_6H_5)_3Sb(OC_6H_5)_2$	138, **31**, *19*
$C_{30}H_{27}O_4P_2Sb$	$(C_6H_5)_3Sb(OP(O)(H)C_6H_5)_2$	189, **42**, *4*
$(C_{30}H_{27}O_4Sb)_n$	$((C_6H_5)_3Sb(O_2CC_6(CH_3)_4CO_2))_n$	188, **41**, *8*
$C_{30}H_{33}Br_2Sb$	$(4\text{-}C_2H_5CH{=}CHC_6H_4)_3SbBr_2$	91
$C_{30}H_{33}O_4Sb$	$(C_6H_5)_3Sb(O_2C_6H_9)_2$	155, **34**, *11*
$C_{30}H_{35}N_2O_2Sb$	$(C_6H_5)_3Sb(ON{=}C(CH_2)_5)_2$	160, **35**, *18*
$C_{30}H_{37}O_4S_2Sb$	$(C_6H_5)_3Sb(O_2CCH_2SC_4H_9)_2$	175, **39**, *6*
$C_{30}H_{39}N_2O_2Sb$	$(C_6H_5)_3Sb(ON{=}C(C_2H_5)C_3H_7)_2$	160, **35**, *17*
$(C_{30}H_{41}N_2Sb)_n$	$((C_6H_5)_3Sb(HN\text{-}(CH_2)_{12}\text{-}NH))_n$	194, **44**, *4*
$C_{30}H_{45}Cl_2O_3Sb$	$(C_{10}H_{15}O)_3SbCl_2$	25
C$_{31}H_{24}NO_2Sb$	$(C_6H_5)_3Sb(2\text{-}OC_6H_4CH{=}NC_6H_4O\text{-}2)$	146, **32**, *23*
$C_{31}H_{26}NO_2Sb$	$(C_6H_5)_3Sb(2\text{-}OC_{10}H_6CH{=}NCH_2CH_2O)$	145, **32**, *20*
$(C_{31}H_{26}N_3OSb)_n$	$((C_6H_5)_3Sb(HN\text{-}C_6H_4\text{-}CONH\text{-}C_6H_4\text{-}NH))_n$	194, **44**, *17*
$C_{31}H_{27}N_2Sb)_n$	$((C_6H_5)_3Sb(HN\text{-}C_6H_4\text{-}CH_2\text{-}C_6H_4\text{-}NH))_n$	194, **44**, *16*

$C_{31}H_{28}F_3O_3Sb$	$(4\text{-}CH_3C_6H_4)_3Sb(OC(CF_3)CH_2C(C_6H_5)O_2) \cdot 0.5\,C_2H_4Cl_2$	147, **32**, *35*
$C_{31}H_{29}N_2O_4Sb$	$(4\text{-}CH_3C_6H_4)_3Sb(ON{=}CHC_4H_3O)_2$	161, **35**, *28*
$(C_{31}H_{29}N_4O_2Sb)_n$	$((C_6H_5)_3Sb(HN\text{-}C_4HN_2(CH_2\text{-}C_6H_3(OCH_3)_2)\text{-}NH))_n$	194, **44**, *8*
$C_{32}H_{20}Cl_3N_2O_8Sb$	$(4\text{-}ClC_6H_4)_3Sb(O_2CC_6H_4NO_2\text{-}4)_2$	182, **40**, *41*
$C_{32}H_{20}Cl_5O_4Sb$	$(4\text{-}ClC_6H_4)_3Sb(O_2CC_6H_4Cl\text{-}4)_2$	182, **40**, *39*
$C_{32}H_{21}F_{24}O_2Sb$	$(C_6H_5)_3Sb(OCH_2(CF_2)_5CHF_2)_2$	134/135
$C_{32}H_{21}N_4O_8Sb$	$(C_6H_5)_3Sb(C_7H_3NO_2\text{-}NO_2)_2$	191, **43**, *10*
$C_{32}H_{22}Cl_3O_4Sb$	$(4\text{-}ClC_6H_4)_3Sb(O_2CC_6H_5)_2$	182, **40**, *38*
$C_{32}H_{23}Cl_2O_4Sb$	$(C_6H_5)_3Sb(O_2CC_6H_4Cl\text{-}2)_2$	181, **40**, *23*
	$(C_6H_5)_3Sb(O_2CC_6H_4Cl\text{-}3)_2$	181, **40**, *24*
	$(C_6H_5)_3Sb(O_2CC_6H_4Cl\text{-}4)_2$	181, **40**, *25*
$C_{32}H_{23}N_2O_2S_2Sb$	$(C_6H_5)_3Sb(C_7H_4NOS)_2$	191, **43**, *11*
$(C_{32}H_{23}N_2O_2Sb)_n$	$((C_6H_5)_3Sb(HN\text{-}C_{14}H_6O_2\text{-}NH))_n$	194, **44**, *6*
$C_{32}H_{23}N_2O_4Sb$	$(C_6H_5)_3Sb(C_7H_4NO_2)_2$	191, **43**, *9*
$C_{32}H_{23}N_2O_8Sb$	$(C_6H_5)_3Sb(O_2CC_6H_4NO_2\text{-}2)_2$	181, **40**, *32*
	$(C_6H_5)_3Sb(O_2CC_6H_4NO_2\text{-}3)_2$	182, **40**, *33*
	$(C_6H_5)_3Sb(O_2CC_6H_4NO_2\text{-}4)_2$	182, **40**, *34*
$C_{32}H_{24}Cl_3N_2O_4Sb$	$(4\text{-}ClC_6H_4)_3Sb(O_2CC_6H_4NH_2\text{-}4)_2$	182, **40**, *40*
$C_{32}H_{25}N_4Sb$	$(C_6H_5)_3Sb(C_7H_5N_2)_2$	191, **43**, *12*
$C_{32}H_{25}O_4Sb$	$(C_6H_5)_3Sb(O_2CC_6H_5)_2$	180, **40**, *22*
	$(C_6H_5)_3Sb(OC_6H_4CHO\text{-}2)_2$	139, **31**, *33*
$C_{32}H_{25}O_6Sb$	$(C_6H_5)_3Sb(O_2CC_6H_4OH\text{-}2)_2$	181, **40**, *26*
	$(C_6H_5)_3Sb(O_2CC_6H_4OH\text{-}4)_2$	181, **40**, *27*
$C_{32}H_{26}Cl_3O_2Sb$	$(4\text{-}ClC_6H_4)_3Sb(OC_6H_4CH_3\text{-}4)_2$	140, **31**, *39*
$C_{32}H_{26}Cl_3O_4Sb$	$(4\text{-}ClC_6H_4)_3Sb(OC_6H_4OCH_3\text{-}4)_2$	140, **31**, *38*
$C_{32}H_{26}NO_2Sb$	$(C_6H_5)_3Sb(2\text{-}OC_6H_4C(CH_3){=}NC_6H_4O\text{-}2)$	146, **32**, *24*
$C_{32}H_{27}N_2O_4Sb$	$(C_6H_5)_3Sb(O_2CC_6H_4NH_2\text{-}2)_2$	181, **40**, *29*
	$(C_6H_5)_3Sb(O_2CC_6H_4NH_2\text{-}3)_2$	181, **40**, *30*
	$(C_6H_5)_3Sb(O_2CC_6H_4NH_2\text{-}4)_2$	181, **40**, *31*
$C_{32}H_{27}O_2Sb$	$(C_6H_5)_3Sb(OCH(C_6H_5)CH(C_6H_5)O)$	145, **32**, *11*
$C_{32}H_{29}N_4O_2Sb$	$(C_6H_5)_3Sb(ON{=}C(NH_2)C_6H_5)_2$	161, **35**, *24*
$C_{32}H_{29}O_2Sb$	$(C_6H_5)_3Sb(OC_6H_4CH_3\text{-}2)_2$	139, **31**, *28*
	$(C_6H_5)_3Sb(OC_6H_4CH_3\text{-}3)_2$	139, **31**, *29*
	$(C_6H_5)_3Sb(OC_6H_4CH_3\text{-}4)_2$	139, **31**, *30*
$C_{32}H_{29}O_4Sb$	$(C_6H_5)_3Sb(OC_6H_4OCH_3\text{-}4)_2$	139, **31**, *27*
$C_{33}H_{29}Cl_2O_2Sb$	$(2\text{-}CH_3C_6H_4)_3Sb(OC_6H_4Cl\text{-}4)_2$	140, **31**, *41*
	$(4\text{-}CH_3C_6H_4)_3Sb(OC_6H_4Cl\text{-}4)_2$	140, **31**, *46*
$C_{33}H_{29}N_2O_6Sb$	$(2\text{-}CH_3C_6H_4)_3Sb(OC_6H_4NO_2\text{-}4)_2$	140, **31**, *42*
	$(4\text{-}CH_3C_6H_4)_3Sb(OC_6H_4NO_2\text{-}4)_2$	141, **31**, *47*
$C_{33}H_{31}O_2Sb$	$(2\text{-}CH_3C_6H_4)_3Sb(OC_6H_5)_2$	140, **31**, *40*
	$(4\text{-}CH_3C_6H_4)_3Sb(OC_6H_5)_2$	140, **31**, *45*
$C_{33}H_{41}N_2O_2Sb$	$(4\text{-}CH_3C_6H_4)_3Sb(ON{=}C(CH_2)_5)_2$	161, **35**, *27*
$C_{34}H_{23}N_2O_4Sb$	$(C_6H_5)_3Sb(C_8H_4NO_2)_2$	191, **43**, *8*
$C_{34}H_{26}Cl_3O_4Sb$	$(4\text{-}ClC_6H_4)_3Sb(O_2CC_6H_4CH_3\text{-}4)_2$	182, **40**, *42*
$C_{34}H_{27}Cl_2O_6Sb$	$(C_6H_5)_3Sb(O_2CCH_2OC_6H_4Cl\text{-}4)_2$	174
$C_{34}H_{28}NO_2Sb$	$(C_6H_5)_3Sb(OC(C_6H_5){=}CHC(CH_3){=}NC_6H_4O\text{-}2)$	146, **32**, *22*
$C_{34}H_{29}N_4O_6Sb$	$(C_6H_5)_3Sb(ON{=}C(CH_3)C_6H_4NO_2\text{-}4)_2$	160, **35**, *22*
$C_{34}H_{29}N_4Sb$	$(C_6H_5)_3Sb(C_7H_4N_2\text{-}CH_3)_2$	191, **43**, *13*

$C_{34}H_{29}O_4S_2Sb$	$(C_6H_5)_3Sb(O_2CCH_2SC_6H_5)_2$	175, **39**, *7*
$C_{34}H_{29}O_4Sb$	$(C_6H_5)_3Sb(O_2CC_6H_4CH_3\text{-}2)_2$	182, **40**, *35*
	$(C_6H_5)_3Sb(O_2CC_6H_4CH_3\text{-}3)_2$	182, **40**, *36*
	$(C_6H_5)_3Sb(O_2CC_6H_4CH_3\text{-}4)_2$	182, **40**, *37*
$C_{34}H_{29}O_6Sb$	$(C_6H_5)_3Sb(O_2CC_6H_4OCH_3\text{-}4)_2$	181, **40**, *28*
	$(C_6H_5)_3Sb(O_2CCH_2OC_6H_5)_2$	174
$C_{34}H_{29}O_8Sb$	$(C_6H_5)_3Sb(OC_8H_7O_3)_2$	142
$C_{34}H_{31}N_2O_2Sb$	$(C_6H_5)_3Sb(ON{=}C(CH_3)C_6H_5)_2$	160, **35**, *21*
$C_{34}H_{31}N_2O_4Sb$	$(C_6H_5)_3Sb(ON{=}C(H)C_6H_4OCH_3\text{-}4)_2$	160, **35**, *20*
$C_{34}H_{31}O_4P_2Sb$	$(C_6H_5)_3Sb(OP(O)(H)CH{=}CHC_6H_5)_2$	189, **42**, *3*
$C_{34}H_{31}O_5Sb$	$(C_6H_5)_3Sb(2\text{-}OC_6H_4OC_2H_4OC_2H_4OC_6H_4O\text{-}2)$	147, **32**, *28*
$C_{34}H_{33}O_2Sb$	$(C_6H_5)_3Sb(OC_6H_4C_2H_5\text{-}4)_2$	139, **31**, *31*
$C_{35}H_{25}N_2O_2Sb$	$(C_6H_5)_3Sb(C_5N_2O_2(C_6H_5)_2)$	196
$C_{35}H_{25}N_6O_{12}Sb$	$(4\text{-}CH_3C_6H_4)_3Sb(C_7H_2NO_2\text{-}(NO_2)_2)_2$	192, **43**, *22*
$C_{35}H_{27}N_4O_8Sb$	$(4\text{-}CH_3C_6H_4)_3Sb(C_7H_3NO_2\text{-}NO_2)_2$	192, **43**, *21*
$C_{35}H_{29}Cl_2O_4Sb$	$(2\text{-}CH_3C_6H_4)_3Sb(O_2CC_6H_4Cl\text{-}4)_2$	183, **40**, *44*
	$(3\text{-}CH_3C_6H_4)_3Sb(O_2CC_6H_4Cl\text{-}4)_2$	183, **40**, *49*
	$(4\text{-}CH_3C_6H_4)_3Sb(O_2CC_6H_4Cl\text{-}4)_2$	184, **40**, *54*
$C_{35}H_{29}N_2O_2S_2Sb$	$(4\text{-}CH_3C_6H_4)_3Sb(C_7H_4NOS)_2$	192, **43**, *23*
$C_{35}H_{29}N_2O_4Sb$	$(4\text{-}CH_3C_6H_4)_3Sb(C_7H_4NO_2)_2$	192, **43**, *20*
$C_{35}H_{29}N_2O_8Sb$	$(2\text{-}CH_3C_6H_4)_3Sb(O_2CC_6H_4NO_2\text{-}4)_2$	183, **40**, *46*
	$(3\text{-}CH_3C_6H_4)_3Sb(O_2CC_6H_4NO_2\text{-}4)_2$	183, **40**, *51*
	$(4\text{-}CH_3C_6H_4)_3Sb(O_2CC_6H_4NO_2\text{-}4)_2$	184, **40**, *56*
$C_{35}H_{31}N_4Sb$	$(4\text{-}CH_3C_6H_4)_3Sb(C_7H_5N_2)_2$	192, **43**, *24*
$C_{35}H_{31}O_4Sb$	$(2\text{-}CH_3C_6H_4)_3Sb(O_2CC_6H_5)_2$	183, **40**, *43*
	$(3\text{-}CH_3C_6H_4)_3Sb(O_2CC_6H_5)_2$	183, **40**, *48*
	$(4\text{-}CH_3C_6H_4)_3Sb(O_2CC_6H_5)_2$	183, **40**, *53*
$C_{35}H_{33}N_2O_4Sb$	$(2\text{-}CH_3C_6H_4)_3Sb(O_2CC_6H_4NH_2\text{-}4)_2$	183, **40**, *45*
	$(3\text{-}CH_3C_6H_4)_3Sb(O_2CC_6H_4NH_2\text{-}4)_2$	183, **40**, *50*
	$(4\text{-}CH_3C_6H_4)_3Sb(O_2CC_6H_4NH_2\text{-}4)_2$	184, **40**, *55*
$C_{35}H_{35}O_2Sb$	$(2\text{-}CH_3C_6H_4)_3Sb(OC_6H_4CH_3\text{-}4)_2$	141, **31**, *44*
	$(4\text{-}CH_3C_6H_4)_3Sb(OC_6H_4CH_3\text{-}4)_2$	141, **31**, *49*
$C_{35}H_{35}O_4Sb$	$(2\text{-}CH_3C_6H_4)_3Sb(OC_6H_4OCH_3\text{-}4)_2$	140, **31**, *43*
	$(4\text{-}CH_3C_6H_4)_3Sb(OC_6H_4OCH_3\text{-}4)_2$	141, **31**, *48*
$C_{36}H_{27}Br_2O_3Sb$	$(4\text{-}C_6H_5OC_6H_4)_3SbBr_2$	88
$C_{36}H_{27}Br_2Sb$	$(2\text{-}C_6H_5C_6H_4)_3SbBr_2$	91
	$(4\text{-}C_6H_5C_6H_4)_3SbBr_2$	91
$C_{36}H_{27}Cl_2O_3Sb$	$(4\text{-}C_6H_5OC_6H_4)_3SbCl_2$	54
$C_{36}H_{27}Cl_2Sb$	$(4\text{-}C_6H_5C_6H_4)_3SbCl_2$	57/8
$C_{36}H_{27}I_2O_3Sb$	$(4\text{-}C_6H_5OC_6H_4)_3SbI_2$	100, **22**, *6*
$C_{36}H_{27}I_2Sb$	$(4\text{-}C_6H_5C_6H_4)_3SbI_2$	100, **22**, *12*
$C_{36}H_{27}N_2O_2Sb$	$(C_6H_5)_3Sb(OC_9H_6N)_2$	142
$C_{36}H_{27}SSb$	$(4\text{-}C_6H_5C_6H_4)_3SbS$	203
$C_{36}H_{29}O_2Sb$	$(2\text{-}C_6H_5C_6H_4)_3Sb(OH)_2$	129
	$(4\text{-}C_6H_5C_6H_4)_3Sb(OH)_2$	129
$C_{36}H_{29}O_4Sb$	$(C_6H_5)_3Sb(O_2CCH{=}CHC_6H_5)_2$	177
$C_{36}H_{33}N_4Sb$	$(C_6H_5)_3Sb(C_7H_4N_2\text{-}C_2H_5)_2$	191, **43**, *14*
$C_{36}H_{33}O_4S_2Sb$	$(C_6H_5)_3Sb(O_2CCH_2SCH_2C_6H_5)_2$	175, **39**, *1*
$C_{36}H_{33}O_6Sb$	$(C_6H_5)_3Sb(O_2C_9H_9O)_2$	156, **34**, *13*

$C_{36}H_{33}O_8Sb$	$(C_6H_5)_3Sb(O_2CCH_2OC_6H_4OCH_3\text{-}2)_2$	174
	$(C_6H_5)_3Sb(O_2CCH_2OC_6H_4OCH_3\text{-}3)_2$	174
$C_{36}H_{37}O_4Sb$	$(C_6H_5)_3Sb(O_2C_6H_5(CH_3)_2C)_2$	155, **34**, *10*
$C_{36}H_{65}O_8Sb$	$(C_4H_9)_3Sb(O_2CCH{=}CHCO_2(CH_2)_7CH_3)_2$	176
$C_{36}H_{73}O_4Sb$	$(C_4H_9)_3Sb(O_2C(CH_2)_{10}CH_3)_2$	165
$C_{37}H_{29}N_2O_2Sb$	$(C_6H_5)_3Sb(C_5N_2O_2(CH_2C_6H_5)_2)$	196
$C_{37}H_{29}N_2O_4Sb$	$(4\text{-}CH_3C_6H_4)_3Sb(C_8H_4NO_2)_2$	192, **43**, *19*
$C_{37}H_{33}Cl_2O_6Sb$	$(4\text{-}CH_3C_6H_4)_3Sb(O_2CCH_2OC_6H_4Cl\text{-}4)_2$	175
$C_{37}H_{35}N_4O_6Sb$	$(4\text{-}CH_3C_6H_4)_3Sb(ON{=}C(CH_3)C_6H_4NO_2\text{-}4)_2$	161, **35**, *30*
$C_{37}H_{35}N_4Sb$	$(4\text{-}CH_3C_6H_4)_3Sb(C_7H_4N_2\text{-}CH_3)_2$	192, **43**, *25*
$C_{37}H_{35}O_4Sb$	$(2\text{-}CH_3C_6H_4)_3Sb(O_2CC_6H_4CH_3\text{-}4)_2$	183, **40**, *47*
	$(3\text{-}CH_3C_6H_4)_3Sb(O_2CC_6H_4CH_3\text{-}4)_2$	183, **40**, *52*
	$(4\text{-}CH_3C_6H_4)_3Sb(O_2CC_6H_4CH_3\text{-}4)_2$	184, **40**, *57*
	$(4\text{-}CH_3C_6H_4)_3Sb(O_2CCH_2C_6H_5)_2$	175
$C_{37}H_{37}N_2O_2Sb$	$(4\text{-}CH_3C_6H_4)_3Sb(ON{=}C(CH_3)C_6H_5)_2$	161, **35**, *29*
$C_{38}H_{29}F_6O_6Sb$	$(C_6H_5)_3Sb(OC(CF_3)CH_2C(C_6H_5)O_2)_2$	144, **32**, *9*
$C_{38}H_{35}N_2O_4Sb$	$(C_6H_5)_3Sb(CH_3COCHCONHC_6H_5)_2$	135
$C_{38}H_{37}O_4Sb$	$(C_6H_5)_3Sb(O_2C_{10}H_{11})_2$	156, **34**, *12*
$C_{38}H_{41}O_2Sb$	$(C_6H_5)_3Sb(OC_6H_4(C_4H_9\text{-}t)\text{-}4)_2$	139, **31**, *32*
$C_{38}H_{53}O_4S_2Sb$	$(C_6H_5)_3Sb(SCH_2CO_2C_8H_{17})_2$	205, **47**, *6*
$C_{39}H_{27}Br_8O_8Sb$	$(4\text{-}CH_3C_6H_4)_3Sb(O_2CC_6Br_4CO_2CH_3\text{-}2)_2$	184, **40**, *60*
$C_{39}H_{27}Cl_8O_8Sb$	$(4\text{-}CH_3C_6H_4)_3Sb(O_2CC_6Cl_4CO_2CH_3\text{-}2)_2$	184, **40**, *59*
$C_{39}H_{35}O_4Sb$	$(4\text{-}CH_3C_6H_4)_3Sb(O_2CCH{=}CHC_6H_5)_2$	177
$C_{39}H_{35}O_8Sb$	$(4\text{-}CH_3C_6H_4)_3Sb(O_2CC_6H_4CO_2CH_3\text{-}2)_2$	184, **40**, *58*
$[C_{39}H_{39}As_2O_2Sb]^{2+}$	$[(CH_3)_3Sb((C_6H_5)_3AsO)_2]^{2+} \cdot 2[ClO_4]^-$	107, **24**, *4*
$C_{39}H_{39}N_4Sb$	$(4\text{-}CH_3C_6H_4)_3Sb(C_7H_4N_2\text{-}C_2H_5)_2$	192, **43**, *26*
$[C_{39}H_{39}O_2P_2Sb]^{2+}$	$[(CH_3)_3Sb((C_6H_5)_3PO)_2]^{2+} \cdot 2[ClO_4]^-$	107, **24**, *3*
$C_{39}H_{39}SbSi_2$	$(CH_3)_3Sb(Si(C_6H_5)_3)_2$	210
$C_{39}H_{55}O_4Sb$	$(4\text{-}CH_3C_6H_4)_3Sb(O_2C(CH_2)_7CH_3)_2$	169
$C_{41}H_{43}O_4Sb$	$(C_6H_5CH_2)_3Sb(O_2C_{10}H_{11})_2$	155, **34**, *8*
$C_{42}H_{30}Br_5N_{12}O_{18}Sb$	$(2,4\text{-}(CH_3)_2\text{-}5\text{-}((NO_2)_3C_6H_2NH)\text{-}6\text{-}Br\text{-}C_6H)_3SbBr_2$	91
$[C_{42}H_{35}O_2S_2Sb]^{2+}$	$[(C_6H_5)_3Sb((C_6H_5)_2SO)_2]^{2+} \cdot 2[ClO_4]^-$	107, **24**, *7*
$C_{42}H_{39}N_2Sb$	$(C_6H_5)_3Sb(C_{12}H_{12}N)_2$	191, **43**, *15*
$C_{42}H_{49}O_4Sb$	$(C_6H_5)_3Sb(O_2\text{-}4(CH_3)_2CHC_6H_4(CH_3)_2C)_2$	156, **34**, *14*
$C_{42}H_{65}S_2Sb$	$(C_6H_5)_3Sb(SC_{12}H_{25})_2$	205, **47**, *5*
$C_{44}H_{35}N_2O_2Sb$	$(C_6H_5)_3Sb(ON{=}C(C_6H_5)_2)_2$	161, **35**, *23*
$C_{44}H_{35}N_2O_4Sb$	$(C_6H_5)_3Sb((ON(C_6H_5)COC_6H_5))_2$	162
$C_{45}H_{39}O_2Sb$	$(4\text{-}CH_3C_6H_4)_3Sb(OC_6H_4C_6H_5\text{-}4)_2$	141, **31**, *50*
$C_{45}H_{71}O_3S_2Sb$	$(4\text{-}CH_3OC_6H_4)_3Sb(SC_{12}H_{25})_2$	205, **47**, *8*
$C_{46}H_{55}N_2O_2Sb$	$(C_6H_5)_3Sb(OC_{28}H_{40}N_2O)$	147, **32**, *27*
$C_{47}H_{41}N_2O_2Sb$	$(4\text{-}CH_3C_6H_4)_3Sb(ON{=}C(C_6H_5)_2)_2$	161, **35**, *31*

$\mathbf{C}_{48}H_{73}O_4S_2Sb$	$(C_6H_5)_3Sb(SCH_2CH_2CO_2C_{12}H_{25})_2$	205, **47**, *7*
$\mathbf{C}_{51}H_{79}O_4S_2Sb$	$(4\text{-}CH_3C_6H_4)_3Sb(SCH_2CH_2CO_2C_{12}H_{25})_2$	205, **47**, *9*
$\mathbf{C}_{52}H_{65}O_6Sb$	$(C_6H_5)_3Sb(O_2C(CH_2)_2C_6H_2(C_4H_9\text{-}t)_2\text{-}3,5\text{-}OH\text{-}4)_2$	175
$[\mathbf{C}_{54}H_{45}As_2O_2Sb]^{2+}$	$[(C_6H_5)_3Sb((C_6H_5)_3AsO)_2]^{2+} \cdot 2[ClO_4]^-$	107, **24**, *9*
$C_{54}H_{45}N_2P_2Sb$	$(C_6H_5)_3Sb(N{=}P(C_6H_5)_3)_2$	191, **43**, *17*
$[C_{54}H_{45}O_2P_2Sb]^{2+}$	$[(C_6H_5)_3Sb((C_6H_5)_3PO)_2]^{2+} \cdot 2[ClO_4]^-$	107, **24**, *8*
$C_{54}H_{45}O_4SbSi_2$	$(C_6H_5)_3Sb(O_2Si(C_6H_5)_3)_2$	156, **34**, *15*
$C_{54}H_{45}SbSi_2$	$(C_6H_5)_3Sb(Si(C_6H_5)_3)_2$	210
$C_{54}H_{63}N_2P_2Sb$	$(c\text{-}C_6H_{11})_3Sb(N{=}P(C_6H_5)_3)_2$	190, **43**, *5*
$C_{54}H_{85}O_4Sb$	$(C_6H_5)_3Sb(O_2C(CH_2)_{16}CH_3)_2$	169
$\mathbf{C}_{57}H_{51}N_2P_2Sb$	$(4\text{-}CH_3C_6H_4)_3Sb(N{=}P(C_6H_5)_3)_2$	192, **43**, *27*
$\mathbf{C}_{59}H_{51}O_{12}Sb$	$(C_6H_5CH_2)_3Sb((O_2C_6H_5)_3C)_2$	155, **34**, *7*
$\mathbf{C}_{62}H_{67}O_6Sb$	$(C_{10}H_7)_3Sb(O_2CCH_2C_6H_2(C_4H_9\text{-}t)_2\text{-}3,5\text{-}OH\text{-}4)_2$	175

Ligand Formula Index

The ligands containing carbon atoms can be used to locate a compound in this volume. These ligands are listed in the Ligand Formula Index by number of carbon atoms. The number of identical ligands in a compound is not taken into consideration. Thus several compounds may be listed at one position. Compounds having two or more different carbon-containing ligands occur at more than one position. The variable organic ligands are placed in the first two columns, while nonorganic ligands such as OH, halogen, chalcogen, etc., appear in the third column.

Page references are printed in ordinary type, table numbers in boldface, and compound numbers within the tables in italics.

C_2H_5	–	CO_3	111
C_2H_5	–	Cl	20/1
C_2H_5	–	F	3
C_2H_5	–	HO	116
C_2H_5	–	I	94/5
C_2H_5	–	NO_3	109
C_2H_5	–	O	115/6
C_2H_5	–	O_4S	110
C_2H_5	–	S	200
C_2H_5	–	Se	209
C_2H_5	$C_2H_3O_2$	–	165
C_2H_5	C_2H_5O	–	132
C_2H_5	C_2O_4	–	186
C_2H_5	$C_3H_2O_4$	–	186
C_2H_5	C_3H_6NO	–	159, **35**, *9*
C_2H_5	C_4H_8NO	–	159, **35**, *10*
C_2H_5	C_4H_9O	–	132
C_2H_5	$C_4H_9O_2$	–	132
			155, **34**, *4*
C_2H_5	$C_6H_5NO_2S$	–	196, **46**, *1*
C_2H_5	C_6H_5O	–	138, **31**, *18*
C_2H_5	$C_7H_4ClO_2$	–	180, **40**, *13*
C_2H_5	$C_7H_4NO_4$	–	180, **40**, *17*
C_2H_5	$C_7H_5O_2$	–	179, **40**, *12*
C_2H_5	$C_7H_6NO_2$	–	180, **40**, *14*
			180, **40**, *15*
			180, **40**, *16*
C_2H_5	$C_7H_7NO_2S$	–	196, **46**, *2*
C_2H_5	$C_8H_4O_4$	–	186
C_2H_5	$C_8H_7O_2$	–	180, **40**, *18*
			180, **40**, *19*
C_2H_5	C_8H_8NO	–	160, **35**, *11*
C_2H_5O	CH_3	–	132
C_2H_5O	C_2H_5	–	132
C_2H_5O	C_4H_9	–	132
C_2H_5O	C_6H_5	–	134
$C_2H_5O_2$	C_6H_5	–	134
C_2H_5S	CH_3	–	205, **47**, *2*
C_2H_6OS	CH_3	–	107, **24**, *1*
C_2H_6OS	C_6H_5	–	107, **24**, *5*
C_2O_4	CH_3	–	186
C_2O_4	C_2H_5	–	186
C_2O_4	C_4H_9	–	186
C_2O_4	C_6H_5	–	186
$\mathbf{C}_3F_6O_2$	C_6H_4Cl	–	147, **32**, *30*
$C_3F_6O_2$	C_6H_5	–	144, **32**, *7*
$C_3F_6O_2$	C_7H_7	–	147, **32**, *34*
$C_3H_2NOS_2$	C_6H_5	–	191, **43**, *7*
$C_3H_2NO_2$	CH_3	–	171, **37**, *9*
$C_3H_2O_4$	CH_3	–	186

$C_3H_2O_4$	C_2H_5	–	186
$C_3H_2O_4$	C_6H_5	–	186
$C_3H_3O_2$	C_6H_5	–	176
C_3H_5			
$CH_2{=}C(CH_3)$	–	Br	76
$CH_2{=}C(CH_3)$	–	Cl	28
$CH_2{=}C(CH_3)$	–	F	5
$CH_2{=}C(CH_3)$	–	I	96
$CH_2{=}CHCH_2$	–	Br	76
$CH_2{=}CHCH_2$	–	Cl	28
$CH_3CH{=}CH$	–	Br	75/6
$CH_3CH{=}CH$	–	Cl	27/8
$CH_3CH{=}CH$	–	F	5
$CH_3CH{=}CH$	–	I	96/7
$C_3H_5NO_2$	C_4H_9	–	112, **26**, *6*
$C_3H_5O_2$	CH_3	–	164
$C_3H_5O_2$	C_6H_5	–	168
$C_3H_5O_2S$	C_6H_5	–	175, **39**, *2*
C_3H_6NO	CH_3	–	158, **35**, *1*
C_3H_6NO	C_2H_5	–	159, **35**, *9*
C_3H_6NO	C_6H_5	–	160, **35**, *13*
C_3H_6NO	C_7H_7	–	161, **35**, *25*
$C_3H_6NS_2$	CH_3	–	207/8
C_3H_7			
n-C_3H_7	–	Br	70
n-C_3H_7	–	Cl	21
n-C_3H_7	–	I	95
n-C_3H_7	–	O	116/7
n-C_3H_7	–	S	200
n-C_3H_7	–	Se	209
n-C_3H_7	$C_2H_3O_2$	–	165
n-C_3H_7	$C_5H_3F_8O$	–	132
n-C_3H_7	$C_6H_5NO_2S$	–	196, **46**, *3*
n-C_3H_7	$C_7H_5O_2$	–	180, **40**, *20*
n-C_3H_7	$C_7H_7NO_2S$	–	196, **46**, *4*
$(CH_3)_2CH$	–	Br	71
$(CH_3)_2CH$	–	Cl	22
$(CH_3)_2CH$	–	F	3
$(CH_3)_2CH$	–	I	95
C_3H_8ClSi	–	Cl	24
C_3H_9OSi	CH_3	–	162
C_3H_9OSi	C_6H_5	–	162
C_3H_9Si	C_6H_5	–	210
C_4 H_2O_4	C_6H_5	–	187, **41**, *2*
	C_6H_5	–	187, **41**, *3*
$C_4H_3N_5O$	C_6H_5	–	194, **44**, *20*
$C_4H_3N_5O_2$	C_6H_5	–	194, **44**, *7*
C_4H_3O	–	Cl	58
C_4H_3S	–	Br	91
C_4H_3S	–	Cl	58

C_4H_3S	–	O	129
$C_4H_4Cl_2$	–	Cl	30/1
$C_4H_4NO_2$	CH_3	–	190, **43**, *1*
$C_4H_4NO_2$	C_6H_5	–	191, **43**, *6*
$C_4H_4NO_2$	C_7H_7	–	192, **43**, *18*
$C_4H_4N_2S_4Zn$	C_6H_5	–	194, **44**, *18*
$C_4H_4O_4$	C_6H_5	–	186
$C_4H_4O_4S$	C_6H_5	–	187, **41**, *1*
$C_4H_5NO_4$	C_6H_5	–	187
$C_4H_5O_2$			
$O_2CCH{=}CHCH_3$	CH_3	–	176
$O_2CCH{=}CHCH_3$	C_6H_5	–	177
$O_2CCH{=}CHCH_3$	C_7H_7	–	177
$O_2CC(CH_3){=}CH_2$	C_6H_5	–	177
$C_4H_7OS_2$			
$SC(S)OC_3H_7$	CH_3	–	206, **48**, *1*
$SC(S)OCH(CH_3)_2$	CH_3	–	206, **48**, *2*
$C_4H_7O_2$	CH_3	–	164
$C_4H_7O_2$	C_6H_5	–	168
$C_4H_7O_2$	C_7H_7	–	169
$C_4H_7O_2S$	C_6H_5	–	175, **39**, *3*
C_4H_8NO	CH_3	–	158, **35**, *2*
C_4H_8NO	C_2H_5	–	159, **35**, *10*
C_4H_8NO	C_6H_5	–	160, **35**, *14*
C_4H_8NO	C_7H_7	–	161, **35**, *26*
C_4H_9			
$n\text{-}C_4H_9$	–	Br	71
$n\text{-}C_4H_9$	–	Cl	22
$n\text{-}C_4H_9$	–	CNO	103, **23**, *2*
$n\text{-}C_4H_9$	–	COS_2	112, **26**, *3*
$n\text{-}C_4H_9$	–	CO_3	112, **26**, *1*
$n\text{-}C_4H_9$	–	CS_3	112, **26**, *5*
$n\text{-}C_4H_9$	–	F	3
$n\text{-}C_4H_9$	–	HO	117
$n\text{-}C_4H_9$	–	I	95
$n\text{-}C_4H_9$	–	O	117
$n\text{-}C_4H_9$	–	S	200/1
$n\text{-}C_4H_9$	–	Se	209
$n\text{-}C_4H_9$	$C_2H_3O_2$	–	165
$n\text{-}C_4H_9$	$C_2H_3O_2S$	–	172
$n\text{-}C_4H_9$	C_2H_5O	–	132
$n\text{-}C_4H_9$	C_2O_4	–	186
$n\text{-}C_4H_9$	$C_3H_5NO_2$	–	112, **26**, *6*
$n\text{-}C_4H_9$	$C_7H_4ClNO_2$	–	112, **26**, *9*
$n\text{-}C_4H_9$	$C_7H_5NO_2$	–	112, **26**, *8*
$n\text{-}C_4H_9$	$C_7H_5O_2$	–	180, **40**, *21*
$n\text{-}C_4H_9$	$C_9H_{17}NO_2$	–	112, **26**, *7*
$n\text{-}C_4H_9$	$C_{12}H_{12}O_4$	–	186
$n\text{-}C_4H_9$	$C_{12}H_{19}O_4$	–	176
$n\text{-}C_4H_9$	$C_{12}H_{23}O_2$	–	165
$(CH_3)_2CHCH_2$	–	Br	72

$(CH_3)_2CHCH_2$	–	Cl	22
$(CH_3)_2CHCH_2$	–	CNO	103, **23**, *3*
$(CH_3)_2CHCH_2$	–	COS_2	112, **26**, *4*
$(CH_3)_2CHCH_2$	–	CO_3	112, **26**, *2*
$(CH_3)_2CHCH_2$	–	I	96
$(CH_3)_2CHCH_2$	–	O	117
$(CH_3)_2CHCH_2$	$C_2H_3O_2$	–	165
$(CH_3)_2CHCH_2$	$C_7H_5NO_2$	–	112, **26**, *10*
C_4H_9O	CH_3	–	132
C_4H_9O	C_2H_5	–	132
C_4H_9O	C_6H_5	–	134
$C_4H_9O_2$			
$OOC(CH_3)_3$	CH_3	–	155, **34**, *1*
$OOC(CH_3)_3$	C_2H_5	–	155, **34**, *4*
$OOC(CH_3)_3$	C_6H_5	–	155, **34**, *9*
$OOC(CH_3)_3$	C_6H_{11}	–	155, **34**, *5*
$OOC(CH_3)_3$	C_7H_7	–	155, **34**, *6*
$OCH_2CH_2OC_2H_5$	C_2H_5	–	132
$OCH_2CH_2OC_2H_5$	C_6H_5	–	134
$C_4H_{10}O_2Tl$	C_6H_5	–	156, **34**, *16*
$C_4H_{10}PS_2$	CH_3	–	209
$C_4H_{11}Si$	–	Br	73
$C_4H_{11}Si$	–	Cl	24
$C_4N_2S_2$	CH_3	–	208
$C_4N_2S_2$	C_6H_5	–	208
C$_5H_2N_5$	C_6H_5	–	194, **44**, *10*
$C_5H_3Br_2O_4$	C_6H_5	–	177
$C_5H_3Cl_2O_4$	C_6H_5	–	177
$C_5H_3F_8O$	C_3H_7	–	132
$C_5H_4NO_2$	C_6H_5	–	160, **35**, *19*
$C_5H_4NO_2$	C_7H_7	–	161, **35**, *28*
$C_5H_5F_3O_3$	C_6H_4Cl	–	147, **32**, *31*
$C_5H_5F_3O_3$	C_6H_5	–	144, **32**, *8*
$C_5H_5N_3$	C_6H_5	–	194, **44**, *22*
$C_5H_5N_5O$	C_6H_5	–	194, **44**, *21*
$C_5H_5O_4$	C_6H_5	–	177
$C_5H_6O_4$	C_6H_5	–	186
$C_5H_7O_3$	C_6H_5	–	174
$C_5H_8NS_2$	CH_3	–	208
C_5H_9	–	Br	72
$C_5H_9OS_2$	CH_3	–	206, **48**, *3*
$C_5H_9O_2$	–	O	118
$C_5H_9O_2$	CH_3	–	164
$C_5H_9O_2S$	C_6H_5	–	175, **39**, *4* 175, **39**, *5*
$C_5H_{10}NO$			
$ON{=}C(CH_3)C_3H_7$	CH_3	–	159, **35**, *3*
$ON{=}C(CH_3)C_3H_7$	C_6H_5	–	160, **35**, *15*
$ON{=}C(C_2H_5)_2$	CH_3	–	159, **35**, *4*
$ON{=}C(C_2H_5)_2$	C_6H_5	–	160, **35**, *16*

C_5H_{11}	–	Br	72
C_5H_{11}	–	Cl	23
C_5H_{11}	–	F	3
C_5H_{11}	–	I	96
C_5H_{11}	–	NO_3	109
C_5H_{11}	–	O	117/8
C_5H_{11}	–	O_4S	110
C_5H_{11}	–	S	201
C_5H_{11}	–	Se	209
$C_6Cl_4O_2$	C_6H_5	–	145, **32**, *15*
C_6Cl_5	–	Br	87
C_6Cl_5	–	Cl	52
C_6Cl_5O	C_6H_5	–	138, **31**, *23*
C_6F_5	–	Br	87
C_6F_5	–	Cl	51
C_6F_5	–	ClO_4	107
C_6F_5	–	F	8
C_6F_5	–	NO_3	109
C_6F_5	CH_3O	–	135
C_6F_5	C_2F_6NO	–	162
$C_6H_3Br_2O$	C_6H_5	–	138, **31**, *25*
$C_6H_3Cl_2$	–	Cl	52
$C_6H_3F_2$	–	F	8
C_6H_4Br	–	Br	87
C_6H_4Br	–	Cl	52/3
C_6H_4Br	–	F	9
C_6H_4Br	–	I	100, **22**, *2*
$C_6H_4BrNO_2S$	C_6H_5	–	197
C_6H_4BrO	CH_3	–	137, **31**, *5*
C_6H_4BrO	C_6H_4Cl	–	139, **31**, *36*
C_6H_4BrO	C_6H_5	–	138, **31**, *24*
C_6H_4Cl	–	Br	87
C_6H_4Cl	–	Cl	51/2
C_6H_4Cl	–	F	8
C_6H_4Cl	–	I	100, **22**, *1*
C_6H_4Cl	$C_2HCl_3O_2$	–	147, **32**, *29*
C_6H_4Cl	$C_3F_6O_2$	–	147, **32**, *30*
C_6H_4Cl	$C_5H_5F_3O_3$	–	147, **32**, *31*
C_6H_4Cl	C_6H_4BrO	–	139, **31**, *36*
C_6H_4Cl	C_6H_4ClO	–	139, **31**, *35*
C_6H_4Cl	$C_6H_4NO_3$	–	140, **31**, *37*
C_6H_4Cl	C_6H_5O	–	139, **31**, *34*
C_6H_4Cl	$C_7H_4ClO_2$	–	182, **40**, *39*
C_6H_4Cl	$C_7H_4NO_4$	–	182, **40**, *41*
C_6H_4Cl	$C_7H_5O_2$	–	182, **40**, *38*
C_6H_4Cl	$C_7H_6NO_2$	–	182, **40**, *40*
C_6H_4Cl	C_7H_7O	–	140, **31**, *39*
C_6H_4Cl	$C_7H_7O_2$	–	140, **31**, *38*
C_6H_4Cl	$C_8H_7O_2$	–	182, **40**, *42*
$C_6H_4ClNO_2S$	C_6H_5	–	197

C_6H_5S	CH_3	–	205, **47**, *4*
$C_6H_6N_2$	C_6H_5	–	194, **44**, *11*
$C_6H_6O_2P$	CH_3	–	189, **42**, *2*
$C_6H_6O_2P$	C_6H_5	–	189, **42**, *4*
$C_6H_8O_4$	C_6H_5	–	186
$C_6H_9O_2$	C_6H_5	–	155, **34**, *11*
$C_6H_{10}NO$	CH_3	–	159, **35**, *5*
$C_6H_{10}NO$	C_6H_5	–	160, **35**, *18*
$C_6H_{10}NO$	C_7H_7	–	161, **35**, *27*
$C_6H_{10}NS_2$	CH_3	–	208
C_6H_{11}	–	Br	73
C_6H_{11}	–	Cl	23
C_6H_{11}	–	O	118
C_6H_{11}	–	S	201
C_6H_{11}	–	Se	209
C_6H_{11}	$C_2H_2O_3$	–	144, **32**, *4*
C_6H_{11}	$C_4H_9O_2$	–	155, **34**, *5*
C_6H_{11}	$C_{18}H_{15}NP$	–	190, **43**, *5*
$C_6H_{11}O_2S$	C_6H_5	–	175, **39**, *6*
$C_6H_{12}NO$	C_6H_5	–	160, **35**, *17*
$C_6H_{12}O_2$	C_6H_5	–	145, **32**, *10*
$C_6H_{14}N_2$	C_6H_5	–	194, **44**, *3*
C$_7H_2N_3O_6$	C_7H_7	–	192, **43**, *22*
$C_7H_2O_6$	C_6H_5	–	188, **41**, *9*
$C_7H_3F_{12}O$	C_6H_5	–	134/5
$C_7H_3N_2O_4$	C_6H_5	–	191, **43**, *10*
$C_7H_3N_2O_4$	C_7H_7	–	192, **43**, *21*
$C_7H_4ClNO_2$	C_4H_9	–	112, **26**, *9*
$C_7H_4ClO_2$			
2-$ClC_6H_4CO_2$	C_6H_5	–	181, **40**, *23*
3-$ClC_6H_4CO_2$	C_6H_5	–	181, **40**, *24*
4-$ClC_6H_4CO_2$	CH_3	–	179, **40**, *2*
4-$ClC_6H_4CO_2$	C_2H_5	–	180, **40**, *13*
4-$ClC_6H_4CO_2$	C_6H_4Cl	–	182, **40**, *39*
4-$ClC_6H_4CO_2$	C_6H_5	–	181, **40**, *25*
4-$ClC_6H_4CO_2$	C_7H_7	–	183, **40**, *44*
			183, **40**, *49*
			184, **40**, *54*
C_7H_4NOS	C_6H_5	–	191, **43**, *11*
C_7H_4NOS	C_7H_7	–	192, **43**, *23*
$C_7H_4NO_2$	C_6H_5	–	191, **43**, *9*
$C_7H_4NO_2$	C_7H_7	–	192, **43**, *20*
$C_7H_4NO_4$			
2-$O_2NC_6H_4CO_2$	CH_3	–	179, **40**, *6*
2-$O_2NC_6H_4CO_2$	C_6H_5	–	181, **40**, *32*
3-$O_2NC_6H_4CO_2$	CH_3	–	179, **40**, *7*
3-$O_2NC_6H_4CO_2$	C_6H_5	–	182, **40**, *33*
4-$O_2NC_6H_4CO_2$	CH_3	–	179, **40**, *8*
4-$O_2NC_6H_4CO_2$	C_2H_5	–	180, **40**, *17*

$C_6H_5CH_2$	$C_{10}H_{11}O_2$	–	155, **34**, *8*
$C_6H_5CH_2$	$C_{19}H_{15}O_6$	–	155, **34**, *7*
2-$CH_3C_6H_4$	–	Br	88
2-$CH_3C_6H_4$	–	Cl	54/5
2-$CH_3C_6H_4$	–	F	9
2-$CH_3C_6H_4$	–	HO	128
2-$CH_3C_6H_4$	–	I	100, **22**, *10*
2-$CH_3C_6H_4$	–	O	128
2-$CH_3C_6H_4$	C_2H_2ClNO	–	195, **45**, *8*
2-$CH_3C_6H_4$	C_2H_3NO	–	195, **45**, *5*
2-$CH_3C_6H_4$	$C_2H_3O_2$	–	169
2-$CH_3C_6H_4$	C_6H_4ClO	–	140, **31**, *41*
2-$CH_3C_6H_4$	$C_6H_4NO_3$	–	140, **31**, *42*
2-$CH_3C_6H_4$	C_6H_5O	–	140, **31**, *40*
2-$CH_3C_6H_4$	$C_7H_4ClO_2$	–	183, **40**, *44*
2-$CH_3C_6H_4$	$C_7H_4NO_4$	–	183, **40**, *46*
2-$CH_3C_6H_4$	$C_7H_5O_2$	–	183, **40**, *43*
2-$CH_3C_6H_4$	$C_7H_6NO_2$	–	183, **40**, *45*
2-$CH_3C_6H_4$	C_7H_7O	–	140, **31**, *44*
2-$CH_3C_6H_4$	$C_7H_7O_2$	–	140, **31**, *43*
2-$CH_3C_6H_4$	$C_8H_7O_2$	–	183, **40**, *47*
3-$CH_3C_6H_4$	–	Br	89
3-$CH_3C_6H_4$	–	Cl	55
3-$CH_3C_6H_4$	–	F	9
3-$CH_3C_6H_4$	–	I	100, **22**, *11*
3-$CH_3C_6H_4$	–	O	128
3-$CH_3C_6H_4$	–	S	203
3-$CH_3C_6H_4$	C_2H_2ClNO	–	195, **45**, *9*
3-$CH_3C_6H_4$	C_2H_3NO	–	195, **45**, *6*
3-$CH_3C_6H_4$	$C_7H_4ClO_2$	–	183, **40**, *49*
3-$CH_3C_6H_4$	$C_7H_4NO_4$	–	183, **40**, *51*
3-$CH_3C_6H_4$	$C_7H_5O_2$	–	183, **40**, *48*
3-$CH_3C_6H_4$	$C_7H_6NO_2$	–	183, **40**, *50*
3-$CH_3C_6H_4$	$C_8H_7O_2$	–	183, **40**, *52*
4-$CH_3C_6H_4$	–	Br	89
4-$CH_3C_6H_4$	–	Cl	55/7
4-$CH_3C_6H_4$	–	CNS	105
4-$CH_3C_6H_4$	–	F	9
4-$CH_3C_6H_4$	–	HO	128
4-$CH_3C_6H_4$	–	I	99/100
4-$CH_3C_6H_4$	–	O	128
4-$CH_3C_6H_4$	–	S	203
4-$CH_3C_6H_4$	CHO_2	–	169
4-$CH_3C_6H_4$	$C_2Cl_3O_2$	–	175
4-$CH_3C_6H_4$	$C_2F_3O_2$	–	175
4-$CH_3C_6H_4$	$C_2HCl_3O_2$	–	147, **32**, *33*
4-$CH_3C_6H_4$	$C_2HF_3O_2$	–	147, **32**, *32*
4-$CH_3C_6H_4$	$C_2H_2BrO_2$	–	175
4-$CH_3C_6H_4$	C_2H_2ClNO	–	195, **45**, *10*
4-$CH_3C_6H_4$	$C_2H_2ClO_2$	–	175
4-$CH_3C_6H_4$	C_2H_3NO	–	195, **45**, *7*

2-$CH_3OC_6H_4$	C_6H_5	–	139, **31**, *28*
3-$CH_3OC_6H_4$	–	Br	88
3-$CH_3OC_6H_4$	–	Cl	53
3-$CH_3OC_6H_4$	–	I	100, **22**, *4*
3-$CH_3OC_6H_4$	CH_3	–	137, **31**, *12*
3-$CH_3OC_6H_4$	C_6H_5	–	139, **31**, *29*
4-$CH_3OC_6H_4$	–	Br	88
4-$CH_3OC_6H_4$	–	Cl	53
4-$CH_3OC_6H_4$	–	I	100, **22**, *7*
4-$CH_3OC_6H_4$	–	NO_3	109
4-$CH_3OC_6H_4$	–	O	127
4-$CH_3OC_6H_4$	CH_3	–	137, **31**, *13*
4-$CH_3OC_6H_4$	C_6H_4Cl	–	140, **31**, *39*
4-$CH_3OC_6H_4$	C_6H_5	–	139, **31**, *30*
4-$CH_3OC_6H_4$	C_7H_7	–	140, **31**, *43*
4-$CH_3OC_6H_4$	C_7H_7	–	140, **31**, *44*
4-$CH_3OC_6H_4$	C_7H_7	–	141, **31**, *49*
4-$CH_3OC_6H_4$	$C_{12}H_{25}S$	–	205, **47**, *8*
$C_7H_7O_2$			
2-$CH_3OC_6H_4O$	CH_3	–	137, **31**, *9*
4-$CH_3OC_6H_4O$	CH_3	–	137, **31**, *10*
4-$CH_3OC_6H_4O$	C_6H_4Cl	–	140, **31**, *38*
4-$CH_3OC_6H_4O$	C_6H_5	–	139, **31**, *27*
4-$CH_3OC_6H_4O$	C_7H_7	–	140, **31**, *43*
			141, **31**, *48*
C_7H_7S	CH_3	–	205, **47**, *3*
$C_7H_8N_2O$	C_6H_5	–	194, **44**, *12*
C_7H_{15}	–	I	96
C_7H_{15}	–	O	118
C$_8H_2Cl_2O_4$	C_6H_5	–	187, **41**, *6*
$C_8H_3BrO_4$	C_6H_5	–	187, **41**, *4*
$C_8H_3NO_6$	C_6H_5	–	187, **41**, *5*
$C_8H_4MnO_3$	–	Br	77
$C_8H_4NO_2$	C_6H_5	–	191, **43**, *8*
$C_8H_4NO_2$	C_7H_7	–	192, **43**, *19*
$C_8H_4O_4$			
2-$O_2CC_6H_4CO_2$	C_2H_5	–	186
2-$O_2CC_6H_4CO_2$	C_6H_5	–	186
3-$O_2CC_6H_4CO_2$	C_6H_5	–	186
4-$O_2CC_6H_4CO_2$	C_6H_5	–	186
$C_8H_6ClO_3$	C_6H_5	–	174
$C_8H_6ClO_3$	C_7H_7	–	175
C_8H_7	–	Cl	57
$C_8H_7BrNO_2$	–	Br	89
$C_8H_7BrNO_2$	–	O	129
$C_8H_7N_2$	CH_3	–	190, **43**, *3*
$C_8H_7N_2$	C_6H_5	–	191, **43**, *13*
$C_8H_7N_2$	C_7H_7	–	192, **43**, *25*
$C_8H_7N_2O_3$	CH_3	–	159, **35**, *7*
$C_8H_7N_2O_3$	C_6H_5	–	160, **35**, *22*

$C_{10}H_{14}N_2$	C_6H_5	–	194, **44**, *13*
$C_{10}H_{15}O$	–	Cl	25
$C_{10}H_{19}O_2S$	C_6H_5	–	205, **47**, *6*
$\mathbf{C}_{11}H_{11}NO_2$	C_6H_5	–	145, **32**, *21*
$C_{11}H_{13}NO_2$	C_6H_5	–	145, **32**, *19*
$\mathbf{C}_{12}H_8FeO_4$	C_6H_5	–	188, **41**, *10*
$C_{12}H_9$			
$2\text{-}C_6H_5C_6H_4$	–	Br	91
$2\text{-}C_6H_5C_6H_4$	–	HO	129
$4\text{-}C_6H_5C_6H_4$	–	Br	91
$4\text{-}C_6H_5C_6H_4$	–	Cl	57/8
$4\text{-}C_6H_5C_6H_4$	–	HO	129
$4\text{-}C_6H_5C_6H_4$	–	I	100, **22**, *12*
$4\text{-}C_6H_5C_6H_4$	–	S	203
$C_{12}H_9O$	–	Br	88
$C_{12}H_9O$	–	Cl	54
$C_{12}H_9O$	–	I	100, **22**, *6*
$C_{12}H_9O$	CH_3	–	137, **31**, *16*
			138, **31**, *17*
$C_{12}H_9O$	C_7H_7	–	141, **31**, *50*
$C_{12}H_{10}NOP$	C_6H_5	–	198
$C_{12}H_{10}N_2$	C_6H_5	–	194, **44**, *19*
$C_{12}H_{10}N_2O_4S_2$	CH_3	–	194, **44**, *2*
$C_{12}H_{10}N_2O_4S_2$	C_6H_5	–	194, **44**, *9*
$C_{12}H_{10}OS$	C_6H_5	–	107, **24**, *7*
$C_{12}H_{12}N$	C_6H_5	–	191, **43**, *15*
$C_{12}H_{12}O_4$	C_4H_9	–	186
$C_{12}H_{12}O_4$	C_6H_5	–	188, **41**, *8*
$C_{12}H_{17}O_2$	C_6H_5	–	156, **34**, *14*
$C_{12}H_{19}O_4$	C_4H_9	–	176
$C_{12}H_{23}O_2$	C_4H_9	–	165
$C_{12}H_{25}S$	C_6H_5	–	205, **47**, *5*
$C_{12}H_{25}S$	C_7H_7O	–	205, **47**, *8*
$C_{12}H_{26}N_2$	C_6H_5	–	194, **44**, *4*
$\mathbf{C}_{13}H_9NO_2$	CH_3	–	143, **32**, *3*
$C_{13}H_9NO_2$	C_6H_5	–	146, **32**, *23*
$C_{13}H_{10}NO$	C_6H_5	–	161, **35**, *23*
$C_{13}H_{10}NO$	C_7H_7	–	161, **35**, *31*
$C_{13}H_{10}NO_2$	C_6H_5	–	162
$C_{13}H_{11}NO_2$	C_6H_5	–	145, **32**, *20*
$C_{13}H_{11}N_3O$	C_6H_5	–	194, **44**, *17*
$C_{13}H_{12}N_2$	C_6H_5	–	194, **44**, *16*
$C_{13}H_{14}N_4O_2$	C_6H_5	–	194, **44**, *8*
$\mathbf{C}_{14}H_8N_2O_2$	C_6H_5	–	194, **44**, *6*
$C_{14}H_{10}BrN_4O_6$	–	Br	91
$C_{14}H_{11}NO_2$	C_6H_5	–	146, **32**, *24*
$C_{14}H_{12}O_2$	C_6H_5	–	145, **32**, *11*

C$_{15}H_{29}O_2S$	C_6H_5	–	205, **47**, *7*
$C_{15}H_{29}O_2S$	C_7H_7	–	205, **47**, *9*
C$_{16}H_{13}NO_2$	C_6H_5	–	146, **32**, *22*
$C_{16}H_{16}O_5$	C_6H_5	–	147, **32**, *28*
$C_{16}H_{23}O_3$	$C_{10}H_7$	–	175
C$_{17}H_{10}N_2O_2$	C_6H_5	–	196
$C_{17}H_{25}O_3$	C_6H_5	–	175
C$_{18}H_{15}AsO$	CH_3	–	107, **24**, *4*
$C_{18}H_{15}AsO$	C_6H_5	–	107, **24**, *9*
$C_{18}H_{15}NP$	C_6H_5	–	191, **43**, *17*
$C_{18}H_{15}NP$	C_6H_{11}	–	190, **43**, *5*
$C_{18}H_{15}NP$	C_7H_7	–	192, **43**, *27*
$C_{18}H_{15}OP$	CH_3	–	107, **24**, *3*
$C_{18}H_{15}OP$	C_6H_5	–	107, **24**, *8*
$C_{18}H_{15}O_2Si$	C_6H_5	–	156, **34**, *15*
$C_{18}H_{15}Si$	CH_3	–	210
$C_{18}H_{15}Si$	C_6H_5	–	210
$C_{18}H_{35}O_2$	C_6H_5	–	169
C$_{19}H_{14}N_2O_2$	C_6H_5	–	196
$C_{19}H_{15}O_6$	C_7H_7	–	155, **34**, *7*
C$_{28}H_{40}N_2O_2$	C_6H_5	–	147, **32**, *27*

Table of Conversion Factors

Following the notation in Landolt-Börnstein [7], values which have been fixed by convention are indicated by a bold-face last digit. The conversion factor between calorie and Joule that is given here is based on the thermochemical calorie, cal_{thch}, and is defined as 4.1840 J/cal. However, for the conversion of the "Internationale Tafelkalorie", cal_{IT}, into Joule, the factor 4.1868 J/cal is to be used [1, p. 147]. For the conversion factor for the British thermal unit, the Steam Table Btu, Btu_{ST}, is used [1, p. 95].

Force	N	dyn	kp
1 N (Newton)	1	10^5	0.1019716
1 dyn	10^{-5}	1	1.019716×10^{-6}
1 kp	9.8066**5**	$9.8066\mathbf{5} \times 10^5$	1

Pressure	Pa	bar	kp/m^2	at	atm	Torr	lb/in^2
1 Pa (Pascal) = 1 N/m^2	1	10^{-5}	1.019716×10^{-1}	1.019716×10^{-5}	0.986923×10^{-5}	0.750062×10^{-2}	145.0378×10^{-6}
1 bar = 10^6 dyn/cm^2	10^5	1	10.19716×10^3	1.019716	0.986923	750.062	14.50378
1 kp/m^2 = 1 mm H_2O	9.8066**5**	$0.98066\mathbf{5} \times 10^{-4}$	1	10^{-4}	0.967841×10^{-4}	0.735559×10^{-1}	1.422335×10^{-3}
1 at = 1 kp/cm^2	$0.98066\mathbf{5} \times 10^5$	0.98066**5**	10^4	1	0.967841	735.559	14.22335
1 atm = 76**0** Torr	$1.0132\mathbf{5} \times 10^5$	1.0132**5**	1.033227×10^4	1.033227	1	760	14.69595
1 Torr = 1 mm Hg	133.3224	1.333224×10^{-3}	13.59510	1.359510×10^{-3}	1.315789×10^{-3}	1	19.33678×10^{-3}
1 lb/in^2 = 1 psi	6.89476×10^3	68.9476×10^{-3}	703.069	70.3069×10^{-3}	68.0460×10^{-3}	51.7149	1

Work, Energy, Heat	J	kWh	kcal	Btu	MeV
1 J (Joule) = 1 Ws = 1 Nm = 10^7 erg	1	2.778×10^{-7}	2.39006×10^{-4}	9.4781×10^{-4}	6.242×10^{12}
1 kWh	$3.\mathbf{6} \times 10^6$	1	860.4	3412.14	2.247×10^{19}
1 kcal	4184.**0**	1.1622×10^{-3}	1	3.96566	2.6117×10^{16}
1 Btu (British thermal unit)	1055.06	2.93071×10^{-4}	0.25164	1	6.5858×10^{15}
1 MeV	1.602×10^{-13}	4.450×10^{-20}	3.8289×10^{-17}	1.51840×10^{-16}	1

1 eV/mol ≙ 23.0578 kcal/mol = 96.473 kJ/mol

Power	kW	PS	kp m/s	kcal/s
1 kW = 10^{10} erg/s	1	1.35962	101.972	0.239006
1 PS	0.73550	1	**75**	0.17579
1 kp m/s	$9.80\mathbf{665} \times 10^{-3}$	0.01333	1	2.34384×10^{-3}
1 kcal/s	4.18**40**	5.6886	426.650	1

References:

[1] A. Sacklowski, Die neuen SI-Einheiten, Goldmann, München 1979. (Conversion tables in an appendix.)
[2] International Union of Pure and Applied Chemistry, Manual of Symbols and Terminology for Physicochemical Quantities and Units, Pergamon, London 1979; Pure Appl. Chem. **51** [1979] 1/41.
[3] The International System of Units (SI), National Bureau of Standards Spec. Publ. No. 330 [1972].
[4] H. Ebert, Physikalisches Taschenbuch, 5th Ed., Vieweg, Wiesbaden 1976.
[5] Kraftwerk Union Information, Technical and Economic Data on Power Engineering, Mülheim/Ruhr 1978.
[6] E. Padelt, H. Laporte, Einheiten und Größenarten der Naturwissenschaften, 3rd Ed., VEB Fachbuchverlag, Leipzig 1976.
[7] Landolt-Börnstein, 6th Ed., Vol. II, Pt. 1, 1971, pp. 1/14.
[8] ISO Standards Handbook 2, Units of Measurement, 2nd Ed., Geneva 1982.

Key to the Gmelin System of Elements and Compounds

System Number	Symbol	Element
1		Noble Gases
2	H	Hydrogen
3	O	Oxygen
4	N	Nitrogen
5	F	Fluorine
6	**Cl**	**Chlorine**
7	Br	Bromine
8	I	Iodine
8a	At	Astatine
9	S	Sulfur
10	Se	Selenium
11	Te	Tellurium
12	Po	Polonium
13	B	Boron
14	C	Carbon
15	Si	Silicon
16	P	Phosphorus
17	As	Arsenic
18	Sb	Antimony
19	Bi	Bismuth
20	Li	Lithium
21	Na	Sodium
22	K	Potassium
23	NH_4	Ammonium
24	Rb	Rubidium
25	Cs	Caesium
25a	Fr	Francium
26	Be	Beryllium
27	Mg	Magnesium
28	Ca	Calcium
29	Sr	Strontium
30	Ba	Barium
31	Ra	Radium
32	**Zn**	**Zinc**
33	Cd	Cadmium
34	Hg	Mercury
35	Al	Aluminium
36	Ga	Gallium

HCl

$ZnCl_2$

System Number	Symbol	Element
37	In	Indium
38	Tl	Thallium
39	Sc, Y La—Lu	Rare Earth Elements
40	Ac	Actinium
41	Ti	Titanium
42	Zr	Zirconium
43	Hf	Hafnium
44	Th	Thorium
45	Ge	Germanium
46	Sn	Tin
47	Pb	Lead
48	V	Vanadium
49	Nb	Niobium
50	Ta	Tantalum
51	Pa	Protactinium
52	**Cr**	**Chromium**
53	Mo	Molybdenum
54	W	Tungsten
55	U	Uranium
56	Mn	Manganese
57	Ni	Nickel
58	Co	Cobalt
59	Fe	Iron
60	Cu	Copper
61	Ag	Silver
62	Au	Gold
63	Ru	Ruthenium
64	Rh	Rhodium
65	Pd	Palladium
66	Os	Osmium
67	Ir	Iridium
68	Pt	Platinum
69	Tc	Technetium[1]
70	Re	Rhenium
71	Np, Pu...	Transuranium Elements

$CrCl_2$

$ZnCrO_4$

Material presented under each Gmelin System Number includes all information concerning the element(s) listed for that number plus the compounds with elements of lower System Number.

For example, zinc (System Number 32) as well as all zinc compounds with elements numbered from 1 to 31 are classified under number 32.

[1] A Gmelin volume titled "Masurium" was published with this System Number in 1941.

A Periodic Table of the Elements with the Gmelin System Numbers is given on the Inside Front Cover